The Composting Toilet System Book

Version 1.2, Updated

A Practical Guide to Choosing, Pla
Maintaining Composting Toilet S,ms,
a Water-Saving, Pollution-Preventing Alternative

By David Del Porto and Carol Steinfeld

Library of Congress Catalog Card Number 98-88826
ISBN 0-9666783-0-3

✲

First Edition, Second Printing with Additions
Printed in the United States of America

Copyright © 2000 by David Del Porto and Carol Steinfeld. All rights reserved.

Published by
The Center for Ecological Pollution Prevention (CEPP)
P.O. Box 1330
Concord, Massachusetts 01742-1330

For information about permission to reprint selections from this book,
write to the Center for Ecological Pollution Prevention

Printed on recycled-content paper

Del Porto, David.
The composting toilet system book : a practical guide to choosing, planning and maintaining composting toilet systems, a water-saving, pollution-preventing alternative / by David Del Porto and Carol Steinfeld. -- 1st ed.
p. cm.
LCCN: 98-88826
ISBN: 0-9666783-0-3

1. Toilets. 2. Compost. 3. Water conservation.
I. Steinfeld, Carol. II. Title.

TD775.D45 1998 628'.744
QBI98-1628

Notes & Acknowledgements

We hope you find this book a valuable tool for considering your wastewater and toilet system options, reducing your water and wastewater costs and helping to protect public and environmental health.

Thanks to a number of factors, there is a growing demand for information about composting toilet and graywater systems. At the same time, composting toilet technology appears to be "ready for prime time" and progressing fast. David Del Porto, co-author of this book, has sold thousands of composting toilet systems of different manufacturers and designs since 1973 as the principal of Ecos and EcoTech, and designed several for homeowners in the United States and projects in developing countries through his ecological engineering firm, Sustainable Strategies. Says David: In that time, I've gone from naively enthusiastic about the early systems, and on through the valley of harsh reality, as I faced disappointed owners, malfunctioning or failed systems and all sorts of issues of odors, capacity, and so forth. Throughout North America, Europe and developing countries, I've seen enthusiastic but inexperienced promoters install inappropriate, ill-sited technologies and then provide little or no information and training to the users about the operation and maintenance of these systems. I've learned a lot, and I've used what I learn to develop design performance standards for composting toilets. There's been a lot of change in this field, and I expect to see a lot more, as these technologies become more widely accepted as wastewater management solutions.

✲ ✲ ✲ ✲ ✲ ✲ ✲

We thank all of the people who contributed information and experiences to this book, especially the many composting toilet owners/operators who sent us photos, often at their own expense. These folks are true ecological heroes and innovators who walk their talk. Thanks, too, to the manufacturers and designers who supplied information and references for their systems.

Special thanks to Elva Del Porto, Dan Harper, Kevin Ison (whose "International Composting Toilet News" web site made it easier to access some European companies as well as Leigh Davison, whom we also thank), Dave Rapaport, George Anna Clark of Espacio de Salud, Clint Elston of AlasCan, Uno Winblad...and Real Goods, Jade Mountain, Chelsea Green Publishing Co. and others for being patient with us.

—David Del Porto and Carol Steinfeld

This book was written and designed solely by the authors, with editorial help and review from Patti Nesbitt and Deborah Dwyer.

We hope to continually update and expand this book, so we invite information, updates and comments from those designing and living with these systems.

Contact:
The Center for Ecological Pollution (CEPP)
P.O. Box 1330
Concord, MA 01742-01330 USA
Email: EcoP2@hotmail.com
Worldwide Website: http://www.cepp.cc and http://www.ecological-engineering.com/cepp/ctbook.html

Check the web site for new contact information, book updates and news about bulletins and newsletters on ecological engineering topics from CEPP.

Cover:
Clockwise from top left: Patti Nesbitt adds bark mulch to her Carousel composting toilet system (photo: Carol Steinfeld); the Pioneer, a self-contained composting toilet by Vera Miljø (photo: Vera Miljø); a urine-diverting toilet from Vera Miljø (photo: Vera Miljø); David Del Porto's graywater-irrigated greenhouse with fountain (photo: David Del Porto); and a twin-bin net toilet system on the island of Yap (photo: David Del Porto).

CONTENTS

Adapted from the cover illustration of The Integral Urban House (Sierra Books, out of print). Used with permission by Sierra Books.

CHAPTER ONE

Shift Happens!

Composting Toilet Systems Emerge as Viable Wastewater Alternatives

This book focuses on composting toilets (also known as dry, waterless and biological toilets and non-liquid saturated systems) because, among wastewater treatment technologies, they are one of the most direct ways to avoid pollution and conserve water and resources. Of course, most people who install composting toilets do so simply because they need to have a toilet system where a septic system cannot be installed.

Long used by developing countries, parks, off-the-grid homeowners, and cottage owners around the world, composting toilet systems are now making their way into mainstream year-round homes, for many reasons:

- Flush toilets are increasingly used with composting systems, making these systems more socially acceptable.
- More graywater (washwater) systems are emerging and getting approved.
- Increasingly, service contracts are available for maintaining composting toilet systems.
- Water shortages threaten at least one-third of the world. Some estimates place it at one-half.
- Many states are tightening on-site wastewater system standards, so that many of the United States' millions of septic systems are now considered inadequate, and therefore in noncompliance. As a result, many property owners are seeking ways to supplement their septic systems so they can avoid installing new ones. Diverting excrement and flush water from the flow removes more than 90 percent of the pollution, leaving only graywater to manage.
- Population densities are increasing in cities and coastal areas, intensifying the challenge of managing human waste.
- More people are converting vacation homes into year-round residences. These homes are often in remote and environmentally sensitive natural areas, such as seacoasts, lakes and mountains, with limited capacity for wastewater disposal.
- Individuals and institutions are increasingly interested in sustainable technologies, as the public's awareness of sustainability issues grows.
- A sewerless society? According to the United States Environmental Protection Agency and the United States Census Bureau, on-site systems are increasingly chosen over central sewer systems by property owners and municipalities because they cost less than a central sewer system. (USEPA, "Response to Congress on Use of Decentralized Wastewater Treatment Systems")
- Public health specialists at development agencies worldwide are promoting effective and ecological on-site waste treatment systems that save water and help prevent the spread of disease.

At the same time, the acceptance of composting toilet systems as a technology has grown tremendously. They are far more efficient, refined and proven. Every year, more states change laws and regulations to permit them. Even researchers at Harvard University have decided that this is the technology of the future, and have developed a high-tech prototype "smart" composting toilet with solid-state sensors and microchips that control the process.

Composting toilet systems are in place, have improved and are increasingly used worldwide.

The Big Flush

Every day, most of us use an average of 3.3 gallons (12.5 liters) of drinking-quality water to flush a toilet just once. At 5.2 flushes per day, the average American uses 6,263 gallons (23,705 liters) of drinking water each year to flush away 1,300 pounds of excrement. By the year 2000, that means 1,615,590,954,000 gallons (6 trillion liters) of water will be flushed away daily in the United States!

However, until recently, there has been a dearth of information about how to choose a system and how to maintain it. Meanwhile, regulations that pertain to these systems change monthly, as regulators learn more about separated blackwater and graywater systems. And decision makers have simply been unaware of the breadth of the wastewater problem, as much of the information about the relationship of nutrients and pathogens in excrement to disease and dying waters is buried in scholarly papers in scientific journals.

Now composting toilet technology and its regulatory and market climates are changing. The challenge of designing composting toilets is providing adequate control of the composting process—temperature, moisture, exhaust, perhaps mixing, etc.—at affordable prices. These costs are coming down. At the same time, the availability of service contracts makes this more of a user-friendly technology. In the future, it is likely that owner/operators will not maintain their own composting toilet systems unless they elect to do so. The U.S. Environmental Protection Agency and regulators worldwide are recommending the formation of on-site management districts in response to poorly maintained or inadequate conventional on-site systems. These would involve a central organization that manages a district's on-site systems, so no matter what system you had, an agency would be accountable for its performance. This also would allow on-site systems to receive the federal funds and financing that were once provided only for central wastewater treatment plants.

Thanks to these developments, composting toilets—long considered appropriate only for remote applications—may soon be widely viewed as a conventional wastewater treatment technology with obvious advantages for the present and the future.

Reference

USEPA, "Response to Congress on Use of Decentralized Wastewater Treatment Systems," 832-R-97-001b (Washington, D.C.: USEPA Office of Wastewater Management, April 1997)

Peter Rogers, Gordon MacKay Professor of Environmental Engineering; Professor of City Planning, Environmental Sciences; Harvard University, Cambridge, Mass.

The flush of the future? This family uses two micro-flush toilets that flow to a composter in their upscale house in Connecticut.

A Composting Toilet System Saves a House

When a western Massachusetts homeowner took steps to sell his home, he triggered a town review process that ultimately condemned the house.

His health agent discovered that his septic system was located in high groundwater and in a town water supply district—a clear pollution threat.

The Board of Health wouldn't allow him to install just a holding tank, which can be expensive to pump out (hence holding tanks have a history of "developing" holes of suspicious origin).

The only acceptable solution was a zero-discharge system, one that used up the home's effluent. And they had one in mind: an integrated composting toilet and wastewater system. In this case, an engineered garden system that manages household washwater would be enclosed in a greenhouse; built off the living room, it would offer the added benefit of passive solar heat. The home's three toilets would be replaced with micro-flush toilets that flowed to a composting reactor.

The homeowner ultimately got his asking price from a couple who were intrigued by the system. "The prospective owners love the idea," says David Del Porto, whose firm, Sustainable Strategies, designed the system. "They didn't realize that this was possible. They're really enthusiastic."

Carol Steinfeld, "Composting Toilets Emerge as Viable Alternatives," *Environmental Design and Construction,* June-July 1998)

A Brief History of Dry and Composting Toilet Systems

In the beginning, one simply excreted away from common paths. As societies formed and population densities increased, the unpleasant odors associated with excreta led to a managed approach. Pits in the ground that were used and then covered with earth were the earliest form of management.

At the dawn of the agricultural era, observation that plants grew faster where excrement was deposited on the soil led to adding it to soils to improve the yield of plants. In arid communities, one excreted in the fields where it would do the most good. It wasn't long afterward that the concept of portable nutrients, a.k.a. fertilizer, was born. As the value of the excrement as fertilizer was established, a market for this valuable commodity developed in every organized agricultural society. Technologies employed various means of collecting, managing and transporting the excrement-derived fertilizer.

The Chinese piled collected excrement, mixed it with straw, and inserted into the piles bamboo pipes pierced with aeration holes. The pipes aerated the piles and promoted composting. Composting transformed the excrement into a portable dry form that was easier to move to the field (and made the nutrients more "available" to the plants). Odors were reduced, making the whole process of collecting and transporting excreta a more efficient and pleasant task.

An early composting and drying system was developed in Syria more than 1,000 years ago. The urine was evaporated and the feces dried, and the somewhat decomposed fecal matter was collected and sold. There was even a currency based on the trade value of the excrement.

In China, elaborate roadside privy buildings were constructed to attract travelers, with the objective of collecting the resulting fertilizer with the least cost to the farmer.

The Excreta-Disease Link

It was not until the late 1700s that a connection between disease and feces was made and the concept of "sanitation" was born. Following that revelation, inventors, engineers and entrepreneurs moved to provide the public with all manner of devices to capitalize on this new industry.

Manufactured Composting Toilet Systems

Perhaps one of the first manufactured composting/moldering toilets was the "earth closet," a wooden toilet commode (see the sidebar on page 6) developed in England. Often made of polished woods, it was also a marvelous piece of furniture housing a collecting container into which a "flush" mechanism in back of the system released some soil onto each deposit. When it was full, the collection chamber was taken outside. Its inventor, Henry Moule, observed that the collected material resembled earth in little time. He likely did not understand the specifics of the composting process, and certainly this system could have been improved to encourage faster processing. However, it was a successful system that competed in the mid-1800s with the outhouse, until the ability of the water closet to immediately move wastes out of the house won out.

In the 20th century, the use of aerobic biological decomposition as a method for treating human excreta has been widely practiced. One of the earliest recorded applications was in India, where Sir Albert Howard developed theories of composting in the 1930s. Later, a double-chamber decomposing toilet system called the Gopuri was invented by Appasheb Patwardhan in northwest India in the 1940s. Another two-chamber system was used widely in rural areas in Vietnam in the 1960s. Aeration chimneys were added to both of these designs to remove odors. However, the chimneys also promoted composting, and the twin-chamber system was renamed "the Ventilated Improved Privy," or "VIP" privy. Thousands of these have since been built in Asia, Central America and Mexico.

In the 1930s, a Swedish engineer, Rickard Lindstrom, developed the Clivus Multrum, a concrete

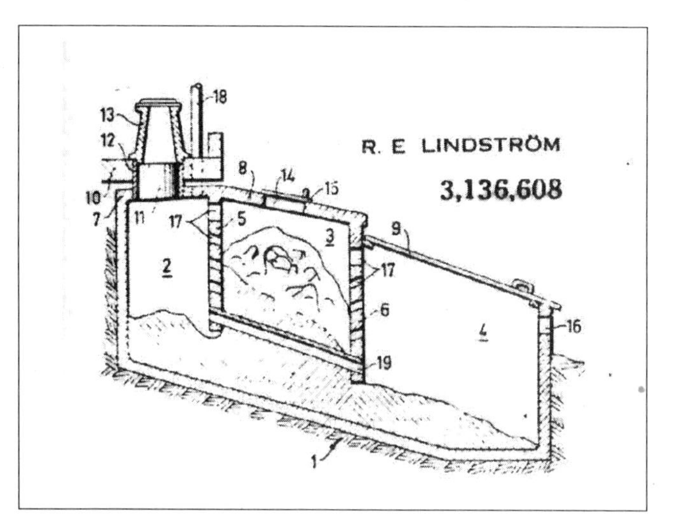

The original Clivus Multrum (Stop the Five-Gallon Flush, *1980)*

The Development of Domestic Wastewater Management Systems

This is a very general diagram of the evolution of types of wastewater systems.
The actual timeline and development sequence vary from location to location in the world.

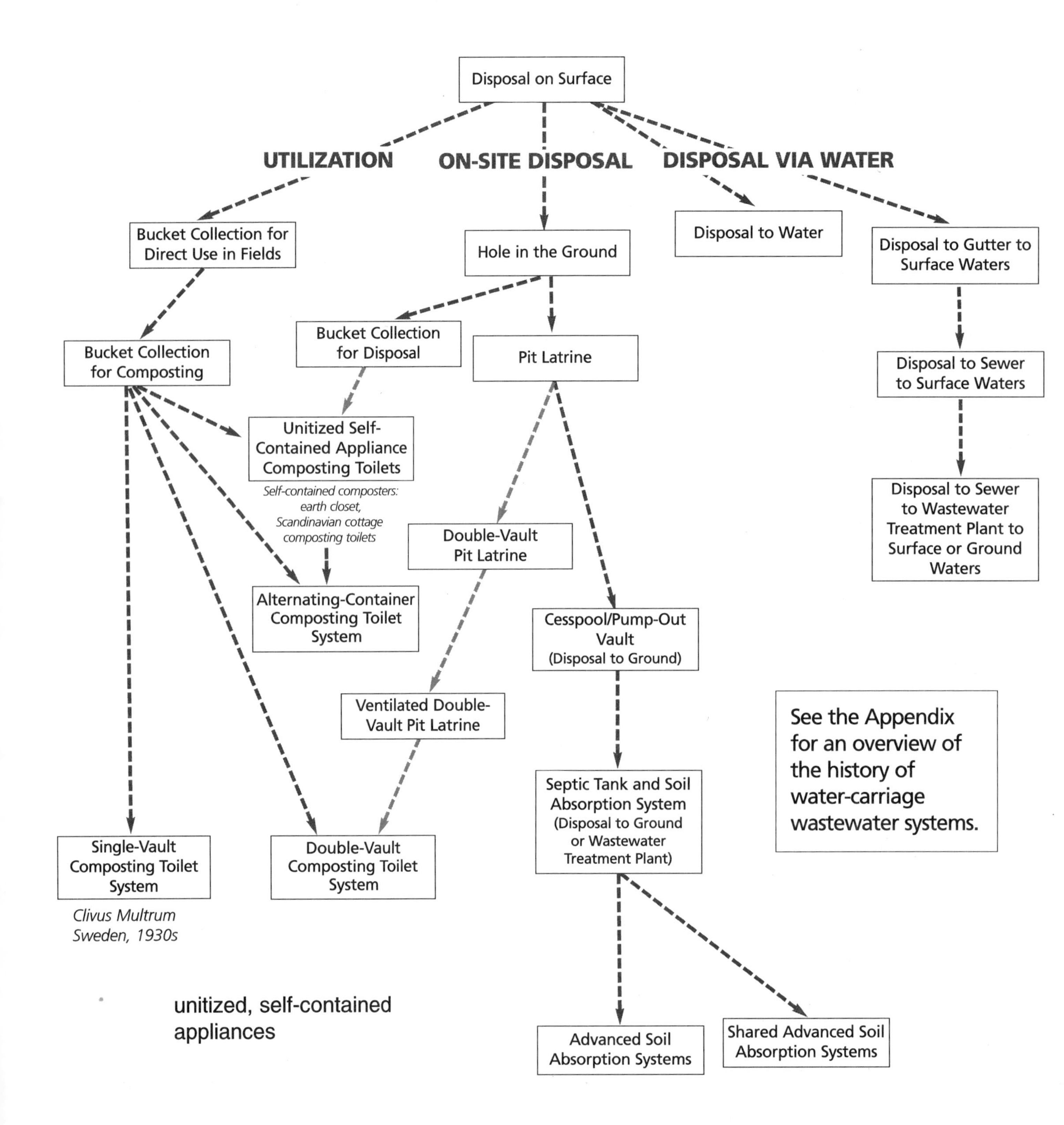

sloped-bottom single-chamber composter containing two baffles and air ducts. He patented his system in 1962, and in 1964, he fabricated out of fiberglass his first unit for market. Lindstrom first came upon the idea of this system in his youth, when cleaning out the stables at his family's farm. One day, his cart of manure tilted and spilled some of its contents onto a large angled rock in the sun. In the following weeks and months, he noticed that the manure slowly moved down the rock as it dried. In a month or two, it appeared to have transformed into soil.

Subsequently, several composting toilet systems appeared on the market in Scandinavia, where rocky soils, seashores, lakeshores and islands abound, making septic systems infeasible in many places. Considered an improvement on the alternating bucket system, many of these composting toilets were either alternating-bucket systems or heated and aerated chambers with some kind of mixing device (such as BioLet, early Sun-Mars, etc.). The Crown Prince of Sweden even held a contest for the most innovative nonpolluting technology. This generated several zero-discharge toilet system designs, among them one upon which several of today's cottage composting toilets are now based.

The Clivus Multrum came to the United States in the 1970s soon after an article about it appeared in *Organic Gardening* magazine. Abby Rockefeller, a member of the renowned Rockefeller banking family and a champion of ecological systems, licensed the technology and patents from Lindstrom, and established a North American manufacturing facility. Well-funded, promoted and marketed, Clivus Multrum came to be the best-known composting toilet in the United States. For many, this brand name still is synonymous with "composting toilet."

An Industry Gets Underway

By the late 1970s, composting toilets were enjoying a relative heyday in the United States. The passage of the Clean Water Act raised awareness of water pollution issues, and composting toilet systems were advertised by manufacturers as the quintessential pollution avoidance solution. Their literature suggested that there was little or no maintenance required, no odors, total destruction of pathogens, no liquids to manage, and only infrequent removal of a crumbly dry, earthlike material that would make flowers bloom and vegetable crops yield enormous bounty. Thousands were installed in the 1970s and 1980s.

However—as with all technologies early in their evolution—there were still some wrinkles to iron out, and not all of the systems were ready for the market. The manufacturers, knowingly or unknowingly, had oversold the product to their distributors. The capacities of systems were often overstated. Manufacturers and dealers withheld information about their maintenance aspects for fear that sales would suffer if people knew what was truly involved. Some design aspects needed to be fine-tuned. Sometimes the devices themselves were of poor quality, often with metal components that quickly corroded—electronic parts, such as switches and motors, failed prematurely in this "challenging" chemical environment. Sometimes the housing of the devices was not even watertight or strong enough for use.

At the same time, the public may not have been ready for these systems. Composting toilets represented a change in approach that required some social marketing, as well as support by the public and regulators. Having excrement in the house represented downward mobility to many. Some owners abandoned common sense and installed their systems in cold places or in nearly inaccessible locations.

In one instance, they were sold through government programs with no service information or support to remote Inuit tribes in Canada, where they were simply used as very expensive holding containers, then put outside when full.

Some owners experienced the reality of large single-chamber systems going anaerobic or compacting. Designs were developed by well-meaning nonprofit organizations and distributed widely to rural back-to-the-landers, who, with their bright-eyed willingness to deal with any problems that came up, suffered through unprocessed human excrement in the finished-product removal chambers, leachate draining through cracks, the presence of flies and other pests—and generally a most unsanitary condition.

As a consequence, a rash of "failures" came to light, often due to improper installations and inadequate maintenance (often through no fault of the user—this information was simply not provided). These failures were experienced throughout the world, wherever these products were distributed.

Then, in 1981, an evaluation by the California Department of Health of systems installed in homes in California reported and documented these problems. The problems illuminated by this report, often referred to as "The Enferadi Study," made it clear that, while this technology had exceptional possibilities of resolving on-site sanitation challenges, there were still issues of design, quality, installation and maintenance information

Manufactured Composting Toilet Systems Aren't So New: The Story of the Earth Closet

Adapted from a report by Wayne Collins

An earth closet is a dry toilet in which dry earth is used to cover excreta. Until about 100 years ago, the traditional "place of easement" for people living in the country was either a privy with a cess pit, or an earth closet. Because the earth closet was shallow, the decomposition process in it was aerobic, allowing the composting process to occur. This is likely the first manufactured composting toilet.

In Britain, Queen Victoria used an earth closet at Windsor Castle, although many types of water closet were available. For many years, the earth and water closets were rival systems, with champions and detractors on both sides.

Henry Moule, the vicar of Fordington in Dorset, championed the earth closet in the1800s. In 1861, he produced a 20-page pamphlet titled "National Health and Wealth, Instead of the Disease, Nuisance, Expense, and Waste Caused by Cess-Pools and Water-Drainage." "The cess-pool and privy vault are simply an unnatural abomination," he thundered. "The water closet... has only increased those evils." And he went on to describe his own amazing discovery.

In the summer of 1859, he decided his cesspool was intolerable and a nuisance to his neighbors. So he filled it in, and instructed all his family to use buckets. At first, he buried the sewage in trenches in the yard, one foot deep, but he discovered by accident that in three or four weeks "not a trace of this matter could be discovered." So he put up a shed, sifted the dry earth beneath it, and mixed the contents of the bucket with this dry earth every morning. "The whole operation does not take a boy more than a quarter of an hour. And within 10 minutes after its completion neither the eye nor nose can perceive anything offensive."

Then he discovered that he could recycle the earth and use the same batch several times, and he began to grow lyrical. "Water is only a vehicle for removing it out of sight and off the premises. It neither absorbs nor effectively deodorises.... The great...agent...is dried surface earth, both for absorption and for deodorising offensive matters." And, he said, he no longer threw away valuable manure, but obtained a "luxuriant growth of vegetables in my garden."

He backed up this last point with scientific experiments showing that potatoes nurtured with earth manure grew one-third bigger than those given only superphosphate.

According to Moule, doctors said that if his scheme could be generally adopted, "much more would be effected by it for the prevention and check of disease and sickness, and for the improvement of health, than Jenner has effected by the discovery of vaccination."

In about 1850, some people in England brought the earth closet inside the house, and various patented systems appeared. Moule produced a commode with a bucket below the seat and a hopper behind it containing fine dry earth or ashes. When you had finished, you pulled a lever to release a measured amount of earth into the bucket and cover the contents. Parker's Patent "Woodstock" Earth Closet had similar automatic mechanisms triggered either by the release of pressure on the seat—so that it "flushed" when you stood up—or by pressing a foot lever.

Moule tried hard to get government support, maintaining, "There can never be a National Sanitation Reform without active intervention by central government. That active intervention can never take place under the Water Sewerage System, without a large increase of local taxation. But let the Dry-Earth System be enforced... and with a vast improvement in health and comfort, local taxation may be entirely relieved."

The medical journal *The Lancet* reported that 148 of his dry-earth closets were used at the Volunteer encampment at Wimbledon—40 or 50 of them used daily by not less than 2,000 men—without the slightest annoyance in terms of aesthetics.

In 1865 the Dorset County School, with 83 boys, changed from water closets to earth closets, and cut annual maintenance costs dramatically. At the same time, odors and diarrhea were eliminated.

For some decades in the second half of the 19th century, the earth closet and the water closet were in hot competition. Almost everything Moule said was true, and much the same arguments are used today by the champions of composting toilets. The environmental considerations have not changed; using water closets is expensive, and merely shifts the problem downstream.

Henry Moule died in 1880, but to the end, he tried to persuade the British government that the earth closet was the system of the future, and he nearly succeeded. Nevertheless, in rich countries, because it rapidly removes the sewage from the house, the water closet has won the battle—so far...

Rev. Moule's Earth Closet, perhaps the first manufactured composting toilet. (Photo:Robert Forsberg)

to be addressed before this technology was truly ready for a broader market. (At the same time, however, this report held these composting toilet systems to a standard that many conventional systems could not meet.)

Composting toilets also received undeserved criticism and opposition from many plumbers, who mistakenly thought dry sanitation would rob them of a good portion of business. In addition, regulators, and even environmentalists, opposed these zero-discharge systems, which they thought would open the door to development in areas considered until then unbuildable due to soils that would not support septic systems (see sidebar).

To be sure, thousands of successful composting toilet systems and their happy owners remain from that time. But this history reminds us that any new technology that represents a major shift in a sometimes culturally sensitive and private part of everyday life requires a strategic introduction, with stress tests, education and regulatory support.

A Technology Comes of Age

More than a decade of evolution later, the performance of composting toilet systems is more commensurate with system designers' and manufacturers' claims. And more and more composting toilets are appearing in mainstream bathrooms, with both dry and flush toilets. The goal for the next millenium is to set up their users for success, which the field is now well equipped to do.

Composting toilet systems are a cost-effective, resource-efficient and increasingly accepted technology—an affordable sanitation system that can be transferred to the world's population at a cost that can be borne by the poorest members of the developing world, yet embraced by affluent North American homeowners who seek a solution to rising wastewater treatment costs.

Surprising Detractors: Environmentalists

Despite their ability to conserve and protect water resources, composting toilet systems have been opposed by some unlikely figures: environmentalists.

In many states, whether a lot is buildable or not depends on the land's capacity to handle wastewater disposed into the ground (a septic system). Some call this "toilet zoning." Many not-so-ecologically-minded developers embraced zero-discharge wastewater treatment methods, which could theoretically allow them to build on all of their wetland lots and other "undevelopable" properties.

Many environmentalists were—and still are—opposed to resource-conserving technologies because they could possibly allow building in environmentally sensitive areas. These environmentalists' well-placed concern, however, reminds us that there is a mechanism for restricting land use: zoning, wetland laws and conservation restrictions. Unfortunately, changing zoning to reflect the unsuitability of lands for high-impact building is opening a can of worms in the minds of planners, who are aware of the long legal battles that changing zoning can bring. While their concerns deserve anyone's sympathy, the continued practice of using polluting technologies is clearly not the solution.

A case in point is the state of New Hampshire: In the mid-1980s, manufacturers and local distributors convinced state regulators that composting toilets with reduced septic tanks and leachfields for graywater would work in environmentally sensitive areas. So, developers subdivided properties into small lots with small tanks and leaching areas. They told the buyers that once they got occupancy permits, they could take out the composters and install standard flush toilets. In fact, some actually put in the plumbing for the flush toilet and concealed it with a self-contained composting toilet! Later, the resulting hydraulic overload by the standard flush toilets and the solids and pollutants in feces, toilet paper and wastewater caused failures of these systems throughout the state. That chagrined regulators, and hardened their positions requiring full-sized septic tanks and leaching fields in all new construction, unless an engineered plan was approved. At the same time, the system plan is tied to the deed of the property.

Today, the increasingly recognized need for zero-discharge wastewater systems is prompting municipalities to creatively counter the potential machinations of unscrupulous developers through a variety of controls and restrictions. They realize that forbidding the use of water-conserving technologies is like "throwing out the baby with the bathwater"—and not the way to limit development.

CHAPTER TWO

The Big Picture: Managing Wastewater Sustainably

We can thank sewers, central wastewater treatment plants and septic systems for allowing us to live in relative cleanliness and for lowering the risk of disease in high-population areas. But their success is a matter of degree: They still fall short in their charge to protect public and environmental health. And they are getting expensive.

Essentially, we are using a valuable resource—drinking water treated and delivered at significant expense—to dilute and dispose of another potentially valuable resource, human excreta. To this we add industrial and household chemicals and stormwater drainage, then we pay a very high cost to transport this combined effluent to a facility which attempts to separate all of those constituents, clean them up to a degree mandated by federal law, and discharge the remaining water back into the environment—usually rivers, oceans and the ground. In most cases, the same nutrients and toxic chemicals that went into the wastewater mix are still present in what leaves the treatment plant. Growing realization of the effects of this is prompting regulators to mandate further treatment—and that's making this "combine-dilute-treat-and-dispose" approach very expensive.

However, even with ever-tightening requirements for advanced wastewater treatment, many of our waterways do not meet the goals of swimmable and fishable quality set by the 1972 Clean Water Act. And, while much progress has been made over the last 25 years, there are many who ask if there might not be a better way.

Every year we are learning more about the longer-term effects of our present approach to wastewater clean-up: partial treatment and disposal. The 1972 Clean Water Act only requires the reduction of suspended solids, biological oxygen demand (BOD)* and fecal coliform bacteria. But now, responsible regulators worldwide are mandating the removal of nutrients, toxic chemicals, parasites, viruses, radioactive wastes and other constituents.

At the same time, planners are asking: In a world where drinking water is increasingly expensive and scarce, can we use this valuable resource for flushing toilets?

It is clear that better ways are needed; the good news is that many are here and more are emerging. However, the answer is not merely a matter of more clean-up at the end of the sewage pipe. A larger, more strategic solution is called for, based on a broad approach:

- ***Prevent*** pollution
- ***Conserve*** water
- ***Separate/Divert*** constituents at their sources, and then
- ***Recycle, Reuse and Utilize*** it.

Plants in an indoor graywater bed in New Mexico. (Photo: Solar Survival Architecture)

Water, Water Everywhere?

Nearly half a billion people around the world face shortages of fresh water, and that number is expected to swell to 2.8 billion people by 2025 as the world population grows, according to a report by The Johns Hopkins University School of Public Health.

By 2025, one in every three of the world's projected 8 billion people will live in countries short of freshwater, the report said.

Associated Press, Aug. 26, 1998

*BOD is a measure of biodegradable carbon. The volume of BOD in water is a measure of the amount of oxygen that is going to be removed from a body of water in order to biologically degrade the carbon therein. The removed oxygen will no longer be available to higher-order animals, such as fish, which will die without enough dissolved oxygen.

Better Ways Emerging: From Disposal to Utilization

More regulations requiring better treatment and increasing costs for treatment, clean-up and water are prompting a reframing of the wastewater issue. In the current system, we create "wastes" that we want "disposed of." A better strategy is to put these outputs to use, just as they are in nature's model. In balanced ecosystems there is no waste: The outputs of one organism are the inputs of another.

Increasingly, constructed natural systems, such as artificial wetlands and sequenced aquatic technologies, are proving to be more reliable and cost-effective ways of managing wastewater and removing or utilizing its unwanted constituents.

A new discipline, ecological engineering, is developing and designing these systems—optimized, controlled, contained and monitored versions of natural systems—which prevent wastewater pollution by using potential pollutants as nutrients in constructed eco-systems, thus preventing wastewater pollution.

For treatment, the advantage of constructed natural systems is that they offer far more complex physical, biological and chemical processes than any of our present technologies. In natural systems, most of the "treatment" occurs in the plant root mass, or "rhizosphere." The rhizosphere serves as a dynamic bioplex, transforming the nutrients and complex organic compounds in wastewater into simpler forms the plants and microscopic animals use for energy and growth. These systems are robust and adaptive, adjusting to the changing nutrient and strength levels of effluent and ambient environmental factors.

"You can't solve a problem with the same kind of thinking that created it in the first place." *—Albert Einstein*

Most conventional central systems offer only secondary treatment: reduction of suspended solids and BOD with optional disinfection. Most on-site systems offer primary treatment: separating solids from liquids and some biological activity. Natural systems often perform that, as well as a third stage, or "tertiary" waste treatment: removing, sequestering or dissimilating toxic materials and using up the nutrients, so what flows out the other end is significantly cleaner. How clean is a function of

Top: A Living Machines Solar Aquatics system in South Burlington, Vermont, manages wastewater for 1,300 residential users.
Left: Specially selected plants in large aerated tanks provide a dynamic filter for microbial treatment—far more complex than mechanical filters—as well as nutrient removal. What's more, this facility is an attractive place to be.

These shrubs and ornamental plants are part of a Washwater Garden system that manages the water from a home's washing machine. (Photo: David Del Porto)

retention time, what and how many living organisms are at work, temperature and the constituents of the effluent.

Some systems use up the effluent entirely, so nothing at all leaves the system except plants or compost. These zero-discharge approaches are typically installed where sensitive environmental conditions and geological conditions prevent disposal solutions.

Ecological = Economical

Planners are also finding that natural systems can provide welcome amenities to their communities. In Arcata, California, a series of artificial wetland marshes provide tertiary wastewater treatment for this coastal community's wastewater treatment plant. They also serve as wildlife habitat and an enjoyable place for bird-watching along the walking trails between the marshes. In South Burlington, Vermont, a greenhouse filled with several tanks of a variety of plants and fish serves as a waste treatment system for 1,600 residents. In these systems, microbes and plants clean up the water.

These are both large central collection operations with integrated natural systems; on-site options include composting toilets and natural graywater systems that use up the effluent on site. In a Vancouver office building and in a Swedish apartment building, composting toilet systems process toilet wastes, while graywater is filtered and used to irrigate the landscapes around the buildings. In homes with composting toilet systems in Toronto and in Massachusetts, graywater is utilized by water-loving plants in planter beds and greenhouses, which are integrated into the homes.

Why Haven't We Changed Sooner?

Progress in innovation and the use of alternative technologies for wastewater treatment has been slow. Factors include:

☐ **Out of Sight...**

Our wastewater has been out of sight and often out of mind. As long as the public health bureaucracy said that what we were doing was good enough, there was little impetus for change. Now, high costs and health concerns are bringing wastewater back to the public's consciousness.

☐ **Prescriptive versus Performance Standards**

Septic systems have been the only on-site wastewater systems regulators would permit—the easiest approach from a regulatory standpoint. (Recycling wastewater, in fact, has been illegal in most states.) Now, regulations are increasingly establishing treatment performance standards. Technologies that can meet those standards will be permitted.

☐ **"Toilet Zoning"**

In many states, if a property's soils are not right for a septic system, you cannot legally build on it, unless there is sewer access. Although the intent of this is to protect the environment, some town planners use such regulations as surrogate zoning bylaws to control growth. Now, with the advent of zero-discharge wastewater technologies, community planners see this control removed, and they fear that development will run unchecked. One answer is national land-use planning—*not* forbiding nonpolluting wastewater treatment methods.

☐ **Cheap Drinking Water**

Until now, drinking water was inexpensive. Since our water sources were perceived to be relatively free of contaminants, the only costs for supplying water were for transmitting it and filtering it. But now, the new Federal Safe Drinking Water Act's more rigorous standards for water cleanliness require most communities to disinfect water. Chlorine has been the disinfectant of choice, because it is cheap. However, chlorine causes its own problems: Its byproducts, such as dioxin, and trihalomethanes, such as chloroform, are known carcinogens. Also, some organisms, such as cryptosporidium, are resistant to it, which has resulted in massive outbreaks of illness, including fatal instances in Cleveland and Detroit. These concerns are prompting some water treatment plants to switch to ozone and ultraviolet disinfection, and that is raising the price of water.

☐ **A Powerful Wastewater Industry**

Since the 1972 passage of the Federal Clean Water Act, a very powerful construction, engineering, manufacturing and government bureaucracy complex perpetuates the centralized collection and treatment model.

"Almost everything ends up in [water] sooner or later.... It's fundamentally moronic to let the water get dirty and then clean it up, as opposed to just keeping it clean... There are many reasons why water pollution is bad. It can make you sick right away, it can kill you right away, it can kill you slowly, it can make you sick slowly. And then, it can do all these same things to the environment."

—Peter Lehner, senior attorney and director of the Natural Resources Defense Council's Clean Water Program

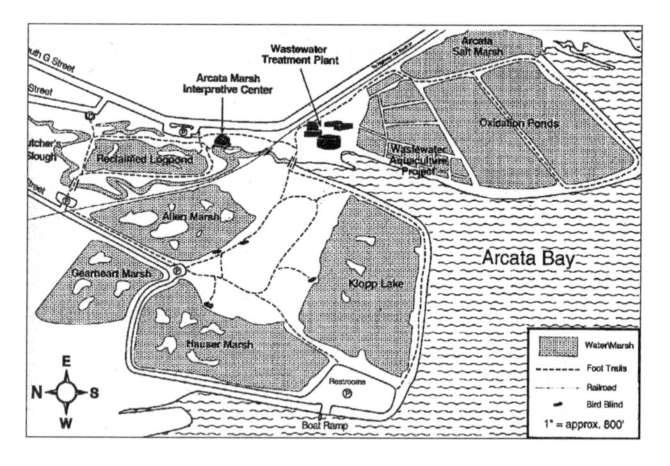

A Vision for...the Present

Perhaps one day, gardens, greenhouses and stands of trees will replace sewage outfall pipes and elevated mound systems. Imagine using our sewage in ecological tree plantations that provide fuel, fiber, construction materials, wildlife habitat and beautiful landscapes.

Picture integrated systems: A small pig farmer's wastewater irrigates a tree farm instead of polluting a river. Envision a community with both homes and commercial buildings in which graywater irrigates plants in a greenhouse, and blackwater is managed by composting toilets or is periodically taken to a central composting facility. All of this is financed, installed and maintained by community management districts—at a substantial savings over the costs of operating a sewage plant and cleaning up polluted waters.

The beauty of this future scenario is that it saves money and does a better job of protecting public and environmental health.

The solutions are here: Some do not easily slip into the place of septic systems or huge central wastewater treatment plants. Some do.

Are composting toilet systems the final answer to all of these problems and issues? Not necessarily. Composting toilet systems are a component of an integrated water-wise strategy. Nor are composting toilets in their current manufactured forms the end of their evolutionary line. When you think of them as aerobic biological systems, you see greater possibilities for this technology and realize this is about a method, not about strange-looking toilets used in cottages.

Top: A map of the Arcata Marsh and Wildlife Sanctuary shows the various marshes and treatment ponds that double as sanctuary and a component of Arcata's wastewater treatment plant.
Left: Walking trails weave between the marshes and are popular with joggers and birdwatchers.

A New Design Protocol for Better Water *Eco*-nomics

A better way to manage wastewater is a five-pronged approach that prevents pollution and treats effluents as potentially valuable resources.

1. Prevent pollution

If you don't create the problem, you don't have to pay for its solution. Avoiding pollution has better "eco-nomics" than waste treatment and disposal.

Examples: Use alternatives to toxic and troublesome chemicals, such as oxygen-providing peroxide-based bleach instead of toxic chlorine-based bleach.

2. Conserve water

Reduce the volume of wastewater by using water more efficiently.

Examples: Install low-flow faucets, showerheads, low-flush and waterless toilets, waterless urinals, high-efficiency clothes- and dishwashing machines and water-pressure-reducing valves. Irrigate with graywater. Use point-of-use water heating. Collect rainwater for nonpotable water use.

3. Divert/Separate wastewater components

Don't mix it all together to create a problem substance, wastewater, that has to be pulled apart at great expense later! Separate it at the source, not at the end of the pipe.

Examples: Keep excrement separate from graywater, and toxic household chemicals out of the mix entirely.

4. Pretreat wastewater

Prepare it at the source for use.

Examples: Filter laundry water to remove non-biodegradable lint and solids. Install grease/oil interceptors for kitchen effluents, hair/particle filters for bathing and washing, and composting toilet systems for excrement and organic residues.

5. Recycle, Reuse, Utilize

When possible, recycle, reuse or utilize effluents. Using them strategically, such as for landscape irrigation, helps save both water supply and wastewater treatment costs—and prevents effluents from becoming a problem.

Examples: Collect "warm-up" water while waiting for bathing water to warm up and use it to flush toilets or for laundry. Use condensate from air conditioners and dehumidifiers for the wash water in the laundry. Use final rinse water from the washing machine in the following first wash cycle. Use disinfected wash water for groundwater recharge. Utilization examples: Use effluents to irrigate landscapes or for evaporative cooling after pretreatment.

Just as "Reduce, Reuse and Recycle" has become the credo of responsible solid waste management, this five-pronged strategy will become more obvious and important to the world's population than can presently be imagined. As diminishing sources of quality drinking water and food are shared with an increasing populations, wasting any of the water and nutrients will become a moral as well as an economic imperative.

A graywater-fed planter bed in a New Mexico house. (Photo: Solar Survival Architecture)

CHAPTER THREE

What Is a Composting Toilet System And How Does It Compost?

What Is a Composting Toilet System?

Composting toilet systems (sometimes called biological toilets, dry toilets and waterless toilets) contain and control the composting of excrement, toilet paper, carbon additive, and, optionally, food wastes. Unlike a septic system, a composting toilet system relies on unsaturated conditions (material cannot be fully immersed in liquid), where aerobic bacteria and fungi break down wastes, just as they do in a yard waste composter. Sized and operated properly, a composting toilet breaks down waste to 10 to 30 percent of its original volume. The resulting end-product is a stable soil-like material called "humus," which legally must be either buried or removed by a licensed septage hauler in accordance with state and local regulations in the United States. In other countries, humus is used as a soil conditioner on edible crops.

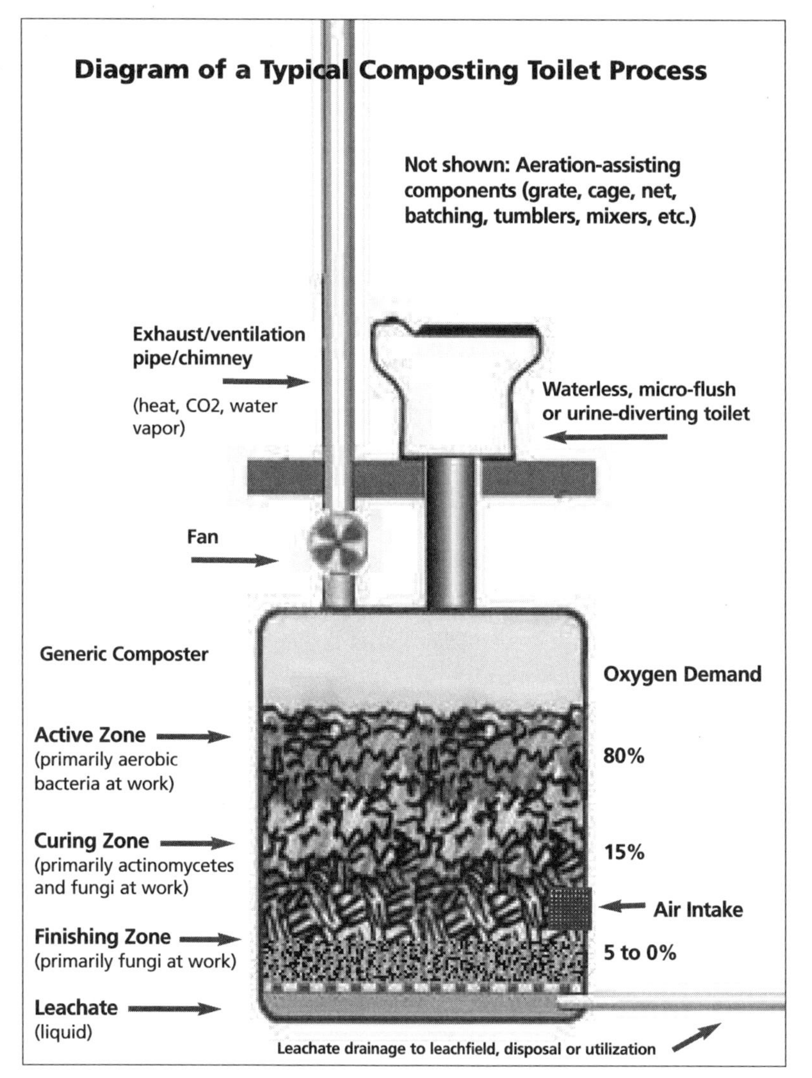

Composting Toilet as Biological Filter

One should consider the composting toilet a biologically active filter that diverts matter out of the wastewater stream and reduces it by microbial digestion. When combined with gray-water filters, such as grease traps for the kitchen, lint filters for the clothes washer and hair traps for bathing, the total integrated system replaces the separating functions performed by a septic tank or a municipal treatment plant. It also removes more than 95 percent of the total solids and most of the biological oxygen demand (BOD). Nitrogen is reduced by more than 80 percent (it's in the urine) and pathogens are all but eliminated.

The primary objective of the composting toilet system is to contain, immobilize or destroy organisms that cause human disease (pathogens), thereby reducing the risk of human infection to acceptable levels without contaminating the immediate or distant environment and harming its inhabitants.

This should be accomplished in a manner that

- is consistent with good sanitation (minimizing both human contact with unprocessed excrement and exposure to disease vectors, such as flies).
- produces an inoffensive and reasonably dry end-product that can be handled with minimum risk.
- minimizes odor.

A secondary objective is to tranform the nutrients in human excrement into fully oxidized, stable plant-available forms that can be used as a soil conditioner for plants.

The main components of a composting toilet are:

- a composting reactor connected to one or more dry or micro-flush toilets;
- a screened exhaust system (often fan-forced) to remove odors, carbon dioxide, and water vapor (the by-products of aerobic decomposition);
- a means of ventilation to provide oxygen (aeration) for the aerobic organisms in the composter;
- a means of draining and managing excess liquid and leachate;
- process controls, such as mixers, to optimize and manage the process; and
- an access door for removal of the end-product.

The composting reactor should be constructed to separate the solids from the liquids and produce a stable, humus end-product with less than 200 most probable number (MPN) per gram of fecal coliform bacteria.

General Types of Composting Toilet Systems

Composting toilet systems can be classified in several ways:

Self-Contained versus Centralized

Composting toilet systems are either *self-contained,* whereby the toilet seat and a small composting reactor are one unit (typically small cottage models), or *centralized* or *remote,* where the toilet connects to a composting reactor that is somewhere else.

Manufactured versus Site-Built

One can either purchase a *manufactured* composting toilet system or have a *site-built* composting toilet system constructed (however, the latter can be difficult to get permitted by local health agents).

Batch (Multiple-Chamber) versus Continuous (Single-Chamber)

Most composting toilet systems use one of two approaches to manage the composting process: either *single-chamber continuous* composting or *multi-chamber batch* composting processes.

A continuous composter (including Clivus Minimuses and such brands as CTS, Clivus Multrum, Phoenix, BioLet, Sun-Mar) features a single chamber into which excrement is added to the top, and the end-product is removed from the bottom. (Systems that feature drawer for removing the end-product at the bottom of a single chamber are only batch systems if the drawer's contents are not exposed to further urine, leachate or unprocessed fecal matter, and instead allow the material to process without new additions.)

A batch composter (such as the EcoTech Carousel, all Vera systems, BioLet NE, and many site-built composters) utilizes two or more interchangeable composting reactors. One is filled at a time, then allowed to cure while another reactor fills, just as with two- and three-bin yard composters.

Proponents of continuous composting maintain that it is simple (takes place in one fixed reactor), allows urine to constantly moisten the process, and allows the center of the mass to heat up through uninterrupted microbial activity.

Advocates of the batch-composting approach say that by not continually adding fresh excrement and urine to older, more advanced material, the material composts more thoroughly, uninterrupted by the added nutrients, pathogens, salts and ammonia in fresh excrement. Also, by dividing the material, it can have more surface area, and thus better aeration. Batch composting also offers an opportunity for unlimited capacity, as one simply adds compost reactors to the process to add capacity.

Batch systems require monitoring the level of the composter to determine when a chamber has filled and a new one must be moved into place. However, because there is more surface area and the material is divided, there is often less or no mixing and raking of the material.

What is true is that complete composting needs time to ecologically cascade all the by-products of the myriad organisms in the composting food web until all of the organic matter is finally transformed into safe, stable humus.

In continuous composters, urine or flush water can leach fresh excrement into the finished compost removal

A self-contained composting toilet

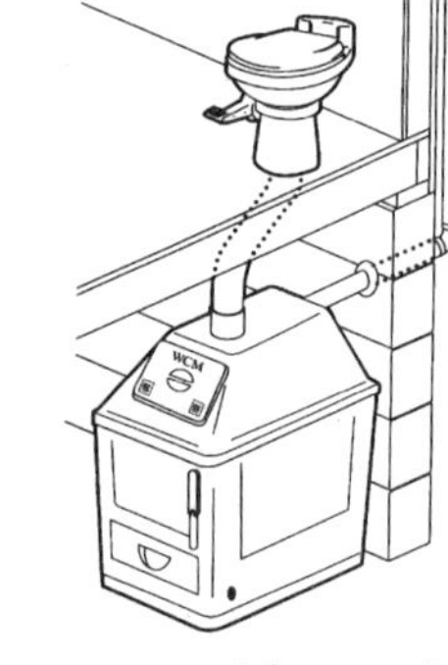

A central (remote) composting toilet

(Illustration: Sun-Mar Corp.)

The Cast of Characters in a Composting Toilet System

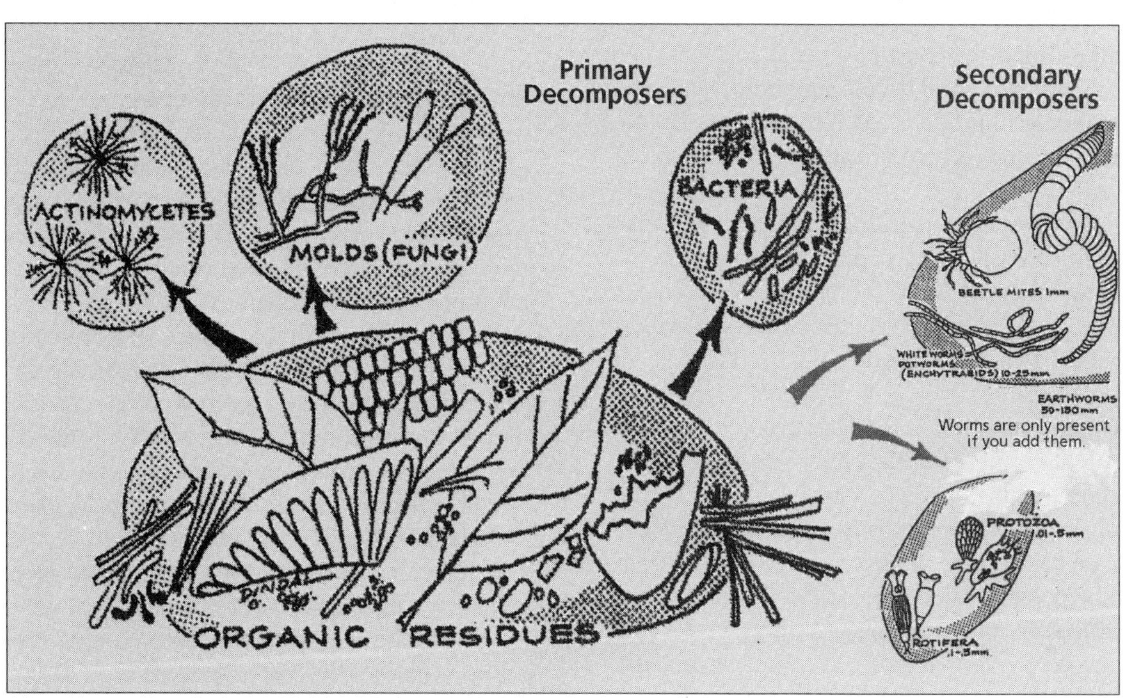

Adapted from an illustration by D.L. Dindal, Soil Ecologist, SUNY College of Environmental Science and Forestry, Syracuse, NY.

This illustration shows the key players in the composting process, as well as some secondary composting organisms and helpers that may or may not be present in a composting toilet system.

area. Segregating the material into batches reduces the risk of having living disease organisms in the finished product.

(One of the authors [Del Porto] has designed, sold and serviced both types of composters. He prefers batch composting-based systems, because they are more forgiving and users seem to be happier with them. However, he would never say that single-chamber continuous systems do not work. They do!)

It is difficult to generalize about which process affords the greatest opportunity for complete processing and minimizes the potential for pathogen survival. In a batch system, a finite supply of nutrients is cycled and recycled through microbe populations until the nutrients, both the free ones and those bound in microbial protoplasm and cell walls, are ultimately converted to stable, fully oxidized forms, and the fungi have performed their work on the remaining lignin and cellulose compounds, releasing antibiotics in the process.

Definitive research is needed in this area.

Active versus Passive

As with solar energy systems, composting systems are usually either *passive* or *active*. Passive systems are usually simple moldering reactors in which ETPA (excrement, toilet paper and additive) is collected and allowed to decompose in cool environments without active process control (heat, mixing, aeration).

Active systems may feature automatic mixers, pile-leveling devices, tumbling drums, thermostat-controlled heaters, fans, and so forth. The trend in the composting of municipal solid waste (garbage and trash), sewage sludge and yard and agricultural residues is toward active systems. By making the process active, the size of the composter can be reduced, because composting efficiency is speeded up (and the volume of the material reduced faster).

Passive systems are designed to optimize the process by design, not mechanical action, allowing only time, gravity, ambient temperature and the shape of the container to control the process. Passive composters are often referred to as moldering toilets, as the process at work is natural uncontrolled decay at cool in-ground temperatures at or below 68°F (20°C). In this cool environment, molds (fungi and actinomycetes) are the primary biological decomposers, because it is a bit too cool for the faster-acting mesophilic and thermophilic bacteria.

What Is Composting?

Composting is the controlled aerobic (atmospheric oxygen-using) biological decomposition of moist organic* solid matter to produce a soil conditioner. Because it requires molecular oxygen, it cannot be immersed in water (saturated).

The emphasis is on "controlled." This sets it apart from the uncontrolled decomposition that occurs in the natural environment: A leaf falls from a tree branch to the forest floor, and microbes transform it into a nutrient form that the tree can consume. The biochemical process is the same.

In a composting toilet, the objective is to transform potentially harmful residuals—mostly human excrement—into a stable, oxidized form.

The primary microorganisms responsible for composting are bacteria, actinomycetes and fungi. However, algae, mixomycetes (slime molds), viruses, lichens and mycoplasmas are other organisms present in the compost process. Soil animals, such as protozoa, amoeba, nematodes, earthworms and arthropods, also perform major roles by degrading surface litter, consuming bacteria and assisting in aeration.

**Biologically derived carbon-containing*

To Get More Technical...

Composting microorganisms produce enzymes that are extracellular—like a chemical aura outside the organism's body. They transform molecules of organic matter into less complex chemicals for microbial growth and energy. All organisms have intracellular enzymes to manage the diverse complexity of the life process. Enzymes are unstable proteins or protein-containing compounds that, when present in small amounts, promote a chemical reaction. The enzymes—such as amylase, cellulase, lipase and protease—are a few of the catalysts responsible for decomposition.

The solid, liquid and gaseous environment of the bioplex that forms a composting system provides a continuum of microniches for the organisms, and that affects the rate of biological transformation of organic matter.

How fast organic matter composts is affected by (1) environmental factors, (2) the composition and constituents of materials being composted, (3) the health and number of the organisms which are using the materials as a food source and (4) the management of the process by the operator.

Composting is an "unsaturated" aerobic process. This means that the material being composted cannot be

What's So Good about Compost?

"It is ironic that composting, so lately embraced in many economies, is one of the oldest forms of recycling known to humankind. As societies become reacquainted with this practice, its value as a natural solution to problems, from overflowing landfills to anemic soils, will become apparent. Then, with the proper institutional and economic incentives, composting could become as commonplace as the recycling of cans, newspapers, or paper is today."

—Gary Gardiner, *Recycling Organic Waste: From Urban Pollutant to Farm Resource*

Organic materials (defined as that which is derived from life processes and have carbon in them) contain minerals, other chemicals and nutrients.

However, applying raw organic materials directly to the soil is not the best way to use this organic matter and its nutrients. That's because when nutrients in materials are decomposing slowly, they are not available for use by plants. Also, decaying organic matter can tie up soil nitrogen that would otherwise fuel plant growth. That's the value of composting: It converts organic material into a stabilized product that builds soils and releases nutrients in a steady way.

Composting can take place in a matter of hours or years—it depends on the process factors described later. Essentially, bacteria and other organisms feast on carbon-rich matter and digest it, producing humus, a rich, stable medium in which roots thrive. Worked into soils, humus builds soil structure and provides a productive environment for plants and essential soil organisms. For our purposes, however, the primary value of composting is its ability to reduce waste

It also builds soils. Because it is riddled with pores, the humus in compost shelters nutrients and provides extensive surface area to which nutrients can bond; indeed, humus traps three to five times more nutrients, water and air than other soil constituents do. These characteristics also help retain nutrients that could otherwise be leached or eroded away. In this way, adding organic matter to soils further reduces the need for additional nutrient applications.

Compost has also been found to aid in suppression of plant diseases, often reducing or eliminating the need for fungicides. Compost releases plant nutrients gradually—consider it a time-release vitamin pill. It holds moisture. And it's cheap.

The nutrient value of composted yard wastes is modest, amounting to about 10 percent of the nutrients applied as fertilizer. The nutrient content of composted human waste, which is highly nitrogenous thanks to urine, is equal to that of many chemical fertilizers.

immersed in liquid, as that would fill the void spaces (the pores) in the composting mass, and prevent oxygen-carrying air from reaching the organisms digesting the food source. If the material becomes saturated, soon the remaining dissolved oxygen in the liquid is consumed. When there is no dissolved or free oxygen, that condition is said to be "anoxic." When the anoxic condition persists, "anaerobic" organisms, which cannot use molecular oxygen, will take over. This process is called "anaerobiosis," and is typified by offensive odors, such as that of rotten eggs, from sulfides, amines and mercaptans, and flammable methane gas produced by anaerobic bacteria.

Composted human waste from a BioLet NE chamber is emptied to be trenched into a flower garden in Sweden.

Success Factors for Composting Toilet Systems

The successful operation of a composting toilet depends on several factors that must be maintained within certain broad ranges.

Beneficial Organisms and Available Nutrients

In a composting toilet, natural soil organisms act as "Living Machines" (a concept first articulated by Dr. John Todd) to decompose excrement into safe and valuable by-products. So, make sure there is a large population of bacteria, actinomycetes, fungi, yeast, algae, protozoa and other organisms.

To inoculate the composter with soil microbes, add two handfuls of sifted compost from a warm yard compost pile* or successfully composted human waste. Sometimes a scoop of rotting leaves from the forest will do it, too. This will provide all the microbes for start-up. One gram of healthy soil may support 500 million bacteria, 20 million actinomycetes, 900,000 fungi, 100,000 yeast, 500,000 algae and 500,000 protozoa. (*Soils for the Management of Organic Wastes and Wastewaters*)

Illustration: Center for Clean Development

Or, you can purchase compost activators from a number of suppliers, including most garden centers.

How many and how effective they are depend on the environment provided, as well as pH and food supply. Human excrement and other organic material that you put in the toilet feed these microbes. The pH is self-regulating, if all of the other conditions are satisfactory.

Environmental Factors

Set the stage for the microbes to do the job. Variations in environmental conditions in the composter directly affect the various populations of organisms, increasing some and decreasing others.

* Note that adding composted material from outdoors may introduce fly larvae to the system.

The Aerobic Decomposition Agenda

In composting:

- Below 41°F (5°C) is biological zero—little to no active processing takes place.
- From 42° to 67° (6°-20°C), ***psychrophilic*** (moldering) processing takes place.
- From 68° to 112° (21°-45°C), ***mesophilic*** bacteria are dominant. These are the typical bacteria in a composting toilet.
- From 113° to 160° (46°-71°C), ***thermophilic*** bacteria take over, and push the process to the limit.

1. Aeration

Take a deep breath. Let it out. Aahh. Your composter needs to do that, too, as the aerobes require free atmospheric or molecular oxygen. If there is an oxygen deficit (a state called "hypoxia"), the aerobes will die. They will be replaced with anaerobes (organisms that can exist only in the absence of molecular oxygen), which will slow the process and generate odors (thanks to their production of hydrogen sulfide, ammonia and amines) and potentially flammable methane gas.

The ventilation system in a composter should draw sufficient air across and through the ETPA.

A key factor, then, is the surface-area-to-volume ratio of the composting substrate (which includes the microbial population), because surface area allows direct contact with oxygen. If the volume of the composting material is greater than the surface area, then oxygen may not reach the microbes, and the process will be limited by this lack of oxygen. Mixing, tumbling, forced aeration and container design are ways composters provide a good surface-to-volume ratio.

To make the composting process work best, the materials being composted should have a loose texture to allow air to circulate freely within the pile. If the material becomes matted down, compacted or forms too solid a mass, the air will not circulate, and the aerobic organisms will die.

- Add bulking agents, such as wood chips, stale popped popcorn, etc., to increase pore spaces that permit air to reach deep into the biomass and allow heat, water vapor and carbon dioxide to be exhausted. Earthworms also create pores, as well as help break down wastes.
- Maintain adequate air flow through the material by proper ventilation (i.e., pressurized air, using convection or forced air by a fan) and/or by frequently mixing.
- Provide aerators, such as mixers, mesh, grates, air channels or screened pipes to help increase the surface area of the composting mass that is exposed to air.

However, too much air flow can remove too much heat and moisture—make sure your composter is not too cool and dry as a result.

2. Moisture Content

The microbes in the composter need the right amount of moisture to thrive. Too much water (saturated conditions) will drown them, and create conditions for the growth of odor-producing anaerobic bacteria. In optimum conditions, the composting mass has the consistency of a well-wrung sponge—about 45 percent to 70 percent moisture.

When the moisture level drops below 45 percent, it can become too dry for composting. Also, excrement, toilet paper and additive will dry out but not decompose, thus prematurely filling the toilet (a good indicator that the mass is too dry). Dehydrating toilets are becoming popular in developing countries, but they do not stabilize the excrement, and that could lead to trouble.

If the moisture level is higher than 70 percent, leachate will pool at the bottom of the composter. You will have to drain it or evaporate it, because it will drown the microbes.

Urine and/or water from micro-flush toilets contributes most of the moisture in a composter, and may not be distributed evenly over the mass. (Note that urine-separating toilets are now available to reduce the leachate problem.)

Also, in composting toilets with heaters at the bottom, the upper parts of the biomass may become too dry.

In some systems, the drained leachate is resprayed over the top to prevent dehydration. This is not the best practice, as the concentrated salt and ammonia from urine are toxic to the beneficial bacteria and other organisms that are composting. Fresh rainwater (which has little or no dissolved minerals) is best for moisture control, but fresh groundwater from the tap (which contains significant dissolved minerals) will do as well. If the material is too dry, spray the compost mass with water, or add a cup of water periodically. Or, connect the dryer duct exhaust vent to the composter to contribute warm, moist vapor.

3. Temperature

The ambient temperature for acceptable biological decomposition is 68° to 112°F (20° to 45°C). Biological zero is 41°F (5°C), the temperature at which almost no microbial respiration occurs. At this temperature, most microbes cannot metabolize nutrients.

In most composting toilet systems, mesophilic (68° to 112°F/20° to 45°C) composting is at work. The heat generated by these microbes is usually lost through the vent stack, so composting toilets rarely reach thermophilic rates (113° to 160°F/46° to 71°C), which support thermophilic bacteria. This is the hot composting that takes place at the core of active yard waste composters (that's what generates the steam you see rising from the compost on a cold day). Achieving thermophilic composting would require either heating the composter—which could be expensive—or retaining the heat better by insulating or venting it less, which might mean odors and insufficient oxygen. In the highly contained environment of this kind of composter, it's a hard balance to reach.

Moldering toilets support psychrophilic organisms, whose optimum temperature is above 41°F (5°C) and below 68°F (20°C). These are predominately fungi and actinomycetes bacteria such as *Streptomyces griseus*, which produces the antibiotic streptomycin. Moldering systems are sized much larger than mesophilic composting systems to compensate for their reduced processing rate.

Moldering is usually the final phase after mesophilic and thermophilic processes have completed the early work of degrading sugars, fats and proteins. As the process cools, fungi and actinomycetous bacteria slowly digest the cellulose and lignin in plant matter, such as wood chips and toilet paper.

Most small manufactured compost toilets have heaters and thermostats to maintain an internal temperature of 90° to 113°F (32° to 45°C) to support the upper mesophilic composting range, while evaporating excess leachate. Evaporation of leachate tends to drop the temperature of the composter, through evaporative cooling, thus requiring continuous inputs of energy (heat) to maintain acceptable temperatures.

Generally, the rate of processing in a biochemical system is directly proportional to the increase of temperature (within certain limits, the rate doubles with every 18°F/8°C increase). The warmer the process, the more capacity in a composter. The cooler the process, the slower the rate—and more time or volume capacity may be needed for processing.

A composter at or below 41°F/5°C will only accumulate excrement, toilet paper and additive until the temperature rises. That is why composter manufacturers state their capacities at 65°F/18°C (comfortable room temperature of an average human-occupied space).

In the earlier days of the composting toilet field, it was thought that the heat generated by the composting process would be enough to evaporate the leachate. (In fact, some optimistically saw the composter as a household furnace!) Alas, it was not to be, as much of this heat goes up the exhaust pipe.

For more on heat sources, see Chapter 4.

4. Carbon-to-Nitrogen (C:N) Ratio

While an important relationship to remember for aerobic bacteria nutrition is the carbon-to-nitrogen—the "C:N"—ratio, of the food source, its significance in composting toilets is often overstated.

Microorganisms require digestible carbon as an energy source for growth, and nitrogen and other nutrients, such as phosphorous and potassium, for protein synthesis to build cell walls and other structures (in the same way humans need carbohydrates and proteins). When measured on a dry weight basis, an optimum C:N ratio for aerobic bacteria is 25:1.

Primarily due to the high nitrogen (from urea, creatine, ammonia, etc.) content and low carbon (glucose) content, urine has a low C:N ratio (0.8:1). Therefore, if the objective is to oxidize all of the nitrogen urinated into the toilet, this would require adding digestible carbonaceous materials on a regular basis. (See the section on "Urine" in Chapter 4). However, the practical fact is that urine, which contains most of the nitrogen, settles by gravity to the bottom of the composter, where it is drained away or evaporated. In either case, the nitrogen passes through the ETPA and is lost to the process! For that reason, adding large amounts of carbon will not help process the nitrogen, and will just fill up the composter faster.

The primary reason then to add carbon material is to create air pockets in the composting material (that's why some call carbon additive "structure material" or "bulking agent").

Digestible carbonaceous materials include carbohydrates (sugar, starch, toilet paper, popped popcorn), vegetable or fruit scraps, finely shredded black-and-white newsprint, and wood chips. A small handful of dry matter per person per day or a few cups every week is a good rule of thumb to absorb excess moisture, create pores in the composting material, and maintain a helpful C:N ratio. For more on additives, see Chapter 6.

5. Process Control for Management

Process control is how you optimize the composting process by controlling the external variables that affect the process, as described above.

The following are some process controls that may be in a composter:

- Motorized and manual mixing or turning provides aeration and moves microbial communities into contact with unprocessed ETPA.
- Blowers and fans remove odors and gasses that are the by-products of composting, such as carbon dioxide and water vapor. Fan-speed controllers optimize the efficiency of the fan.
- Heaters maintain optimum temperature for the microbes and evaporate the water from leachate.
- Pumps are used to move leachate to a management system. Others are used to spray water over the mass.
- Warning indicators and alarms tell the manager when something needs attention.

Some compost system designers are integrating sensors and microprocessors that trigger and manage these process controls (so you don't have to). These are common in larger processing plants for municipal solid waste, sludge biosolids and agricultural residues; they monitor temperature, carbon dioxide production, and oxygen and moisture content of the composting mass. Some of these are being integrated into AlasCan composting systems. Harvard University designed a prototype of such a composter (but it is still far too expensive for production). Small manufactured composters have yet to become so sophisticated, but it is just a matter of time before these sensors and automatic process controls are part of most commercial composting toilets.

Pathogens and Composting Toilet Systems

Properly designed and maintained, composting toilet systems should contain, immobilize and/or destroy pathogens, organisms that cause human disease.

Still, when planning systems, pathogens and vectors need to be considered. This is particularly a concern in areas with endemic diseases, such as warm regions, where pathogens flourish.

When discussing human excrement systems, the diseases we are concerned about include amebiasis, cholera, cryptosporidiosis, gastroenteritis, infectious hepatitis, parasite-related disease, salmonellosis, shigellosis, typhoid fever,and other diarrheal diseases.

How Pathogens Are Attenuated, Immobilized Or Destroyed in a Composting Toilet System

Whereas conventional wastewater treatment technologies depend on chemical or thermal disinfection to reduce pathogens, in a composting toilet it is accomplished by the following:

1. Containment

Pathogens cannot survive for long once they have left the human host. They have co-evolved over thousands of years with the human race and can thrive only within the narrow chemical and environmental parameters of the human body. Like all organisms, human pathogens have specific lifetimes. An organism's lifetime is shortened in the hostile environment of an aerobic composter. Human pathogens in a composter are like fish out of water: They don't live for long. Containing the excreta for an extended period of time brings about the death of pathogens and reduces the risk of infecting new hosts through ingestion, the primary pathway for enteric pathogen transmission.

2. Competition

The competition among composting organisms for available carbon and other nutrients is intense. Human pathogens become food for the well-adapted aerobic soil organisms that thrive in the composter. When the available nutrients are consumed, the microorganisms begin to consume their own protoplasm to obtain energy for cell maintenance. When this occurs, the microorganisms are said to be in the "endogenous phase." When these organisms die, their protoplasm and cellular matter is digested by other organisms. Eventually, if no new food sources are presented, all of the energy will be released and the matter fully oxidized. The end of this phase results in an end-product that is very stable and safe.

3. Antagonism

Some composting organisms produce toxic substances which harm, inhibit or kill other organisms. For example, the actinomycete *Streptomyces griseus* produces streptomycin, a well-known antibiotic. The soil bacteria *Bdellovibrio bacteriovorus* parasitizes the infamous *Echerichia coli* (*E. coli*), and multiplies within the host cell, eventually killing it. (*Soils for the Management of Organic Wastes and Wastewaters*)

4. Adverse Environmental Factors

Factors such as pH, temperature, moisture and ammonia content also play roles. Note: Temperatures above 131°F (55°C) do kill pathogens in a short time, but composting toilet systems do not attain these temperatures unless the process is artificially heated.

Other: Pasteurization

An even faster and controlled way of killing or reducing pathogens is pasteurization. This method is derived from the work of Louis Pasteur (1822–1895), a French researcher who discovered that pathogens were destroyed by heating the matter in question for one hour at 62°C (143.6°F), then cooling it rapidly to prevent re-inoculation. Pasteurization can be accomplished by a variety of means, but the most common is by applying sufficient heat from external sources such as electric or propane heaters. In some cases, the use of microwave or solar energy is used.

What Are Pathogens?

Pathogens, or pathogenic organisms, are responsible for the transmission of communicable diseases. They are generally bacteria, viruses, amoebae or protozoa and parasites, such as worms that invade the body and cause illness by a variety of means that overwhelm the immune system and damage or destroy living tissue.

Excreta, primarily the feces of human and domestic animals, can contain pathogens. Each pathogen has its own life cycle. For some, that includes a "saprophytic" or non-host period, when it is viable after exiting its human host and transmission from one host to another can occur. This is the same stage in which these pathogens can be killed or immobilized. Long-term survival of pathogens outside a host is rare, because they are outside their best environment. They usually need to be put into an aqueous environment to live longer.

Fortunately for general public health, only a few of the hundreds of pathogens have high enough survival rates in an aerobic environment (like the composter) that they can play a significant role.

Typical Pathogen Survival Rates at 20° to 30°C in Various Environments*

	Survival Time[†] in Days		
Pathogen	**Freshwater and wastewater**	**Crops**	**Soil**
Bacteria			
Fecal coliforms[‡]	<60 but usually <30	<30 but usually <15	<120 but usually <50
Salmonella (spp.)[‡]	<60 but usually <30	<30 but usually <15	<120 but usually <50
Shigella[‡]	<30 but usually <10	<10 but usually <5	<120 but usually <50
Vibrio cholerae[§]	<30 but usually <10	<5 but usually <2	<120 but usually <50
Protozoa			
E. histolytica cysts	<120 but usually <15	<10 but usually <2	<20 but usually <10
Helminths			
A. lumbricoides eggs	Many months	<60 but usually <30	<Many months
Viruses			
Enteroviruses[‡]	<120 but usually <50	<60 but usually <15	<100 but usually <20

*Adapted from Feachem et al. (1983).

[†] Includes polio, echo, and *Coxsackie* viruses

[‡] In seawater, viral survival is less, and bacterial survival is very much less than in fresh water.

[§] *V. Cholerae* survival in aqueous environments is a subject of current uncertainty.

(Ron Crites and George Tchobanoglous, *Small and Decentralized Wastewater Management Systems* (United States: McGraw-Hill, 1998).

How are they transmitted?

Pathogens are carried to new hosts from infected persons' excrement by:

- Direct contact with raw feces or, in rare cases, urine
- Vectors (insects, rats. birds, etc.) that pick up contaminated matter on their feet and deposit it on human food or in washing or drinking water
- Washwater from bathing and laundry of infected individuals
- Indirect contact with floor drains and pipes that are collecting and breeding grounds for bacteria
- Ingesting contaminated meats and vegetables
- Drinking or bathing in contaminated water

Pathogens and Inadequate Sanitation

Pathogens that are transmitted in human excrement are of particular concern in parts of the world that have inadequate or no sanitation. The consequences of failing to provide adequate sanitary conditions—particularly a means of containing and treating human excreta—range from simple diarrhea and stomach cramps to death, especially for children and the elderly.

In the developing world—particularly warmer climates, where pathogens flourish—inadequate sanitation means excreta in water sources. Containing excreta through the use of pit latrines can be a step in the right direction, but these latrines can still contaminate groundwater.

A better sanitation measure is the containment and composting of excrement and washing one's hands with soap and water after defecating. Simple hand washing is now re-emerging as the most important measure in preventing disease transmission. Hand washing breaks the primary connection between surfaces contaminated with fecal organisms and the introduction of these pathogens into the human body. The use of basic soap and water, not exotic disinfectants, when practiced before eating and after defecating, may save more lives than all modern methodologies and technologies combined.

Measuring the Risk of Infection

The likelihood of infection by the various pathogens (called "mean severity" in the public health field, see chart in the Appendix) is important for risk assessment. For example, in northern climates where the temperature drops below freezing, *Ascaris lumbricoides* (roundworm) is virtually nonexistent. However, in warm climates it is a common pathogen excreted by humans, dogs, cats and other animals. The risk, therefore, is lower in the North than in the South.

Most pathogens we should be concerned about are found in the feces of infected people and are transmitted by contact, directly or indirectly, through open wounds on the skin or ingestion of contaminated food and water. Water contaminated with the feces of an infected person is the most common carrier of pathogens. Poor personal hygiene is another significant pathway. If the hands are not thoroughly washed after defecation, transmission of fecal pathogens on the hand to one's mouth or foods is inevitable.

What dose does it take to get sick?

The *infectious dose* is the number of pathogens required to infect a host. It varies from pathogen to pathogen. Ingestion of only one or more cysts or eggs of parasitic worms may cause disease. In contrast, most bacterial or viral illnesses require the consumption of hundreds to thousands of organisms to produce illness.

(See the Appendix for an explanation and listing of wastewater-borne pathogens of concern.)

Ascaris lumbricoides (Roundworm)

Roundworm may be the most ubiquitous parasite of humans, with an estimated one billion people infected worldwide. In some communities, infection rates reach 100 percent. It deserves special attention here, because of its reputation for long-term survival in dry or anoxic environments. However, moist composting is a hostile environment for parasites out of their hosts; in a composter, the nutritional, physical, chemical and thermal conditions fall out of the narrow range needed for their long-term survival. Also, antagonism with other, better, adapted organisms will kill foreign organisms.

Not that it should be downplayed: *Ascaris* is tenacious. One report documented the discovery of eggs that were viable *years* after being painted into the floor of a slaughter house. Ascaris can be the "old man" of pathogens when excreted into a composting toilet. But if you have ascaris in your composter, you probably have it in you, as well as in your pets, yard, animal pens and garden, too! What is key then is to keep this material out of water and the food cycle. The composter is usually part of the solution (containing and processing it), and properly maintained, not likely to contribute to its transmission.

Some research has been conducted in this realm in recent years. In a paper titled "Composting Toilets and Intestinal Parasites," medical scientist Sandie Safton reported on studies of end-product/humus from 16 composting toilet systems—two manufactured and three site-built models—in southeast Australia. She found organisms such as *Blastocystis hominis,*

Dietnamoeba fragilis, Entamoeba coli and *Enterobius vermicularis* (pinworm) in the excrement deposited in the systems. A total of 118 humus/end-product samples from all of the systems tested negative for parasites and commensal organisms. "The complete absence of these organisms in the final product indicates that the systems are in fact working with respect to the destruction of pathogenic parasites and commensals," she wrote. "It may not necessarily be the high temperatures that are associated with composting that are responsible for pathogen destruction, but it may simply be time and adverse environmental conditions. The fact that parasites and commensals that have resistant ex-host (out-of-human) stages are unable to survive in these conditions suggests that bacterial pathogens would also be destroyed. The humus/end- product could therefore, with the exception of viruses, be considered pathogen free."

Zentrum für Angewandte Ökologie Schattweid, a Swiss research center that studies ecological growing and lifestyle practices, found pathogens were destroyed after exposing end-product from composting toilets to temperatures of about 160°F/71°C at least twice over two days.

In Norway, Jordforsk Centre for Soil and Environmental Research found polio virus was killed in composting toilets. In a 1974 study Jordforsk tested human waste inoculated with polio virus and composted in thermophilic conditions (temperatures reaching 140°F/60°C) and mesophilic conditions (reaching 100°F/38°C). Coliform and polio were almost totally destroyed with thermophilic composting. In mesophilic conditions, coliform was reduced by 90 percent, and polio by 99.2 percent. This underlines the need for long-term containment or further processing.

References

Sandie Safton, "Composting Toilets and Intestinal Parasites," presented at the Innovative Approaches to the On-Site Management of Waste and Water conference, Southern Cross University, Lismore, N.S.W., 1996

Karol Enferadi, Cooper, et al., "Field Investigation of Biological Toilet Systems and Gray Water Systems" (USEPA, 1980).

T. Rohrer and C. Jaeber, "Versuchezur Hygienisierung von Fäkalkomposten aus Komposttoiletten in Einzelhaushalten" (Zentrum Für Angewandte Ökologie Schattweid, Switzerland, 1997)

Polio study, Jordforsk, Saghellinga, Ås, Norway

J. Martin and D. Focht, *Soils for the Management of Organic Wastes and Wastewaters,* (Soil Science Society of America, Wisconsin, 1977)

Composting versus Drying

Drying toilets are increasingly popular in the developing world where arid climates can dry excrement fast and where water is scarce. Many of these systems rely solely on dehydration to desiccate the feces. This matter is then put out on fields. Urine is diverted and drained into the nearby soil. (Urine can make an excellent fertilizer, but it is rarely collected from these systems.) Often caustic lime or wood ash is added to control odors, but these additives are a doubled-edged sword: They do aid in odor control, but they can also inhibit or stop the natural biological decomposition process. The result is dried, unstable feces and toilet paper.

While drying toilets can be a less expensive form of excrement management (these systems may require no vent/exhaust chimneys, we are reluctant to encourage dehydration here, as there is the possibility that it would ultimately result in large masses of rehydratable waste and the attendant problems of attracting flies and other vectors.

Further, dehydrating pathogenic bacteria can prompt certain pathogens to produce endospores, essentially a form that can survive for decades until rehydrated, when they would return to their full potency. The eggs of the parasite *Ascaris lumbricoides* have been viable for years in a dry environment.

Also, the nutrients in the dried, unoxidized matter are not available to plants. Dried organic matter will require biological processing before plants can use its nutrients safely, so processing might as well occur in the beginning.

(See the Appendix for a summary of a 1999 study of drying toilets in Viet Nam.)

The Basics of Vectors and Insects

In the field of environmental health, a vector is any organism that conveys disease to another organism. Vectors are an important consideration in the design and installation of composting toilets, because they can be a carrier of excrement.

Many kinds of animals, flies, beetles, mites, and other arthropods will be attracted to fresh human excrement for feeding purposes, if they can access it. Crawling insects and mites are mostly just a nuisance. Of greater concern are excrement-frequenting arthropods, such as flies, and the issue here is whether they can come and go: crawl on excrement and then land on humans or food.

All insects, spiders, mites, crustaceans (crabs, shrimp, wood lice, water fleas, barnacles), centipedes, millipedes and others are members of the phylum Arthropoda, the only major invertebrate phylum with members adapted for life on truly dry land.

Arthropods will be found in abundance on and in all healthy soils and on all healthy living organisms, including the human body. Their presence is indicative of health and ecological integrity, and their presence in a composting system should not be alarming. They only become a concern when they can leave the composting system, as they can potentially be vectors for disease transmission. For example, flies that get into a composting toilet reactor can walk on excrement, then fly out of the composter to a food source, and the excrement on their legs can be transmitted to food.

For that reason, restricting fly access to human excrement is an important aspect of the prevention of enteric disease, as well as an aesthetic issue.

(Keep in mind, though, that overall poor sanitation (handling feces, no washing) and water quality rank first for consideration of direct transmission risk, far outweighing flies. Flies are the next most likely cause of fecal ingestion, and you probably need a lot of flies to do it.)

However, when biologically processed excreta has been transformed to a stable and fully oxidized state (not just desiccated)—in other words, it has been turned into humus—it will no longer attract most of the arthropods of concern, such as flies, as humus offers them no food or suitable breeding ground.

The control of arthropods is critical to the safe operation of these technologies, and is usually not addressed adequately in the promotional materials or the operation and maintenance manuals of most composting systems.

Composting systems can be maintained so that these organisms are not allowed to become a health or nuisance issue (see Chapter 5 for details about how to prevent and get rid of flies).

Some management methods:

- Use a toilet stool with a water-seal trap.
- Use a toilet seat and lid that is gasketed, or a fitted lid for the toilet opening.
- Screen ventilation openings.
- Seal cracks and openings (a smoke test can reveal them).
- Apply environmentally benign insect repellents, such as pyrethrins and diatomaceous earth.
- Avoid putting kitchen scraps into the composter.

In a study by the USEPA and California Health Department (Enferadi et al., 1980), samples were taken from a variety of dry toilets for which the owners were not taking adequate precautions, and the results were dramatic and predictable:

"A wide assortment of flies, beetles, mites, spiders and other arthropods were found to inhabit the dry toilets. Mites were the most abundant kind of arthropods collected, generally occurring in numbers of 1,000 to over 100,000 per liter of solids. Fly larvae frequently occurred in numbers of 10 to 1,000 per liter, along with numerous beetles, moths and various wingless insects. No spiders considered harmful to humans were found in the vaults. The cockroach traps were all negative for cockroaches.

"Insects, mites, and other arthropods can gain entrance into the vaults by several means. They can fly through an open toilet, kitchen or access port, or they will land on the vault and crawl through small cracks or inefficient seals. Many of the gnats and small flies can pass through ordinary window screens. Several kind of mites attach to and 'hitch a ride' on flies and beetles. Another common access of importance is introduction of the eggs and larvae of flying insects, and all stages of flightless insects and mites in bedding material (e.g., forest litter and leaves) or kitchen wastes put into the vaults.

"Once in the vault, many kinds of insects and mites can pass their whole life cycle in the shelter of the vault chamber and increase their numbers rapidly and indefinitely. Examples here are the gnats and small flies including the fruit fly (Drosophila) and most beetles and mites. Others, including the larger domestic flies, must leave the vault to mate and cannot continue to colonize the solids without continuous access to the vault by egg-bearing females.

"The medical or nuisance significance of the arthropods found in the dry toilet vaults will vary according to their habits and to different circumstances of their occurrence. The wingless insects and other unobtrusive arthropods which tend to stay in the vault once they are introduced are of little or no medical or nuisance significance. However, flying insects commonly found in the dry toilet vaults may leave the vault at times and become a nuisance in the home if the dry toilet is installed within the home.... The gnats, small flies (except *Drosophila*), and beetles are of minimal significance as vectors of disease because they do not ordinarily seek out human food, and if emanating from a detached privy house, are not likely to enter the home at all....

"The domestic flies will enter homes even if produced in a detached privy house and will frequent human food. The vector potential of the kinds found here is not considered as great as the common house fly (*Musca domestica*), which was not found to frequent or breed in the vaults. However, if enteric disease agents were shed into the vault, the domestic flies found in this study would be one definite way that such agents could be transmitted to other persons.

"It could not be confirmed that flies were breeding in the composter."

Flying Insects Found in Waterless Toilets in One Study (Enferadi et al., 1980)

The following types of flies were found in improperly installed or maintained composting toilets in a 1980 study by the California Health Department and the USEPA.

Flies (Diptera)
Gnats (Nematocera)
- Psychodidae
- Fungivoridae
- Scatopsidae

Small Flies (Acalyptcrare Muscoid Flies)
- Phoridae
- Milichiidae
- Sphaeroceridae
- Heleomyziidae
- *Drosophila**

Large Flies (Calyprerare Muscoid Flies)
- *Fannia**
- *Muscina**
- *Ophyra**
- *Calliphorida**

Beetles (Coleoptera)
- Nitidulidae
- Staphylinidae
- Histeridae

Moths (Lepidoptera)
- Pyralidea

*Domestic flies

References

J. Martin and D. Focht, *Soils for the Management of Organic Wastes and Waste Waters* (Wisconsin: SCCA, ASA, CSSA, 1977)

Clarence Golueke, *Composting* (Emaus, Pa.: Rodale Press, 1972)

Ron Crites and George Tchobanoglous, *Small and Decentralized Wastewater Management Systems* (WCB/McGraw-Hill, 1998)

Karol Enferadi, Robert Cooper, Storm Goranson, Adam Oliveiri, John Poorbaugh, Malcom Walker, Barbara Wilson, "Field Investigation of Biological Toilet Systems and Grey Water Systems" (State of California Health Department, Department of Health Services and the Water Engineering Laboratory, Office of Research and Development, U.S. Environmental Protection Agency. EPA/600/2-86/069, 1980)

The Elements of Human Excrement: A Summary of Various Studies

Ralf Otterpohl, a water system consultant and professor at the University of Hamburg in Germany, took various European and North American studies of the content of human excrement and averaged them. Note that there is a wide range of variability in the content of excrement from person to person, place to place. Factors include nutrition, climate, health, age and lifestyle (Feachem et al., 1993). For example, vegetarians produce higher quantities of feces with a higher water content than those who are meateaters. A majority of the data is from Scandinavian countries, where many investigations of the quantity and composition of human excrement were carried out (Gotaas, et al.).

The following data are from Europe and North American studies. Note that nutrients may be different in less resource-rich regions. The specifications are in person-per-day quanitities.

In the literature are both volume and measure-referred values. These were converted on the assumption that density equals 1.0 kg/dm3. This can over-state the literature specification for the density of urine of 1.001 to 1.028 (average value 1.015) kg/dm³ (Altmann et al., 1974; Adamsson et al., 1995).

Urine

Parameter	Unit	No. of Samples	Min.	Max.	Standard Deviation	Average	Median	Value
Volume*	l/ppd	16	0.5	2.0	0.26	1.2	1.2	**1.2**
Weight*	g/ppd	16	500	2,000	260	1,200	1,200	**1,200**
Total Solids	g/ppd	7	20	147	32.9	72.4	60.0	**60**
Organic Total Solids	g/ppd	5	65	85	9.58	39.1	45	**45**
Organic Carbon	g/ppd	5	1.8	11.9	2.83	6.6	8.4	**8.5**
BOD_5	g/ppd	2	1.8	13.6	2.17	7.5	7.5	**7.5**
COD	g/ppd	2	5.4	30	4.74	15.1	15.1	**15**
Total Nitrogen	g/ppd	14	3.6	16	2.45	10.4	10.9	**11**
Total Phosphorus	g/ppd	14	0.5	2.5	0.41	1.1	1.0	**1**
Potassium	g/ppd	11	1.0	4.9	0.59	2.3	2.5	**2.5**
Calcium	g/ppd	6	0.15	2.2	1.06	1.3	1.4	**1.4**
Magnesium	g/ppd	4	0.06	0.2	0.14	0.2	0.1	**0.1**
Carbon to Nitrogen C/N	n/a	6	0.4	1.2	0.27	0.8	0.8	**0.8**

*Density = 1.0 kg/dm³ ppd = per person per day g = grams l = liters

Feces

Parameter	Unit	No. of Samples	Min.	Max.	Standard Deviation	Average	Median	Value
Volume*	l/ppd	9	0,07	0.4	0.05	0.18	0.15	**0.15**
Weight*	g/ppd	9	200	400	50	180	150	**150**
Total Solids	g/ppd	6	30	60	8.6	44.7	45.0	**45**
Organic Total Solids	g/ppd	4	26	58	6.2	44.8	42.0	**42**
Organic Carbon	g/ppd	6	13.2	33	4.1	21.4	21.8	**22**
BOD_5	g/ppd	1	6	18	-	11.1	11.1	**11**
COD	g/ppd	1	19	55	-	33.0	33.0	**33**
Total Nitrogen	g/ppd	11	0.25	4.2	0.9	2.0	1.9	**2**
Total Phosphorus	g/ppd	11	0.1	1.7	0.33	0.7	0.6	**0.6**
Potassium	g/ppd	7	0.2	1.3	0.21	0.7	0.6	**0.6**
Calcium	g/ppd	2	0.67	1.4	0.52	1.1	1.1	**1.1**
Magnesium	g/ppd	1	0.12	0.18	-	0.15	0.15	**0.15**
Carbon to Nitrogen C:N	-	5	5	11.3	1.79	8.2	7.5	**7.5**

*Density = 1.0 kg/dm³ ppd = per person per day g = grams l = liters

Feces and Urine without Flush Water

Parameter	Unit	No. of Samples	Min.	Max.	Standard Deviation	Average	Median	Value
Volume*	l/ppd	8	0.4	1.7	0.38	1.25	1,4	**1.5**
Weight*	g/ppd	8	400	1,700	380	1,250	1,400	**1,500**
Total Solids	g/ppd	6	80	130	11.1	109	105	**105**
Organic Total Solids	g/ppd	5	59	118	4.9	92	90	**90**
Organic Carbon	g/ppd	4	19	45	1.9	30	31	**30**
BOD5	g/ppd	2	12	40	14.1	30	30	**30**
COD	g/ppd	3	32	86	21.0	57	51	**50**
Total Nitrogen	g/ppd	13	5.9	18	19.0	18	13	**13**
Total Phosphorus	g/ppd	14	0.6	4.2	0.5	1.9	2.0	**2**
Potassium	g/ppd	7	1.5	6.1	0.8	2.8	2.7	**3**
Calcium	g/ppd	2	3.0	4.5	0.07	3.7	3.7	**3**

*Density = 1.0 kg/dm^3 ppd = per person per day g = grams l = liters

The organic load (organic C, COD, BOD5) are from the feces, whereas in the urine, COD values include oxidizable nitrogen compounds, which standard BOD5 test methods do not differentiate.

More than 80 percent of the nitrogen is found in the urine is in an organic form (urea, creatin, uric acid). Oxidized nitrogen compounds (nitrite, nitrate) are negligible. Half of the organic solids distribution in domestic wastewater (BOD5 50 percent, COD 42 percent) is from human excrement. With the nutrients (nitrogen, phosphorus), the largest proportion (> 90 percent) is found in the blackwater again. Systems for graywater purification must be calculated with this nutrient balance in mind.

References for Excrement Studies

Ralf Otterpohl, University of Hamburg, personal correspondence, 1998.

M. Adamsson, G. Dave, "Toxicity of Human Urine, Its Main Nitrogen Excretory and Degradation Products to Daphnia Magna;" in: J. Staudenmann; A. Schönborn; C. Etnier: *Recycling the Resource* (Envir. Research Forum Vol. 5-6, p 137-144, 1995).

K. Bahlo, G. Wach, *Naturnahe Abwasserreinigung* (Freiburg, Germany: Ökobuch, 1992).

R.G. Feachem, D.J. Bradley, H. Garelick, D.D. Mara, "Sanitation and Disease–Health Aspects of Excreta and Wastewater Management," World Bank, Studies in Water Supply and Sanitation, Vol. 3, 1983.

H.B. Gotaas, "Composting–Sanitary Disposal and Reclamation of Organic Wastes," World Health Orgnisation, Geneva; 1956; zit. in: I. Fittschen, Water Management in the Ecovillage Toarp, Dipl.-arb. Inst.f. Siwawi, Uni Karlsruhe, 1995.

K. Hayward, "Separate Ways," Water Quality International, January/February 1997.

W. Pieper, Das Scheiss Buch, Der grüne Zweig 123, Löhrbach, 1988; zit. in: F. Wissing, Wasserreinigung mit Pflanzen, Verlag Ulmer, 1995.

CHAPTER FOUR

Choosing and Planning a Composting Toilet System

When deciding on a composting toilet system, your key considerations are local regulations, system performance, lifestyle considerations, installation constraints and costs (not necessarily in that order).

Local Regulations

To install a composting toilet in a new home or as part of a renovation of an existing system, you may need a permit from local or state agencies. (Strictly speaking, you should get a permit for any installation.) What your state and local agent will allow may limit your options, and may rule out site-built systems. Some states have approved systems based on their own research, some do it on a case-by-case basis, and some require NSF listing. Check first. (See "Regulations" for some particulars about what this might involve.)

Performance

Performance is the composter's ability to safely manage the excrement for the number of people who will use the system on a daily basis for the number of days it will be in use. Research the various systems and decide what you think will work best for your situation.

Lifestyle

Lifestyle relates to what you can live with, both in terms of maintenance and aesthetics. Are you willing to maintain the system yourself and how often? Or do you prefer to pay for routine service?

Do you mind being able to see the contents of the composter? Or worry about the occasional fly? Then consider a toilet stool with a trap of some kind or at least a view barrier.

Carefully read Chapter 6 ("Operating and Maintaining It"). Be sure you will be able to manage the tasks detailed or be willing to pay a service person to do them. Read between the lines—this book's and any company's O&M instructions cannot present the sensory (sight and smell) experiences associated with dealing with a system up close and personal. Some require manual mixing of the exrement, toilet paper and additive (ETPA) with a rake or a pitchfork while it is composting. Others require turning mixing device handles. Others have automated this process, but that means more components that break and wear out eventually. Fixing an internal mechanism in a composter can be an unpleasant task. If this bothers you, call a professional, such as your local septage pumping and hauling or sewer cleaning service organization, and arrange to have these tasks performed on a regular fee-for-service basis.

Remember that composters come in two basic types:

- Self-contained appliance composters where the toilet seat is attached directly to the composter and the whole unit usually sits on the floor of the toilet room
- Central, also known as *remote* or *bi-level* systems, whereby the toilet is on the floor, but the composter is on a level below or adjacent to, the toilet room

The larger central systems are usually easier to live with, as they provide more processing volume, and thus are more forgiving and allow longer intervals between maintenance tasks. These are mostly used in year-round homes and public facilities. And they allow you to use a micro-flush toilet.

About Non-Electric Cottage Composting Toilets

Don't assume that non-electric composting toilets perform as well as their electrically heated and power-ventilated big brothers. You will be disappointed if you think you can use a non-electric cottage composting toilet for full-time year-round use by a family of three and more. It won't work without some kind of supplementary system, such as urine diversion, leachate drainage and passive heat.

Non-electric systems cannot actively evaporate leachate with only ambient air temperatures and passive chimney effect. Power venting is always recommended, even if it is from a 12-volt battery source. Make sure exhaust pipes are straight and ventilation is as direct as possible.

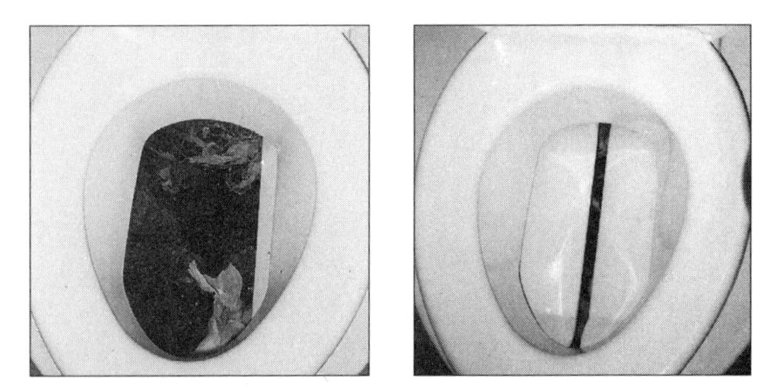

An open BioLet, a self-contained composting toilet, with its view barrier plates open and closed.

Installation Considerations

Consider the particulars of your home or site:

Space for the Composter

Many systems must be installed below the floor underneath the toilet or in a separate building. If you have no basement or sufficient crawl space, your most likely options are to install a vacuum toilet system or to use a self-contained composter (again, if you do not want to manage this, this is not for you).

Other points to remember:

- If you use a waterless toilet, the composter must be positioned directly beneath the toilet. It can be a few floors down, but there should be no angles in the toilet chute (remember: chute straight).
- Be creative: Your composter does not have to be in the basement. Laundry and utility rooms are also potential locations.
- Provide adequate space for making connections to the toilet, exhaust pipes, leachate drains and routine service. It may be helpful to consult with a licensed plumber regarding regulations for installing composters, leachate drains, exhaust pipes and graywater drains.

Space for Maintaining It

When locating the composter, consider how it is going to be maintained and emptied. Try to provide an easy-to-clean area with easy access to outdoors for removing compost. If the system will be maintained by a sewage hauling company, make sure the company's equipment can access the composter (it's best to consult the company first).

Confined Spaces

Public facilities commonly install composters in stand-alone toilet facilities, designed with the smallest possible foundations to minimize construction costs. However, servicing a composter in a tight space is tricky—the air quality may be bad and raking and removing the end-product difficult.

In public facilities, this situation triggers the Occupational Safety and Health Administration's (OSHA) Standard 29 CFR 1910.146 Confined Space rule. A permit may be required, and employers must establish practices and procedures to protect employees from the hazards of working in confined spaces (officially called Permit Required Confined Space, or PRCS). An employer who services composters or owns a composter that will be routinely serviced will need to contact OSHA.

For the homeowner, it may be sufficient to provide a means of pumping fresh air into such a confined space before and during service. Many odorless and invisible gasses, such as carbon dioxide, carbon monoxide, methane (natural gas) and radon, are heavier than air and can fill up a foundation from external sources and displace the oxygen-carrying atmosphere. You cannot survive in such an anoxic environment. If you or a service person climbs down into this room that contains these gasses, you could be in big trouble and not know it until you pass out from lack of oxygen. On August 15, 1996 the American Society of Heating, Refrigerating and Air Conditioning Engineers, Inc. (ASHRAE) and the American National Standards Institute (ANSI) released Standard 62-1989R "Ventilation for Acceptable Indoor Air Quality." Check to see what is an acceptable air-exchange rate for your application. Best, try to avoid this kind of installation.

Be sure to leave plenty of space for maintaining your composter.

Power Access

Most require electricity to operate heating or forced ventilation,exhaust systems (fans). If you do not have adequate power, be it 110- or 220-volt alternating current or 12- or 24-volt direct current, then your option is a passive design (sometimes simply called a "non-electric"). Remember that these are not as predictable performance-wise, and they require exhaust pipes of at least four inches or more in diameter, with no bends or obstructions.

About Locating It Outdoors

Some promotional brochures for composting toilets show the unit under a cottage exposed to the elements. However, after you buy the system, you may see in the operating manual that the unit must be kept at 65°F to operate properly. A frozen composter may never thaw out even during warm summer months, as its insulation prevents the summer heat from penetrating the permafrost in the fiberglass or plastic composter. Even electric heating during the summer may not bring the system to operating temperatures before the cottage is shut down for the season. Then the system is more a holding tank than a composter.

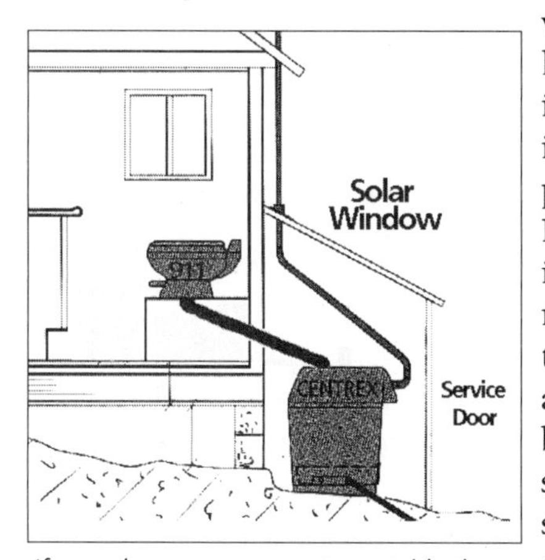

If you place your composter outside the house, keep it sheltered and warm. (Adapted from Sun-Mar)

In your cellar, locate your composter near appliances that give off heat, such as hot water heaters, furnaces and dryers.

Keeping It Warm

As you know, composting happens faster when it is warm. Too-low temperatures will slow down the process. (Some composters are also heated to evaporate leachate.) That means that composters may need to be heated where ambient temperatures are too low for active composting, such as in northern countries.

Since faster composting means faster reduction of the material, heating is a way to increase your composter's capacity. At the same time, the warmer the process, the more you should monitor moisture, as heat dries material.

Approaches to Heating

There are a multitude of ways to get heat to your composter. Be creative, but be safe—don't let your heater get wet if it's not designed to be, and keep it away from flammable materials. Always make sure that a heating system is installed in accordance with state and local building codes.

There are two basic heating approaches:

- Locate your composter in a heated room. Then you do not need to insulate the composter. Keep the ambient temperature at a minimum of 65°F (18°C).
- Heat and insulate the composter itself. If your system doesn't come with a heater, you can buy and install a small heater, such as those used for fish tanks.

To Heat or Not to Heat?

Follow the Q10 Principle to Compost More in Smaller Spaces

When deciding whether you need a heater, consider this principle: By increasing the temperature, you increase the rate of composting.

Increasing temperatures can double the release of energy from the respiration of carbon by aerobic microbes, increasing the rate of processing, within certain limits.

This is known as the Q10 temperature coefficient, which holds that for every 10°C (18°F) rise in temperature, the biochemical rate of reaction is doubled.

In this way, heaters significantly increase the rate of processing, allowing more composting to occur in a smaller space. That means, if you need to increase the capacity of a composter, increase the temperature of the biomass.

Locating the Composter for Warmth

First, be smart about locating your composter. Look for sites that are already warm, such as:

- near a South-facing window
- next to a boiler, furnace, washer, dryer or water heater
- even an interior heated room, such as just under the second floor bathroom

Using Cheap Heat

Then look for cheap or free heat, such as waste heat. Some composters have convenient air intakes that can be connected to a warm air source, such as:

- a heat exchanger connected to the exhaust system of a generator or the flue from a gas hot water heater
- the lint-filtered exhaust from your clothes dryer
- hot discharge air from your air conditioner or dehumidifier
- solar collection (from a collector, window or a greenhouse)
- in the summer, hot air pumped from your attic via ducts
- a connection from an existing space heating system

(More sophisticated systems like heat pumps, while elegant from an engineering point of view, are often cost prohibitive for use with a composter alone.)

Heater Options

• *Conventional Heating Systems*

Electrical heating systems, forced hot water heating systems and forced hot air all are conventional heating options.

You will find compatible heating devices from catalogs and suppliers for recreational vehicles, mobile homes, swimming pools and aquariums. The small submersible heaters in heat-resistant glass used for aquariums make good composter heaters.

In many cases, however, it is best to provide a bottom heater so that the composter is sitting on a warm floor. Consider making a radiant floor heating system just for the composter. Radiant systems are electrical (see self-regulating heaters below), or they are hydronic, whereby a hot liquid, such as water, passes through small-diameter plastic pipes. Have one made by a radiant-savvy plumber, or construct it yourself by making a shallow sandbox into which you place radiant heating tubing.

Self-regulating heaters, similar to those used to keep pipes from freezing, are the most efficient. Self-regulating heating cables consist of two parallel conductors embedded in a heating core made of conductive polymer. The core is radiation-crosslinked to ensure long-term reliability. The heating system automatically adjusts its power output to compensate for temperature changes.

When electricity is applied to cable wires, a current is conducted across microscopic carbon networks. These networks form miniature heater circuits and the cable warms the surrounding area. As the cable itself gets warmer, the plastic expands until most of the carbon circuits are disconnected. Then, if the outside temperature drops, the plastic contracts and the circuits are reconnected. The result is a self-regulating system that delivers just enough heat to keep the composter at a constant temperature without using more power than is necessary.

Some manufacturers put this cable inside the unit so that it warms the leachate, but it is important to be sure that the cables are sewage- and waterproof, and that any connections are made outside the liquid area in accordance with state and local electrical codes. (Available from: Raychem, 300 Constitution Drive, Menlo Park, CA 94025; 800-542-8936)

• *Heated Room*

You can also zone your home heating system or hot water heater to provide heat to your composter. Think about it as heating a very small room in your house. For heating method comparison and conversion purposes, consider the fuel comparison chart below.

These figures are gross fuel data; heating appliances have different efficiency ratings, so ask your local supplier to tell you the conversion factors relevant to your specific application. Electric heating is the lowest cost to install and the highest cost to operate, but it may be the most economical for seasonal use.

Fuel Used	Heat in BTUs
1 kilowatt-hour of electricity	3,412
1 liquid gallon propane	91,600
1 gallon fuel oil	136,000
1 square foot of sunshine/hour (Solar constant)	429.2
1 cord of wood	12,500,000
1 cubic foot, natural gas	1,100

• *Dryer Duct*

A great cheap means of providing heat and humidity is by ducting lint-filtered clothes-dryer exhaust to the composter's air intake. Insulate the dryer duct and composter to prevent heat loss. If un-trapped toilets are used, take care that the dryer blower does not force odors out of the composter system and up through the toilet!

A better way is to disable the dryer blower and allow the composter exhaust fan to draw both the latent and sensible heat from the clothes dryer into the composter and up the exhaust pipe (so the composter fan pulls air out of the dryer). Remember to use a lint filter, or you may have to remove lint from your composter's ducts.

• ***Bread Box Heater***

An incandescent electric lightbulb in a fire-proof box also works, because such a lightbulb produces a lot of heat (in fact, heat production accounts for most of a lightbulb's electrical consumption).

• ***Solar***

For solar heating, you have to have available solar energy, such as a clear, unobstructed south-by-southeast opening unblocked by trees or other buildings. In most North American communities, nine square feet of solar collection area all year long will sufficiently warm a composter in a heated home. Contact your state solar energy association for collector designs.

There are many designs for heating small outhouses with composters in them (see some in the Site-Built Systems section). Some direct solar heat right onto the composting mass through a window, some on the leachate, some on the composter. (Remember that direct solar heat can dry out the surface of the mass, creating a crust that actually insulates the center, so you may actually need more heat. Turning the material and adding water helps.)

This composter is located in a sunny spot next to a sliding glass door, which allows easy removal. In warmer climates, this makes sense.

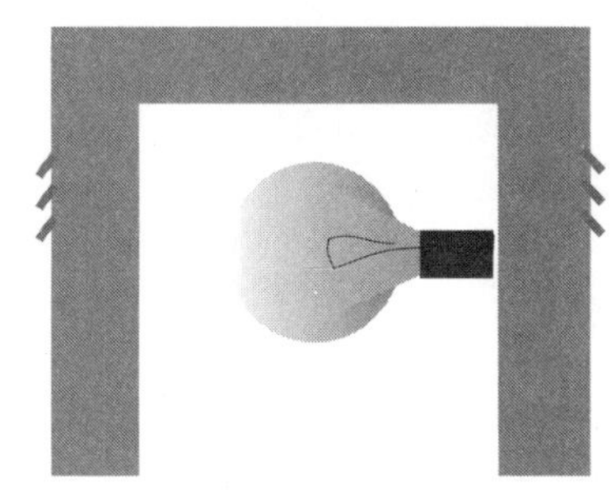

A simple breadbox or paint can heater is simply an incandescent lightbulb in an insulated reflective box. Place it over the composter air-intake to warm incoming air—but be sure the box features vents to allow air in!

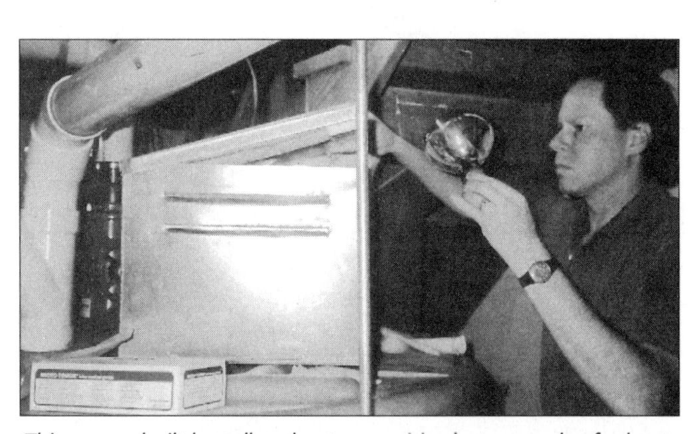

This owner-built breadbox heater provides just enough of a heat boost to keep this basement composter processing well.

Losing Heat

The process of ventilating and exhausting the composting toilet also sends heat out of your composter. Toilet flush water also cools it. Evaporating leachate can cool the process, too, through evaporative cooling. This is most evident in small self-contained composting toilets. Many have heaters and thermostats to keep an internal temperature of 90° to 113°F (32° to 45°C) in order to maintain the upper mesophilic composting range, while evaporating excess liquid. (Sun-Mar, however, only aims for 10°F over ambient.)

So, don't overdo the exhaust system. To minimize heat loss, install a fan speed controller to keep the fan speed at levels just adequate for odor control, but not high enough to cool and dry the material too much. Turn down the fan speed until you smell odor in the bathroom (this does not apply to a trapped toilet), then turn up just a little until the odor goes away. That should be sufficient.

Insulate the composter only if there is more heat inside the composter than outside of it.

Or, place a lit incense stick inside the toilet; if you see smoke coming out, then there is not enough ventilation or exhaust. A rate of 1 to 5 CFM (cubic feet per minute) should control odors and provide sufficient oxygen to the process.

This heat loss also underlines the value of using free and waste heat.

- Insulate the vent pipe, if it passes through cold unheated spaces, to prevent condensation (the vapor from evaporated water and urine, which will run back into the toilet)
- Or paint the vent stack black and put it in its own windowed stack. Solar chimneys help maintain a draft. (This is a little complicated, and only heats the pipe on sunny days.)

Ventilation and Exhaust

Ensuring that air enters (ventilation) and exits (exhaust) the system in the right direction is critical for maintaining composting and preventing odors from entering the home.

Take the Air-Flow Test: If your toilet room is on the side of the house opposite the prevailing winds, remember that wind pressure on the windward side of the house pulls a vacuum on the opposite or leeward side. So, when you open your bathroom window on the leeward side, odor will be pulled from your toilet and into the room. You will know that this is the case if the window curtains blow out. If they blow into the room, then the toilet will be pressurized, and no odor will come into the room. This is not as much of a problem with fan-forced-ventilation/exhaust systems, as the fan can overcome the negative wind pressure. However, when the fan is off, well, that's another story. This is not a problem with systems with seal trap toilets.

Wind turbines work in areas with steady strong winds, but can actually impede air flow at wind speeds of less than 10 to 15 m.p.h., according to a National Forest Service study. A simple plumbing Tee at the top of the pipe both keeps the rain out and provides sufficient venturi action effect (whereby wind sucks air out of the pipe). There are sophisticated extractors, such as the Vacu-stack, which both keeps out the rain and increases the rate of exhaust extraction when wind blows across rigid foils arranged in a globe fashion around the pipe opening (available from wood-stove dealers). (For more information about Vacu-Stack, contact Improved Consumer Products, P.O. Box B, Attleboro Falls, MA 02763.)

Keeping the exhaust chimney warm encourages throughput of the vapors. Top: An oil lamp promotes upward air movement in an exhaust chimney (Vera Miljö). Right: Insulate the pipe, or place it in a warm and sunny location behind a solar window or in a greenhouse, facing South. (Illustration: NCAT)

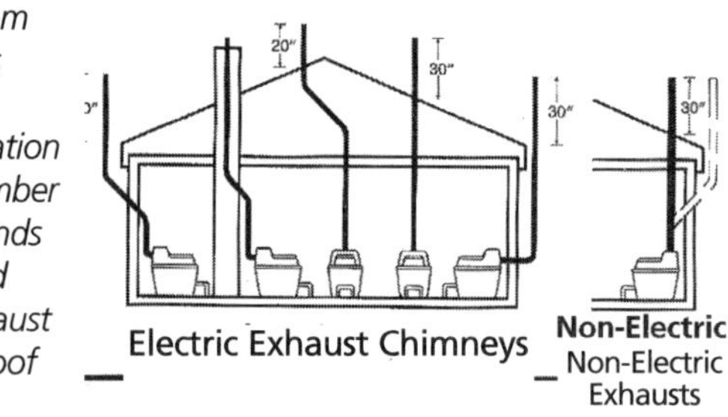

This graphic from Sun-Mar shows some exhaust chimney installation options. Remember to minimize bends and elbows and extend the exhaust pipe over the roof of the house!

A wind turbine from Vera Miljö. These work only at higher wind speeds—otherwise it can hinder air flow through the vent stack.

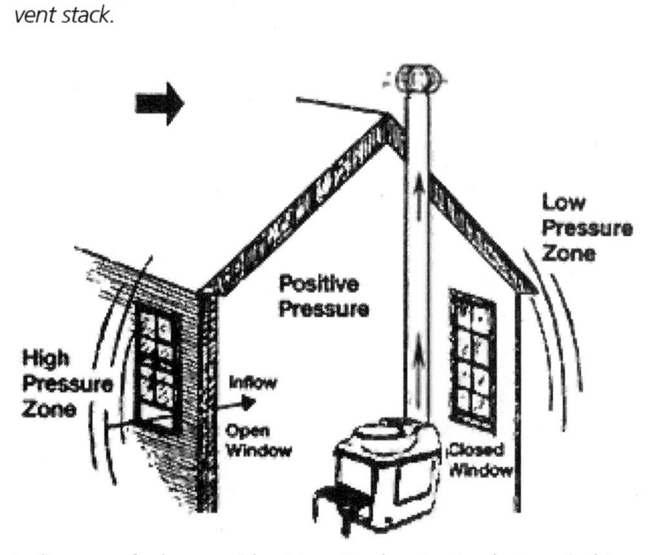

A diagram of a home with a Vacu-Stack extractor (not a wind turbine), which pulls exhaust gasses out of the composter. Unlike a wind turbine, it does not turn. (Graphic adapted from a Vacu-Stack illustration, and is not to scale).

Pulling Odors from the Composter

Some assume that the wider the pipe from the toilet to the composter, the lesser the chance of "skid marks" from stuff going down it. That's probably true, however this connecting pipe can act as a chimney through which odors can back up into the toilet room (if the toilet is not a trapped toilet). In fact, the larger the diameter of the connecting pipe and toilet seat opening, the greater the chance of odor. However, the taller the exhaust pipe, the greater the chimney effect and effective odor exhaust. To reduce this toilet-as-chimney effect, make sure that there is not a negative pressure in the toilet room, so air is not pulled up the connecting pipe.

- Disconnect or reverse the bathroom exhaust fan (the composter's exhaust can take its place).
- Be aware that whole-house attic fans, fireplaces, wood stoves and air conditioners can pull odor from the composter into the building.

Fans are built into many composting toilet systems, and usually their speeds cannot be regulated. These fans will pull air into your composter from either the toilet room or wherever the air intake is. Make sure this is not cold air that will cool your composter. Yes, that might come at a cost, as the composter exhaust fan can suck in as much as 100 cubic feet* of air per minute (cfm) from the building, increasing your heating bill.

That's another reason why a fan-speed controller should be installed with a powerful fan (25 to 100-plus cfm**) so that you can set the minimum amount of fan power to control odor. Again, most composters will need only 1 to 5 cfm*** for adequate aeration for composting, so your primary issue with fan speeds is managing odor.

If the toilet room is cooler than the composter, heat may rise from the composter into the room.

Outside Odors from the Exhaust Pipe

Outside, the odor from the composter will be not normally noticed on the ground if the exhaust pipe terminates at least 12 inches (305 mm) above the peak of the roof. Lower than the roof peak, and odors from the exhaust pipe may be swept to the ground through wind downdrafting.

Odor filters, such as those supplied by Sun-Mar and Orenco, can be installed in the exhaust pipe after the fan and before the pipe terminus. (Orenco Systems, 814 Airway Avenue, Sutherlin, OR 97479-9012; 541/459-4449)

Some manufacturers, such as Sun-Mar, include air diffusers that mix the exhaust gasses with fresh air to dilute the odors and help promote up-draft.

*(2.83 m3)   ** .7-3.0 cmm   ***.03-.14 cmm

Toilet Stool Considerations

(See "Toilet Stools for Composting Systems," Chapter 7, for a more complete discussion of a variety of waterless and flush toilet options and their installation issues.)

In a waterless situation, gravity is the only way to convey ETPA from toilet to composter, so the composter must be almost directly under the toilet, although it can be several floors down. There should be few, if any, angles at all. Do not underestimate the viscosity (think: stickiness) of excrement! If this location won't work, or if you decide that aesthetic and lifestyle issues dictate a barrier between you and the composter, a micro-flush toilet or a vacuum system are your alternatives.

Connecting Pipe Diameter for Dry Toilet Stools

Given the risk of streaking by ETPA down the connecting pipe, it would seem that the bigger the pipe, the better. Not so. If the opening diameter of your toilet is too large, odors are more apt to enter the toilet room, insects have better access to the composter, and pets, toys and infants could fall in (although we have yet to hear of a child falling into a composting toilet system). Too small, and the pipe gets caked with EPTA and may discourage ventilation/exhaust and require frequent cleaning. A good size is eight to 12 inches (203-305 mm).

Remember too that you can purchase the composter from one manufacturer and the toilet from another. Just check the outlet dimensions so that you can get the necessary adapters to connect the toilet to the composter you have selected.

The View from Above

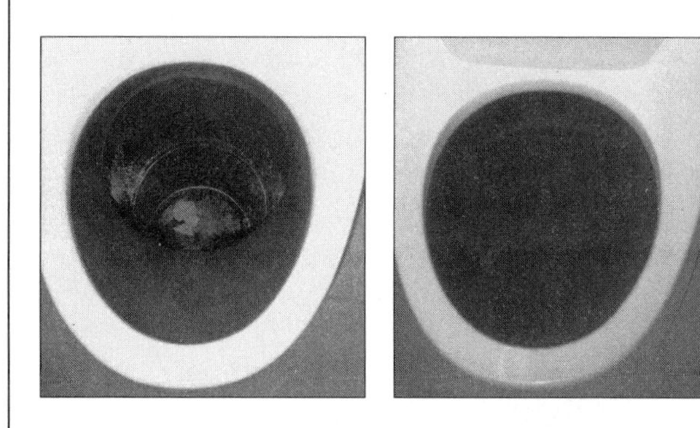

The value of indirect lighting in a bathroom served by a composter is apparent in this bathroom with a Clivus Multrum dry toilet. Left: With overhead lighting, it is possible to see the contents of this Clivus Multrum just three feet below. Right: Not quite the "visual experience" when a different light is on.

Capacity Issues

When deciding what capacity you need, consider:

- How many people will likely use the system every day?
- For large-capacity applications, also ask: How many people will use it every year?

Most manufacturers classify their models by how many people can use the system during a specific period of time and have the system produce an acceptable end-product. This takes into account the composter's ability to contain and compost ETPA, plus its ability to contain or evaporate excess leachate.

It is very important for the buyer to understand the formula that a manufacturer uses. They all differ unless they were tested under a uniform performance standard, such as NSF International's Standard 41 or the Foundation for Product Labeling's Nordic Environmental Labeling Criteria for Closed Toilet Systems.

Seasonal Usage Considerations

Most of the small, self-contained appliance composters were designed for seasonal use only. When a manufacturer states that a model should be used only for seasonal or cottage use, then the number of users stated is usually for 30 to 60 days out of a 365-day year. Manufacturers of seasonal systems usually list a number of users for seasonal periods and a much smaller number for residential (365 days) use.

European manufacturers use the person-day approach (one person over 24 hours). For example, a toilet with 600 person-days capacity, one person could use the system for 600 days or four people for 150 days.

Some composters sized for public facilities are rated by uses per year, as many public facilities measure usage this way. The problem with this is that usage can vary widely, and parks have found their systems rated for 3,000 uses per year cannot manage the fact that 2,000 of the park's uses occur in two months!

Capacity Rating Factors

In a capacity rating, one person is an average adult, averaging the differences between males and females (however, women can use three to four times more toilet paper than men, as women use toilet paper after urinating and men usually do not). Studies have shown that the average adult from a northern European community will produce about 40.6 fluid ounces (1.2 liters) of urine and 20.3 fluid ounces (0.6 liters) of feces daily. In one year then, the same person produces 155.8 gallons (590 liters) of urine and 57.9 gallons (219 liters) of feces. The term "excrement" refers to a combination of urine and feces.

Performance rating organizations take into account "population equivalents" (p.e.), which are the average number of excrement events produced by an average adult person in one 24-hour period. For this standard, one p.e. is defined as 1.2 fecal events and four urine events per person per day.

Day-Use Only versus 24-Hour Use

Remember that the ratio of urine-to-feces volume varies in different settings. In a day-use public facility, there will be a much higher ratio of urine to feces—as high as 10 to 1. However, in a residential setting, a ratio of 3 or 4 to 1 is common. So a predominantly day-use facility will need to consider a system for managing leachate (lots of urine) more than feces.

Is Bigger Better?

The size (volume) of the composter is an important factor, but not the only one. The processing rate can be speeded up by increasing aeration (including through mixing or batching) and raising the temperature.

A composter is not just a storage tank, so remember that buying a larger composter does not mean that you will get the same efficiency of processing that you will get in a smaller one. The larger the chamber, the more important it is to get air to the composting material and avoid compaction. You might even consider installing two smaller ones instead one large one.

Large volume does have advantages: Larger systems serve more users in a fixed time period. They allow longer retention and processing time (for a fixed number of users), which may produce a more thoroughly processed end-product. That can also be achieved by adding composting chambers to a batch composting system.

Cost (Caveat Emptor!)

Cost alone is usually last consideration when choosing a system because the cost factors are driven by the first four categories.

It can be all too easy to choose one of these models based on purchase price alone. It could be a costly mistake.

Of manufactured systems, the small self-contained composting toilets often cost the least. The vast majority of small self-contained composters are purchased for seasonal second homes and cottages. Because they are only used 15 to 60 days per year, they are much smaller and less expensive than large central systems. However, be very careful when selecting any appliance composter for residential use. Yes, it may have been tested for a limited number of persons in continuous use by a small family, but you may soon become weary of the daily and weekly routines required to keep it operating well. Self-contained composters such as BioLet, Sun-Mar, Envirolet, etc., are highly mechanized, but while they are small and lower priced, their operating, maintenance and lifecycle costs can be much higher.

Site-built systems constructed by owners using reused materials and free or very low-cost labor can be the least-cost option. But, again, with site-built systems, you have to consider your ability to get the system permitted by a health official, costs of labor and materials and whether you can live with a system that you might have to fine tune, with no support from a manufacturer.

The size of larger composters may require a significant change in the construction of your house. Consider those costs, too.

Lifecycle Costs

As with any system or appliance, consider long-term and lifecycle costs. Longevity or replacement, costs of parts that wear out, service intervals, installation expenses, management, and power requirements all add up to form the lifecycle cost of a system. Like many products, some seem to have built-in obsolescence, requiring you to replace the system after a few years. This may not be a problem for you if the price is right, but know what you are getting before you purchase. Construction materials, such as painted carbon steel or certain plastics that decompose with time, could be trouble. The more moving parts, heaters, fan motors, etc., the higher the lifecycle cost. The difficulty is in assessing these costs before you make a decision.

Some Additional Considerations when Choosing a Composter:

- ☐ Construction materials: Plastic type (PVC, ABS, polyethylene), fiberglass (resin: epoxy, polyester)
  - Process chamber
  - Exhaust pipe
  - Insulation (if needed)
  - Process control components
- ☐ Plumbing and venting issues
  - Metric or U.S. pipes
  - Exhaust pipe diameter
  - Exhaust pipe insulation diameter
  - Standard plumbing fitting/pipe
  - Location of vent pipe termination
  - Fan-forced exhaust required
  - Supply air and required temperature
  - Exhaust system
  - Leachate drain
- ☐ Power requirements
  - Voltage
  - Amperage
  - AC/DC
  - Nonelectric option
- ☐ Heating Issues
  - Specified temperature for rated performance
  - Can it accommodate supplementary heat
- ☐ Telephone support
- ☐ Warranty issues: Extent and duration
- ☐ Purchasing issues
  - Shipping costs/UPS Common Carrier (big truck)
  - Weight, carton size shipping point (FOB)
  - Factory direct or dealer
- ☐ Service Issues
  - Leachate drain provision
  - Frequency of service at rated capacity
  - Parts available for old models
  - Support for old models
  - Can it be upgraded
  - Maintenance contract availability
  - Access space required for service
  - Are all components for building-code-approved installation furnished in packing carton?
  - Are such components available from local sources
  - Can you use a water-flushed toilet?
- ☐ References of those who have long-term experience (at least two years and 3 or more service/end-product removal cycles).
- ☐ Particular advantages and disadvantages
- ☐ Where are most installations (residential, pubic facilities, vacation homes)
- ☐ Sensitivity during overloading, power outage, extreme temperature
- ☐ Approvals by local agencies and certifications by test laboratories, such as NSF International
- ☐ States approved for
- ☐ Space requirements for installation
  - Footprint (floor space required)
  - Headroom required

Leachate, the Liquid in Your Composting Toilet System

What Is Leachate?

In all composting processes, extra liquid, called "leachate," forms as the result of the accumulation of:

- **urine**
- **water from toilet flushing, toilet bowl washing or condensation**
- **water in cells of organisms and other living matter that is released when they decompose**

As this liquid moves by gravity down through the composting mass, it picks up particles of material and dissolved salt (NACL) and minerals, and accumulates at the bottom of the composter. From there, it should be drained to a planter system, disposed or evaporated.

In composting toilets, leachate can contain significant amounts of fecal microorganisms, and therefore potential pathogens, dissolved nitrogen (primarily in the form of as aqueous ammonia, and nitrates) and salt. Some manufacturers claim that this material is safe, because the high salt and aqueous ammonia in it and its long residence time in the composter kills off pathogens. Some call it "compost tea," a term for leachate from composted plant residues; this oxidized liquid can be used as fertilizer for plants. Leachate from composting toilets, however, is not the same: It is *much* higher in ammonia and salt, and may contain potential pathogens.

★ ***Leachate from a composting toilet can be as dangerous as raw untreated excrement and must be safely managed! It can burn plant roots and pass along pathogens. This is strong stuff.***

A jug of leachate emptied from a BioLet. (photo: BioLet)

This fact was underlined by a report of a study of an AlasCan composter with two micro-flush toilets. It found that the leachate contained almost as many fecal coliform as septic tank effluent (see chart).

One composter, the Phoenix, features an option to pump leachate from the bottom, up and over the top of the composting mass, to keep it hydrated. However, the leachate's plentiful ammonia and salts can disrupt the microbial process. Better to use fresh water for this.

How Much?

As noted earlier in this chapter, studies have shown that the average adult from a northern European community will produce about 40.6 fluid ounces (1.2 liters) of urine and 20.3 fluid ounces (0.6 liters) of feces daily. In one year then, the same person produces 155.8 gallons (590 liters) of urine and 57.9 gallons (219 liters) of feces. That's a lot of liquid.

A wide variety of factors affect natural evaporation (ambient temperature and humidity, plus air flow). If a waterless toilet stool is used, based on typical North American conditions, an estimated 1 to 2 pints of leachate per person daily will not naturally evaporate and will accumulate in the bottom of the composter. Four people could generate 120 to 240 pints or 15 to 30 gallons (57-113 liters) per month.

Removing It

Leachate can be difficult to manage because, as the water evaporates, a sludge forms, which should be removed periodically. It can clog screens, grates and filters in some composters; block outlet holes and openings, drain hoses and pipes and harden on internal mixing equipment, causing maintenance headaches over time.

Some composters have fittings that can be connected to drain pipes. Others, such as the Clivus Multrum, feature an electric sewage pump with a float switch that automatically pumps the leachate to a storage tank.

Analytical Results from Compost Leachate (mg/l)* from an AlasCan Composting Toilet System

Parameter	Date: 8-5-97	12-3-97	Mean
Total Suspended Solids (TSS)	55	214	135
Total Kjeldahl Nitrogen (TKN)	35	113	124
Total Phosphorus (TP)	86	84	85
Biological Oxygen Demand (BOD5)		not tested	
Chemical Oxygen Demand (COD)	731	646	689
Total Organic Carbon (TOC)	151	128	140
Fecal Coliform**	2,600	35,000	18,800

* Fecal Coliform is reported as MPN/100 mls.
** Reported as geometric mean.

The complete report can be seen on the AlasCan web page: http://cloudnet.com/~alascan/testdata.html
"Analysis of Monitoring Results of the Separation and Graywater Treatment System at Chester Woods Park, Olmsted County, Minn.," T. Lee, K. Crawford, and T. Hill. Olmsted County Water Resources Center, 2116 Campus Drive SE, Rochester, MN 55904

A marine bilge pump can be used to manually pump out the leachate from most composters.

Evaporating It

Some composting toilet systems, especially the cottage models, have internal electric heaters to evaporate leachate. They are usually effective, but they require a lot of energy—about 9,000 btu per gallon or approximately 2.7 kwh of electricity per gallon. At 10 cents per kwh, that's about 30 cents a gallon (3.8 liters).

Draining It

Draining it via a drain line lowers your handling of the leachate. Generally, you should drain it to:

- a tank that is pumped out by a septage hauler
- a mini-graywater garden system
- a septic system (septic tank and leachfield)
- a system to be combined with graywater for irrigation of plants
- be diluted with 8 parts water and used to irrigate biologically active soils with plants (with care!)

U.S. regulators usually require that leachate must either be evaporated inside the composter or removed for conventional treatment. One manufacturer suggests using a rock-filled French drain or soak pit—essentially a mini-leachfield—for this purpose. That's fine if you know that your soils can handle it, and it's not going directly into the nearby lake.

Leachate—liquid—collects even in composters with heaters. It is best to provide an automatic means to remove it.

Composter manufacturers rarely mention that leachate can build up, so many people think it only occurs during peak use or power outages. In the authors' experience, it does collect, even in composters with electric heaters. So, it is better to provide an easy and automatic means of removal, such as a gravity drain line or pump.

Leachate can be recombined with filtered graywater for subsurface aerobic irrigation—it is a natural combination of nitrogen and carbon for plants. You can also dilute it with 8 parts water to feed nonedible plants (wear gloves!). Remember: Leachate and graywater contain nutrients *for soil microbes*, which transform them into plant food.

6" mound
Fill
1 to 2" of earth
1 to 2 ft. of gravel
2 ft.
2 ft. of Loose Material
French Drain
ROCK

A French drain for leachate (Sun-Mar)

Top: Leachate tanks and a pump at Walden Pond in Concord, Mass. Above: A mini-Washwater Garden in a five-gallon pail manages leachate from a cottage composting toilet on an island in Maine. It is planted with a cana lily.

Good Drain Line Size

With proper engineering, a composter can drain leachate through a 1- to 2-inch (25.4-50 mm) drain pipe. This large size prevent clogging by solids and allows larger volumes, should a flush toilet be used. Remember that drain lines should be vented. Many codes require separate venting for any drain system, but you may be able to use your composter's vent system.

Install a union (an easily separated connection) between the composter and a vented drain to allow servicing of the drain line. If you can, place the composter on a raised platform so leachate will drain out by gravity (and maintaining the composter will be more convenient).

Urine: Managing It Separately

Several projects worldwide are diverting urine from the wastewater mix and using it. Here's why:

- In wastewater, most of the nutrients—as much as 90 percent of the nitrogen and potassium—is from urine alone.
- Nitrogen that gets into ground and surface waters can create pollution.
- Oxidized and diluted, urine makes a good liquid nutrient for plants.
- Urine is usually sterile in healthy populations.
- Urine is easy to drain away and collect separately.
- Urine mixed with feces produces a malodorous compound—worse than each of its components alone. (Wolgast, 1993)

Several study projects in Sweden have piloted urine collection and use. Urine was collected in homes and apartment buildings in storage tanks, then transported by truck to fields, where it was applied as a fertilizer to crops, such as wheat, corn and oats. In Mexico and Central America and Mexico, some programs collect urine and use it to irrigate plants in containers. (Jonsson, 1997; Winblad, 1997)

However, collecting and using urine presents some challenges:

- The urea in urine—which contains most of the nutrients—degrades rapidly to the gasses ammonia and carbon dioxide unless it is contained or directly utilized. (Sundberg, 1995)
- Urine contained without air will smell bad.

Urine can be diluted and immediately used on plants. When it is used immediately, however, it has not been transformed into a plant-available form—one is relying on the soil microbes in the plant bed to do that. To preserve all of the fertilizer value of urine, it needs to be oxidized or composted (some erroneously call this "fermenting," an anaerobic process). For this, a carbon-urine mixture requires a carbon-to-nitrogen ratio, or C:N, of one part urine to 25 parts carbon to convert all of the nitrogen to nitrates by aerobic soil bacteria. Nitrate is the form of nutrient that is available to plants.

Urine Defined

According to *Van Nostrand's Scientific Encyclopedia:* "Urine is the fluid secreted from the blood by the kidneys, stored in the bladder and discharged by the urethra. In healthy humans, it is amber colored. About 1,250 milliliters (about a third of a gallon or 0.044 cubic feet) of urine are excreted in 24 hours with normal specific gravities of 1.024. Flow ranges from 0.5 to 20 milliliters/minute with extremes of dehydration and hydration. Maximum osmolar concentration is 1,400 compared to plasma osmolarity of 300. In diabetes insipidus, volumes of 4 to 6.6 gallons or 15 to 25 liters per day of dilute urine may be formed. In addition to the substances found in the table below, there are trace amounts of purine bases and methelated purines, glucuronates, the pigments of urochrome and urobilin, hippuric acid and amino acids. In pathological states, other substances will appear: proteins (nephrosis); bile pigments and salts; glucose, acetone, acetoacetic acid and beta-hydrobuttyric acid."

Urine, the Pollutant

The nutrients in urine can cause overgrowth of aquatic plants in lakes, rivers and streams. When they die, they use up oxygen in the water, which suffocates other aquatic life. This is called "hypoxia." When it gets into drinking water supplies, nitrogen can result in such diseases as methaemoglobimaemia, a.k.a. "blue baby syndrome." To protect public health, the Safe Drinking Water Act now requires reducing the nitrogen through tertiary treatment, often involving expensive denitrification equipment.

Keeping It Separate

Our bodies keep urine separate naturally, so we can collect and use it separately. This makes the job of collection and transport easier. (Only in birds and lower organisms are urine and feces combined during excretion.)

Increasing in Scandinavia is the use of convenient urine-diverting toilets, whereby a urine collector and drain are cast into the front of the toilet bowl. (See "Toilet Stools for Composters," Chapter 7.) The urine is collected and mixed with water, and used to irrigate trees and shrubs.

The Fertilizer Value of Urine

The world needs all the nutrients we are flushing away each day in our urine. Given the far-reaching costs of using manufactured fertilizers, utilizing this

A urine-diverting toilet from Vera Miljø

valuable and usually sterile resource deserves more consideration. Urine costs nothing to produce (unless you count the plant and animal protein we eat), but it does have storage and transportation costs, as does commercial chemical fertilizer. Urine is too rich in nutrients to ignore.

And there's lots of it: Two Swedish university studies report that one northern European adult (who consumes plant and animal proteins) produces enough fertilizer in urine to grow 50 to 100 percent of the food requirement for another adult! We excrete these nitrogen-containing compounds as urea, creatine, ammonia and a small amount of uric acid. These nutrients could feed a hungry and growing population at a lower cost than producing more expensive chemical fertilizer. (Sundberg, 1997; Drangert, 1997)

How Much Is Produced?

As the chart on this page shows, a range of 6.7 to 182 grams (.23-6.4 oz.) of nitrogen-containing constituents are excreted by one person in 24 hours. Many studies will give the average daily output for a healthy adult at 11 grams. (Ralf Ottepohl, 1998).

As noted before, the average northern European adult produces about 40.6 fluid ounces (1.2 liters) of urine and 20.3 fluid ounces (0.6 liters) of feces daily. In one year, the same person produces 155.8 gallons (590 liters) of urine and 57.9 gallons (219 liters) of feces. Just one person!

A urine-diverting toilet in Mexico is used with a drying toilet system. Urine is either drained to a rock-filled leachpit or utilized. (Espacio de Salud/César Añorvé)

Why Keep Urine Out of the Composter?

- ☐ Urine adds more water than is necessary to compost feces and toilet paper. The moisture content of feces (66% to 80%), plus the addition of toilet paper, is sufficient for good processing.
- ☐ The nitrogen in urea is quickly converted to ammonia, which is toxic to composting microbes and causes odor.
- ☐ When urine is combined with feces, certain malodorous compounds form, which do not form if they are kept apart. The bacterium, *Micrococcus urea*, is the responsible culprit. (Wolgast, 1993)
- ☐ Although most of the nitrogen is either drained away as leachate or evaporated, it would require significant amounts of carbon (25 parts carbon for one part nitrogen) to provide the optimum ratio to convert the nutrients in urine to a plant-available form. For some systems, that would require wheelbarrows full of additive.
- ☐ The excessive salt (sodium chloride) we put in our foods is excreted in urine. Salt is toxic to living organisms in amounts above their minimum requirements, and can inhibit the composting organisms. Further, it can turn into "salt concrete" in locations where removing it is difficult, and can gum up a composter's mechanisms.
- ☐ The fertilizer value is lost when the ammonia in it volatilizes, and the resulting gas escapes up the exhaust pipe. For that reason, some may overestimate the nutrient value of composted human excrement.

Composition of Daily Urine Normally Produced by an Adult

Substance	Amount
Urea	6.0 - 180 grams (nitrogen)
Creatine	0.3 - 0.8 gram (nitrogen)
Ammonia	0.4 - 1.0 gram (nitrogen)
Uric acid	0.008 - 0.2 gram (nitrogen)
Sodium	2.0 - 4.0 grams
Potassium	1.5 - 2.0 grams
Calcium	0.1 - 0.3 gram
Magnesium	0.1 - 0.2 gram
Chloride	4.0 - 8.0 grams
Bicarbonate	-
Phosphate	0.7 - 1.6 grams (phosphorus)
Inorganic sulfate	0.6 - 1.8 gram (sulfur)
Organic sulfate	0.006 - 0.2 gram (sulfur)

Source: *Van Nostrand's Scientific Encyclopedia*

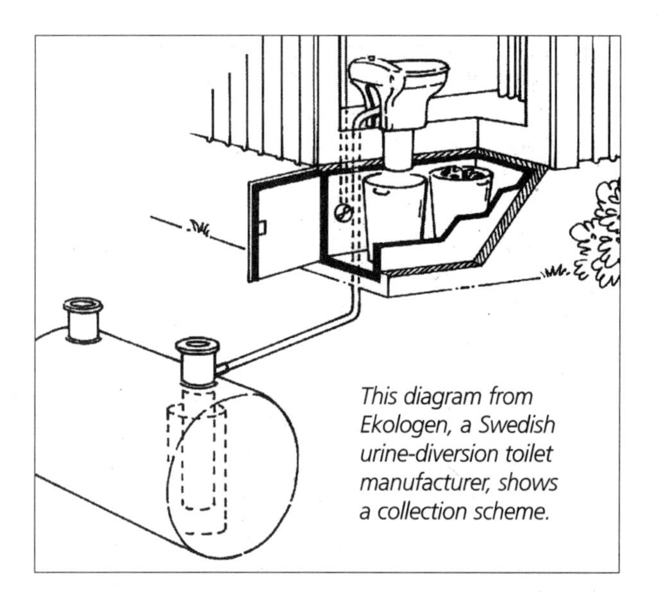

This diagram from Ekologen, a Swedish urine-diversion toilet manufacturer, shows a collection scheme.

What to Do with It

The 115.7 gallons (438 liters) of urine produced by one person in a year is a significant load for any composting toilet system to manage, so dealing with it separately offers some advantages. Draining it for immediate use is the most direct solution. Consider the following:

For one person:

- to evaporate one gallon would require a minimum of 9,000 Btu's (2.6 KwH) of energy (enthalpy of about 1,044 Btu/lb. at 87°F). However, so much is lost in the transfer of heat, you would likely require more Btu's of fuel. (ASHRAE Handbook)
- to store it would require at least two 55-gallon (208-liter) drums
- to soak up one year's worth of urine into a moist carbonaceous form so that it will compost (minimum 50 percent moisture by weight) will require at least 980 pounds (441 kg) of an absorbent, such as bone-dry sawdust. That may require a lot of material. For example: sawdust has a density of 12 pounds (5.4 kg) per cubic foot, so you may need 82.5 cubic feet (2.3 cubic meters) of dry sawdust to soak up 115.7 gallons (438 liters) of urine!

If you plan to use urine for fertilizer, it needs to be oxidized to a plant-available nitrate state first, or diluted and mixed into well-aerated soil, where the aerobic microbes will complete the oxidation ("nitrification") process.

See page 111 for urine composter designs. See the Appendix for notes about inoculants for urine composting.

To utilize the nutrients in urine effectively,

1. Drain urine directly into an aerobic soil system. One way is to drain the urinal/urine-diversion toilet to be combined with the graywater and used on site—no storage necessary.

2. Collect the urine for later use. This requires draining the urinals to storage tanks that can be carried or pumped to the point of use, say, an orchard. Then the urine should be diluted with 8 parts fresh water for every one part urine (because the urea and salt in urine is too concentrated for most plants) and sprinkled on the soil, not on the plants. The soil microbes will oxidize the nitrogen-containing compounds to a form that the plants can best utilize.

Or, make a urine composter. Pour urine mixed with sugar over a bale of hay or into a column, such as an upright 10-inch (254 mm) pipe, packed with peat, sawdust or shredded cardboard. Or pack it with a non-degradable media that provides surface area for aerobic microbes, such as little plastic sponges or coarse sand. Drain the oxidized urine to use as liquid fertilizer. (For more details, see the Appendix.)

The two bacteria mostly responsible for converting urea into a nitrate are *Nitrobacter* and *Nitrosomonas*. These nitrifying bacteria need carbon for their protoplasm, cell walls and the enzymes that convert organic nitrogen molecules to nitrates. For a more specific inoculant, you can grow a *Nitrobacter- Nitrosomonas* culture by mixing urine and sugar, letting it process aerobically, and adding this to the urine barrel/composter. See the Appendix for more specific directions.

The Carbon-Nitrogen Connection

Remember that the C:N ratio of urine—0.8:1—is too low for microbes to use and fully break down into a plant-usable form. To preserve all of the fertilizer value of

Graywater and Urine

Urine is high in nitrogen. Graywater is high in carbon (from soaps, detergents, oils, hair conditioner) and low in nitrogen. Combine the two, and you have a better effluent to use in a planted system. However, both contain a lot of sodium, so be sure to use salt-tolerant plants (see the plant list in "What About Graywater?"). There may be other options; the Center for Ecological Pollution Prevention is presently investigating making a sodium filter out of gypsum.

urine, a carbon-to-nitrogen ratio of 25:1 is needed to convert all of the nitrogen to nitrates by aerobic soil bacteria. (Again, nitrate is the form of nutrient that plants use best.)

But that would require loads of high-carbon materials such as wood chips, peat moss, etc. Most of us are not on a farm that could provide enough straw or leaves from trees for this purpose. And, remember that the primary role of additive is to provide pore spaces for air, not to provide carbon. Instead of adding a room full of sawdust (C:N 400:1) to a year's worth of urine to achieve a C:N of 25:1, consider adding some more concentrated forms of carbon, such as alcohol (as wastewater treatment plants do) or simply sugar.

Why add sugar to a urine composter? Sugars are the most "available" of all the carbohydrates, and can provide a lot of carbon with very little volume compared to straw, leaves and woodchips. These are primarily difficult-to-digest celluloses and lignin, which, in a composter, are decomposed by fungi and actinomycetes, not by bacteria.

Sugar, however, is a simple carbohydrate. Adding some cane sugar may not seem ecologically elegant, but it will surely adjust the C:N ratio, in place of a truckload of sawdust. About a third of a cup of sugar per person per day should do it. (See the Appendix for more details.)

Urine's Variable Reputation

Urine has a bad rap in many cultures due to its association with the smell and disease-causing nature of feces, according to Jan-Olof Drangert of the University of Linkoping in Sweden. In his paper, "Perceptions, Urine Blindness and Urban Agriculture," Drangert reports that this has not always been the case; throughout history, urine has been used for beneficial purposes.

He reports that in Sweden, urine has commonly been used to clean wounds, and to some extent, to drink as therapy. Recently urine has been shown to have a disinfectant property. In the Danish countryside in the 19th century, urine was stored and used as a detergent for washing clothes and yarn dyeing. A century earlier, European artisans collected urine and canine excrement for industrial uses. U.S. military survival training manuals say it is safe to drink one's own urine if no drinking water is available.

Diverting and using urine may seem on the "lunatic fringe" now, but the benefits of doing so are hard to ignore, and it will be a "common sense" practice in the future.

References

Hakan Jonsson, Björn Vinneras, Caroline Höglund, Thor-Axel Stenström, " Source separation of urine" (Wasser & Boden, 51/11, 21-25 ISSN 0043-0951, Berlin, Germany 1999)

Thor-Axel Stenström, Caroline Höglund, Hakan Jonsson, "Evaluation of microbial risks and faecal contamination of urine divertingf sewage systems" (Wasser & Boden. 51/10, 11-14 ISSN 0043-0951, Berlin, Germany 1999)

Caroline Höglund and Thor-Axel Stenström, "Survival of Cryptosporidium parvum in source separated human urine" (Canadian Journal of Microbiology 45: 1-7 1999)

K. Sundberg, "What Does Household Wastewater Contain?" Report 4425 (Stockholm, Sweden: Swedish Environmental Protection Agency, 1995).

Hakan Jonsson, "Assessment of Sanitation Systems and Reuse of Urine," Swedish University of Agricultural Sciences Proceedings of the SIDA Ecological Alternatives in Sanitation Conference, August, 1997.

Uno Winblad, ed., Ecological Sanitation (Stockholm: Sida, 1997).

Soils for the Management of Organic Wastes and Wastewaters, (Madison, Wisconsin: Soil Science Society of America, 1977).

ASHRAE Handbook & Product Directory, 1977 Fundamentals, Psychrometric Tables (New York, NY: American Society of Heating Refrigerating and Air-Conditioning Engineers).

Mats Wolgast, "Rena vatten—om tankar i kretslopp" (Uppsala, Sweden: Creanom HB, 1993).

Jan-Olof Drangert, "Perceptions, Urine Blindness and Urban Agriculture," Ecological Alternatives in Sanitation, Proceedings from Sida Sanitation Workshop, August, 1997.

CHAPTER FIVE

Installing Your System

Some General Installation Guidelines

Proper installation and routine maintenance are critical to the success of composting toilet systems. The following is a general and generic manual to give you an idea of what is involved. Always refer to your system's manual for specifics.

★ *Always follow the manufacturer's instructions so you do not void the warranty!*

Locating the Composter

For easy maintenance and repair, be sure to provide convenient access to the emptying door (if flooding is a concern, it should be at least 18 inches above the floor), as well as the liquid drainage system, the exhaust pipe system and the electrical service connections for the heating system and fan. There are a few reports of owners who installed their systems in tight spaces, and then ultimately discontinued using them simply because servicing them was a hassle.

Check that the support or the floor where the composter will be placed is dry and strong enough to support the load. The composter should be placed on an insulated, level, flat and stable surface, such as concrete, compacted earth, moisture- and rot-resistant lumber or similar material. If the floor is cold (such as in a basement), insulate the composter bottom to prevent heat loss with a minimum of two-inch closed-cell foam insulation board. Foamglas® by Pittsburgh Corning is good because it is an rigid cellular glass foam, will not burn or become unstable, and has constant insulating efficiency and zero moisture permeability. It is very expensive, but you only need a small piece of it.

Access and Drainage Requirements

Consider locating the unit 2 to 18 inches (51-457 mm) above the grade of the floor. This will allow easy access (less stooping) for routine maintenance. Also, an elevated composter creates a grade drop so you can better drain the leachate into another area or container.

A self-contained appliance composter that sits on the toilet room floor has a fixed height from the floor to the toilet seat, but you could place it on a platform to make the removal area easier to get at. Doing this, however, raises the height of the toilet seat. Remember to figure in the space needed to pull out the end-product removal drawer.

Allow yourself plenty of room for maintaining and emptying your composter. (Photo: Vera Miljø)

To move leachate horizontally, you will need to provide a downward pitch of one-quarter inch per linear foot (6.35 mm per linear meter). For example, if the destination is four feet away from the leachate outlet, a one-inch drop is required to drain it by gravity. (Typical pitch for clear, fresh water is one-eighth inch (3.3 mm) per linear foot, but leachate contains suspended solids, and so it requires a steeper pitch.)

Elevating the system may not be necessary if the composter has access doors at knee level or higher, or if there is an integrated leachate collection tank with an automatic pump.

With a self-contained composter that has a drain in the rear, such as the Sun-Mar, it can be helpful to shim (elevate) the front one-eighthto one-quarter inch (but no more) so that liquid drains to the back. This will encourage drainage of the leachate should you need to remove it. Your hardware store can provide you with beveled shims, or you can use cedar roofing or siding shingles. Be sure to keep the unit from tipping from side to side. Of course, you might choose to shim it only just before draining the system. (Draining leachate is mostly an issue with units that do not have power-assisted evaporation systems.)

When in Doubt, Rely on Your Plumber for Installation Assistance

It is impossible to cover all of the installation issues of specific technologies in this book, so let us leave you with one major thought:

When in doubt, rely on your plumber for installation assistance.

Many composting toilet systems merely require installing an exhaust pipe, which carpenters and many homeowners are capable of doing. However, for more complicated installations, such those with micro-flush toilets, a licensed master plumber is the best qualified professional to install your system. No other specialist better understands proper installation of mechanical systems and pipes of any kind. A seasoned plumbing mechanic has worked within the local and state building codes to solve the most difficult drain and ventilation problems. He or she has negotiated all of the walls, floors and roofs of countless buildings to move sanitary effluents and odors safely through the building and perfect the necessary exhaust and make-up air-supply system.

Many plumbers may be unfamiliar with the terms "composting toilet" or "graywater system," but they do understand moving air into and out of any unit process and the drainage or pumping of liquids (leachate or gray water). Give the plumber all of your engineering documents, plans, specifications, and installation and operation manuals. Be sure to refer to the specific approval documents or applicable state code sections. As with regulators, some plumbers are more open to "new" technologies than others. Some might even tell you that certain things cannot be done (because they have not done them before). If that is the case, your options are to either educate the plumber or find another.

For those plumbers who still consider composting toilet systems a funky technology, let them know that the Commonwealth of Massachusetts has included composting toilet systems within its Uniform State Plumbing Code (248 CMR 2.00), and requires that licensed plumbers install plumbing-board-approved composting toilets just as they do water closets (their term for flush toilets). It took many years of lobbying, but the plumbing board ultimately realized that, while these toilets do not require water, they do need proper drainage and venting. And it's a good business opportunity for plumbers.

If you want your plumbing contractor to bid the installation job competitively, you will probably need a Plumbing Specification to furnish to each of the bidding companies.

If you cannot find a plumber to do the installation and want it to be installed according to best plumbing and mechanical practices, you could go to your library and browse through the Uniform Plumbing Code (UPC) published by the International Association of Plumbing and Mechanical Officials (IAPMO) (20001 Walnut Drive South, Walnut, CA 91789-2825; telephone: 714/595-8449). The UPC has been adopted by many of the states in the United States and even such faraway places as the Republic of Palau in the South Pacific. Although the code itself is instructive, the Illustrated Training Manual is rich with information, and provides pictures and diagrams to illustrate all plumbing work, including the exhaust venting, drainage and piping requirement that would apply to a composter installation.

Many of the installation considerations are the same as those for hot water heaters, furnaces and boilers. That is because many of the fundamental issues are the same, such as location, support, length, pitch, rise and clearance of pipes and vent components, methods of supplying make-up air, distances of the access opening above grade (18 inches or 457 mm, for example), seismic bracing, roof flashing techniques, use of flexible pipe, unions and where the end or termination of the exhaust pipe with a vent cap should be relative to open windows, etc.

Installing the Pipes

Always read the installation instructions and plan the installation before making any cuts in the composter, your building or the pipes. To provide more flexibility in locating the pipe (to permit alignment of the toilet-connecting pipe and/or the compost removal door), use flexible synthetic rubber no-hub connectors as unions to facilitate servicing and repairs. They will also serve as vibration-absorbers that reduce noise from fan motors.

Note: In some larger models, the air-intake opening is in the center of the composter lid and it may be covered with a plastic cap, which should be removed. Consider connecting a hot-air supply duct to the air-intake opening, to provide warmed air to the composter. Or you can simply cover it with a fly-screen tube.

Installing a Micro-Flush Toilet

When installing a micro-flush toilet, such as the SeaLand Traveler (see "Toilet Stools for Composting Systems," Chapter 7), with a composter, remember that drainage is very tricky. One pint (half a liter) of water cannot move solids in excrement and toilet paper very far laterally or horizontally unless the pitch is a 35- to 45-degree angle down from the toilet to the composter. The steep downward pitch required for one-pint toilets is far greater than that for conventional toilets, because a well-designed 1.6-gallon flush toilet has a drain-line carry of 59 feet at one-quarter inch per foot pitch in a four-inch PVC pipe.

** 6-liter flush toilet has a drain-line carry of 17.9 meters at 6.35 mm per meter pitch in a 101-mm PVC pipe*

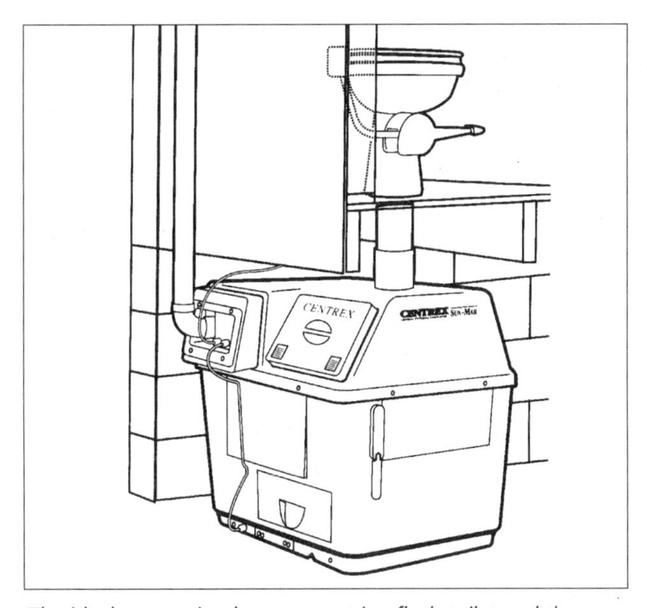

The ideal connection between a microflush toilet and the composter is as direct and vertical as possible. (Graphic: Sun-Mar)

Issues of optimum pipe diameter versus pitch are critical, as a given portion of ETPA will not move *unless* there is sufficient water behind it to push it through or to lift and float it through the pipe. There is an optimum pitch and pipe diameter for every mix of solids and liquids. If the pitch is too shallow, solids will drop to the bottom and the liquids will slowly seep away. If the pitch is too great, then the liquids will precede the solids and the solids will drop out. If it is correct, the flush surge will lift up the solids and carry them down the pipe. If the pipe diameter is too large, then the sidewalls of the pipe will not be scoured by the water, and the energy in the water will be too diffuse to move the solids. You may not know that a problem exists until several portions of solids have blocked the pipe—and then it is too late for a quick and easy solution. For a low-flush installation, consider providing removable clean-out fittings at each end of the horizontal pipe run to provide easy access should a blockage occur.

A licensed plumber should make the connection with an appropriate flange on top of the composter to accommodate the toilet drain line. One innovative plumber we know used a plastic floor-flange that would normally be used to connect a flush toilet to the drain line and to the floor. A union (a threaded plumbing fitting designed to be routinely separated) should serve as a quick disconnect should it be required. Hard connections, such as solvent welding, are too permanent for this purpose.

Look out below!

Some composters may have problems with the impact of the flush water and ETPA falling 30 feet (9 meters) from the second floor toilet to the composter in the basement. It causes major craters and splashing of the processing material. One manufacturer has added splash plates to his design to prevent this. Ask your manufacturer for recommendations.

Installing a Gravity-Drop Connecting Pipe with the Composter Beneath

1. After carefully measuring (twice!), cut a hole in the floor(s) and place the connecting pipe through the hole and into place in the top fitting of the compost container, if it is provided. Remember that the

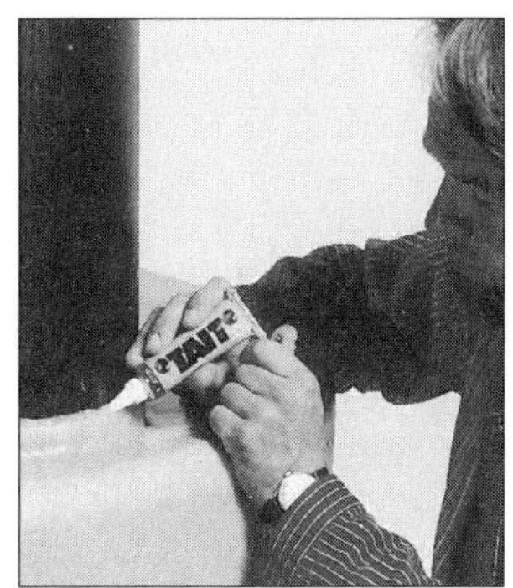
(Photo: Vera Miljø)

composter can be moved to adjust the location. Use a plumb bob from the center of the hole to locate the inlet to the composter.

2. Using a waterproof sealant such as silicone caulk, seal any gap between the connecting pipe and the composter. Unlike solvent-welded joints, silicone perfects a seal, but allows the connection to be broken if you wish to change the location of the composter or pipe in the future.

3. Place the toilet stool atop the connecting pipe. The connecting pipe may be cut to desired height to accommodate your toilet above the floor surface.

4. Position the toilet stool around the pipe (according to the specifications of the toilet stool you have selected).

Connecting the Exhaust Pipe

1. Fasten the top mounting coupling for the vent pipe with screws to the top (or the side) of the composter. Use the flange provided or secure the proper flange for the diameter of the exhaust pipe selected, and after carefully measuring, cut into the top of the composter and caulk it to seal it. Find the best possible place to make the hole on the top of the composter if there is none provided, taking into consideration that the pipe should be as straight as possible. Note: The fewer the elbows, the better the ventilation.

2. If there is no flange available, here is another option: Cut a 2 3/4-inch-long (67 mm) length of the vent pipe and cement it into a coupling. This will leave a short one-inch stub beyond the edge of the coupling. Using the outside diameter of the vent pipe as a guide, draw a circle on the top of the composter where you want the vent pipe to be located.

Carefully cut out a hole in the composter (unless one already exists). Place the stub of the vent pipe in the hole cut into the composter and caulk it thoroughly with silicone to prevent odors from escaping when the system is in use. Or if it is a four-inch (101.6 mm) vent pipe, cement a four-inch closet or toilet-mounting flange over the hole you have cut in the composter and fasten the exhaust pipe to it. Again, remember to insert a union or a no-hub flexible coupling in the line to allow you to remove the composter for service without destroying the exhaust pipe.

3. The fan (if required) should now be installed. (See instructions later in this chapter.)

Easier Pipe Connections

The UniSeal is a simple one-piece inexpensive way to connect pipes to tanks. It is a rubber grommet that is installed in minutes. Use it for pipe diameters from 3/8-to 6-inch. Available from: A.G.S., Inc., 2530 County Road 775, Perrysville, OH 44864; 419/938-7120.

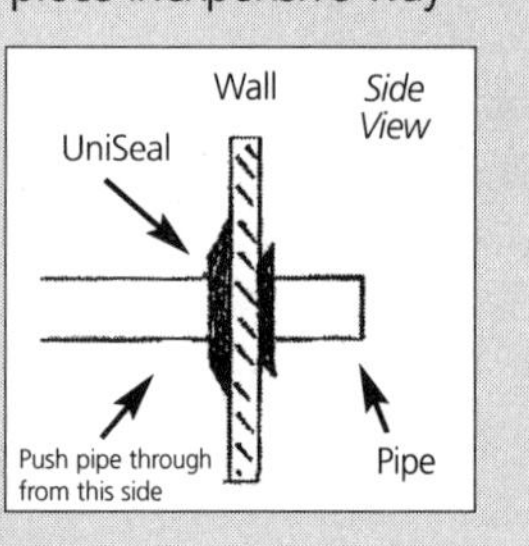

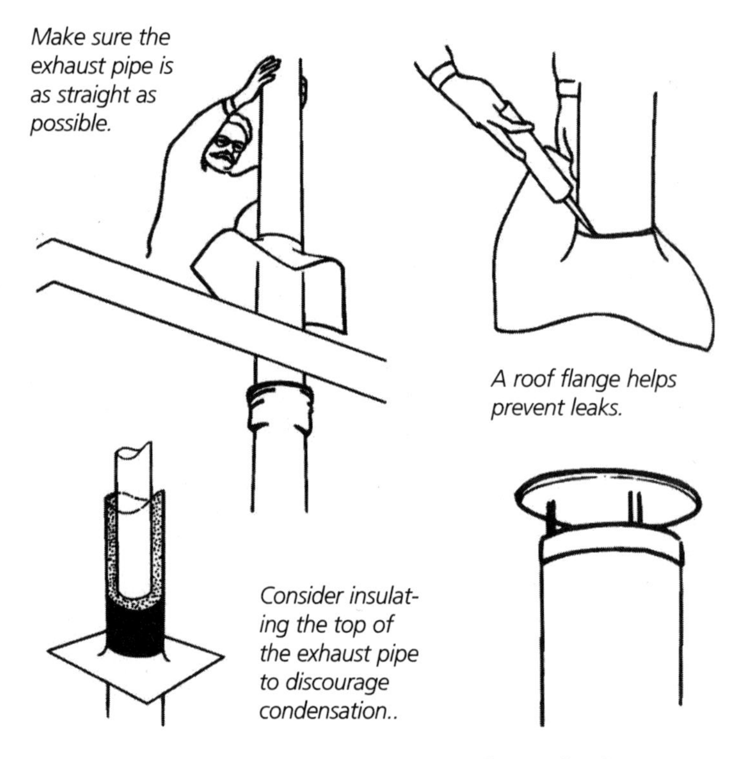
Make sure the exhaust pipe is as straight as possible.

A roof flange helps prevent leaks.

Consider insulating the top of the exhaust pipe to discourage condensation..

Install a rain hat (shown) or pipe Tee on the exhaust pipe terminus.

(Graphics: Vera Miljø)

4. Cut a hole in the roof straight above the exhaust vent pipe. Use a plumb bob to determine exactly where to cut. Push the ventilation pipe through the hole in the roof and down into a roof boot (get the correct one that matches your roof pitch and the pipe diameter) on the roof. Fasten the rubber flashing gasket both to the ventilation pipe and to the roof with the silicone. This prevents rainwater from seeping into the building.

5. Make sure the exhaust vent pipe is as straight as possible and that all joints are sealed with silicone caulk. In colder climates, insulate the vent pipe to prevent condensation and to improve the draft.

6. Wind may cause strain on the pipe where it meets the roof. If you suspect this might cause a problem, install an extra woodstove pipe brace on the roof to prevent damage.

7. Mount an exhaust pipe rain cap (a 90-degree tee coupling will also serve this purpose) and fasten it with the silicone. Cover the outside of the cap or tee coupling with a loose bag of rustproof fly screen. Fasten it around the vertical pipe with corrosion-resistant fasteners, such as stainless-steel wire or hose clamps.

Installing a Leachate Drain Line

Fasten a flexible transparent drain hose (if your plumbing code approves this design) between the composter's leachate drain fitting and the drain pipe or leachate container with clamps. This can function as a leachate-level indicator if the hose is mounted in a vertical position. Better, have your plumber make a permanent drain with hard pipes, with removable unions that flow to a leachate management system.

A leachate line on a 55-gallon net-batch system.

★ Leachate should never reach levels as high as the air intake holes or any nonwatertight doors! Leachate can spill out of the unit and cause serious trouble (odor, damage to floors, and health threat, not to mention an unpleasant mess).

For nonelectric models, the hose can be placed in a horizontal position to continuously drain the liquid to a container or an approved soil absorption system. This can also be done on electric models if the leachate is not evaporating rapidly enough (or to save energy). Periodically check to see if there is excess leachate that should be emptied.

Make sure that all connections are both airtight and watertight. If necessary, use silicone or plastic duct tape to seal joints.

Checking Air Paths

1. Make sure that all connections are airtight and watertight. If necessary, use silicone or plastic duct tape to seal joints. The exhaust system should be constructed so that a minor vacuum is created in the compost container that will draw air into the unit and up the exhaust pipe. Odors may result if this vacuum disappears, due to loose connections or if an emptying door or toilet seat is left open.

2. Ideally, there should be a make-up fresh air supply vent to the wall in the toilet room installed below the level of the toilet stool at the floor. This provides a good air supply and prevents air from being sucked back out of the composter and through the toilet stool, which could result in odors. For this reason, there should be no other exhaust fans in the same room. The composter itself will suck all odors out of the room, so it replaces the customary bathroom vent fan.

Installing a Fan in the Exhaust System

The following instructions are adapted from those provided by FanTech, a popular fan choice of composting toilet system owners. These are for Model FR-100, designed for high-humidity environments. Available in 115 and 220 VAC or 12 VDC. Use a fan-speed controller.

STEP 1

When selecting fan mounting location, consider the following:

- Fan location should allow sufficient access for service. You will want to insert a piece of the exhaust pipe in place of the fan when you are servicing it or the electric power has been turned off. An in-line fan is an obstruction to natural chimney effect draft, so it should only be in-line when it is meant to be continuously running!
- The best location for the fan to be mounted is as close as possible and practical to the termination of the vent pipe. This minimizes the transmission of vibration and motor sounds back to the toilet room.
- In order to prevent bathroom odors, ensure that there are no competing demands for air, such as fireplaces, or bathroom exhaust fans, and open windows on the side of the building *opposite* the direction from which the prevailing wind blows. If necessary, cut one or two inches off the bottom of the bathroom door to supply make-up air to the commode.

STEP 2

Using wood screws, attach the fan motor mounting bracket to a support beam at the selected location. We recommend vertical mounting to reduce condensation build-up. However, if horizontal installation is necessary, either wrap insulation around the fan or drill a 1/4-inch (6.35 mm) hole in the bottom of the housing (along with a threaded insert and drain tubing), allowing condensation to drain out of the fan motor housing.

STEP 3

Attach the fan to the mounting bracket with sheet metal screws. Be careful to ensure that direction of air flow is *up* the exhaust pipe to the roof *from* the composter. Brackets are usually provided with rubber vibration isolation pads or grommets to prevent the transmission of sound through the structure. Do not overtighten. Also, take care not to strip the plastic housing on the fan. Although sheet metal screws are self-tapping, we recommend that pilot holes (no larger than 3/32 inches) be drilled.

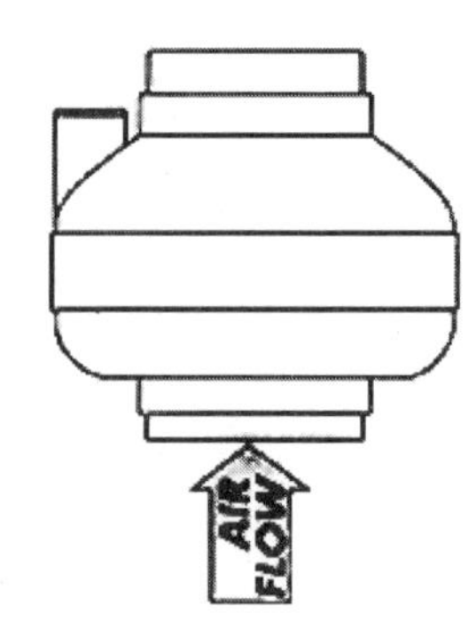

STEP 4

Connect four-inch (101.6 mm0 ID vent pipe (drain, waste and vent pipe recommended) to the inlet and outlet of the fan housing using quick-disconnect rubber couplings such as Fernco flexible couplings. While the worm clamps on the couplings should be snug, take care not to overtighten (60 pounds/27 kg of torque maximum).

NOTE: Steps 2 and 3 may be reversed.

THE ELECTRICAL CONNECTION

Installation work and electrical wiring must be done by qualified persons in accordance with all applicable codes and standards.

DO NOT CONNECT THE POWER SUPPLY until fan is completely installed. Make sure electrical service to the fan is locked in the "OFF" position.

While we recommend that the fan be left running at all times, choose a fan motor suitable for fan speed control use. *Pulling too much air through the composter could cool the process if the temperatures are below the minimum recommended by the manufacturer.* A number of users have found that it is advantageous to have a wall-mounted fan speed controller near the commode so that the fan can be optimally efficient, operated at minimum noise/air velocity, and shut off temporarily while it is being serviced. It is important to note, however, that the fan be turned on immediately following service. NEVER place a switch where it can be reached from a tub or shower.

Bring incoming electrical service through the connector clamp and the knockout on the junction box. Make the necessary connection in accordance with state electrical codes. Ensure a proper ground, unless the fan motor is isolated within a plastic housing, in which case grounding is not necessary.

Secure connections, clamp connector and terminal box screws.

TROUBLESHOOTING

If the fan fails to operate, check the following:

- Is the power on?
- Consult the wiring diagram provided by the fan manufacturer to assure proper connection.
- Check the motor lead wiring and incoming supply leads to assure definite contact.

MAINTENANCE

Since fan bearings are usually sealed and provided with an internal lubricating material, no additional lubrication is necessary. If the fan is to be turned off and left inoperative for more than 10 days, we recommend that it be removed from the vent pipe and stored in a dry environment (and a short section of vent pipe be installed in the fan's position in the vent pipe).

CHAPTER SIX

Operating and Maintaining Your Composting Toilet System

Starting Up Your Composter

To prepare a large composter for use, pour some one-inch bark or wood chips into it to provide a drainage base (this acts much like the drainage material you put on the bottom of a flower pot). On top of that, put a layer of additive, such as horticultural soil-less media (see "Additives," next page), and moisten it so it has the texture of a well-wrung sponge.

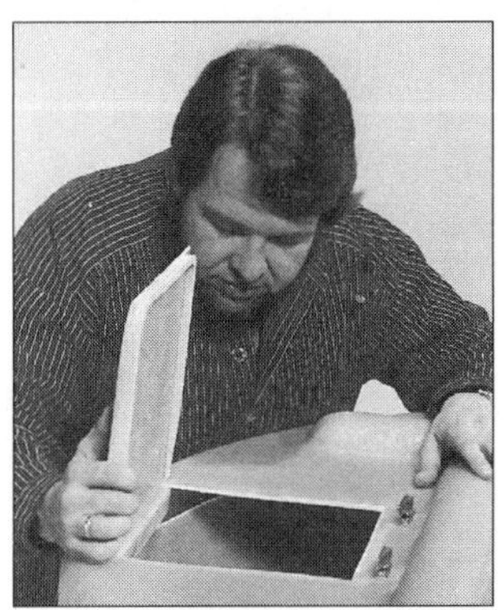
(Photo: Vera Miljø)

The wood chips act as a filter, preventing fine solids from blocking the leachate drain holes or densifying in the end-product removal area. This layer also helps aerate the leachate. For some sloped-bottom composters, the manufacturers recommend filling the composter at least two-thirds full of wood chips. Other manufacturers suggest only two to four inches of wood chips or bark. Don't overfill or you will be removing undecomposed wood chips in a few months!

Quick-Starting Your Composter with Organisms

To get your composter composting soonest, add some compost bacteria—either compost microbes purchased from a garden supply store or a shovelful of well-composted matter from an active manure or yard waste compost pile or perhaps even the forest floor. This process is referred to as "inoculating" the composter. You might think of it as adding yeast to bread dough: Microbes are like packaged yeast, whereas a shovelful of compost is like sourdough starter. Once introduced into the composter, the microorganisms will rapidly multiply to use up the nutrients you have added. The ones that can utilize the nutrients in the ETPA (excrement, toilet paper and additive) will thrive and multiply if you have provided the proper moisture and heat. Organisms that cannot utilize the available nutrients will die.

About Additive (a.k.a. Bulking Agents)

A handful of additive should be added to the composting toilet about every 10th visit (more if you have a small passive system, and wish to cover each deposit).* This will help create air pockets in the material, which aid the composting process. The result will be better processing (see the next page for suggested additives).

The C:N Misperception

Especially in public facilities, composting toilet operators occasionally overemphasize the need to maintain the proper carbon-to-nitrogen ratio of 25:1 in composting toilets. The problem is, you might need a truckload of carbon-containing wood shavings to truly match the nitrogen in a family's composting toilet. In composting toilets, bulking agents are added to soak up the urine and provide porosity to improve aeration, not necessarily to provide the 25:1 C:N ratio.

In composting toilets, bulking agents are added to soak up the urine and provide porosity to improve aeration, not necessarily to provide a 25:1 C:N ratio.

Most of the nitrogen is in urine—which drains to the bottom of the composter, where it is either evaporated or drained away. In either event, the nitrogen is gone and does not need to be properly matched with carbon, unless you want to use the finished product as a fertilizer. In that case, binding up all the nitrogen with carbon so that it composts into humus is worthwhile. However, adding a cup of sugar from time to time would provide a more digestible and volume-effective carbon source than filling up the composter with wood chips (or, as some would suggest, peat moss) that would fill up the composter before it had decomposed.

*Fluidized bed composters, which are engineered systems, do not require bulking agent.

Additives for Composting Toilets

One of the most important rules of good composting is to keep the compost porous and aerated, not compacted.

Biological decomposition requires moisture, oxygen and balanced nutrition, just as humans do. The enzymes that the aerobic bacteria produce to break down larger organic molecules into simple oxidized molecules that can be utilized by plants for growth prefer simple sugars, starches and complex carbohydrates to woody cellulose and lignin found in tree products. Toilet paper, peat moss and wood shavings will also break down, however, it will take longer, and the composter could fill up faster with uncomposted additives as a result

When possible, add dry, starchy kitchen scraps from the house rather than purchasing special materials. Shred or chop additives into 1/4-inch (6 mm) pieces. Leaves tend to mat down, causing loss of air space, so be cautious. Some add fruit and vegetable scraps, too. This is fine for composting, but could draw flies. ***Use plant-originated materials only! Do not add animal products or oils of any kind.***

Adding a cup of additive every 10 visits is a good rule of thumb; some prefer to add something after every visit and some add a cup every two weeks. However, adding a lot of additional matter to the composter can fill it up sooner, reducing its capacity to process excrement.

Suggested Additives

The following is a list of additives in order of their ease of digestion by compostin microbes:

Patti Nesbitt adds a "scoop per poop" of bark mulch to her composter. Less additive works, too.

1. ***Sugars:*** Fruit contains fructose, a fruit sugar, so fruit cores, peelings and bits work well. However, flies like them, too, so you may wish to steer clear of fruit, if flies are a concern. Cut up large pieces. Avoid citrus peels, as they impede decomposition. For carbon alone without the bulk, add a cup of sugar every two weeks.

2. ***Starches:*** These include potatoes (peels, too), yams, pumpkin and squash. Corn is both a sugar and a starch; stale popped corn works best. Vegetables can work well, too, although some may draw flies.

3. ***Complex carbohydrates:*** These include cereals, grains, cooked and dried pastas, fresh and stale bread and biscuits, crackers, popcorn, etc.

Some Additives

Clockwise from upper left: Soil-less growing media, untreated bark, popped popocorn, crushed shells.

Popped popcorn is an ideal additive, as its composition, shape, size, porosity and absorbing qualities creates air pockets, absorbs leachate and breaks down fast.

Houseplant trimmings and leaf mould also work. Avoid grass clippings, which are highly nitrogenous and tend to mat down.

(If yours is a self-contained composting toilet with a grate sifter, remember not to add anything that cannot pass through the grate.)

*4. **Cellulose:*** Good cellulose-rich materials to use include wood shavings, fine shredded bark, leaves and small chips (but not cedar, eucalyptus wood, birch or pine bark or other woods that inhibit biological decomposition). Do not add sawdust, which is too fine and will compact. Toilet paper (preferably white with no perfume), finely chopped newsprint (no colors or glossy paper), plain brown corrugated cardboard, dry wood pulp from a paper mill and wood pellets for stoves can be added, as can chopped dry straw, hay, alfalfa, pea straw, kenaf, and grain hulls, such as rice and buckwheat.

*5. **Minerals:*** Minerals will not break down, but they can help aerate the compost. Horticultural-grade perlite and vermiculite work well as aerating agents.

Packaged Additives

In addition to bark pieces, the professional nursery industry offers a wide variety of soil-less mixes for bedding plants. These work quite well as composting toilet additives.

A wide range of mixes are available. Some of our favorites from Fafard of Canada: Mix #52 consists of 24 percent peat moss, 8 percent perlite, 8 percent vermiculite, and 60 percent bark. Premier Pro-Mix 'BX' includes dolomitic and calcitic limestone as a pH adjuster (neutralizes the slightly acidic urine). Scotts Coir Growing Medium #560 contains processed bark ash and coconut husk fiber. Coconut coir (fiber) is available both loose and in the form of mats, which make excellent first layers.

They are available for about $12 a bale or large bag from Griffin Greenhouse and Nursery Supplies, 1619 Main Street, P.O. Box 36, Tewksbury, MA 01876; 978/851-4346. Or contact your local nursery supplier.

Earthworms

In the last 10 to 15 years, vermicomposting—using earthworms to hasten the breakdown of organic matter—has become popular.

Worms transform organic material in their digestive tracts, so that their fecal matter, called "castings," is rich in nutrients that are ready for plants. Equally important, by virtue of their ability to burrow deep and come to the surface often, earthworms provide deep aeration for soils and prevent compaction.

Earthworms are not very happy at the high temperatures that can occur in composting (over 100°F/38°C), nor can they tolerate low or high moisture levels, highly acidic (low pH) environments, or being in material that is often mixed, tumbled or chopped. So they will not be happy in many of the small composters. In large composters, they should help, especially in continuous composters, because of the compaction problems in these systems. Keep in mind that worms prefer kitchen scraps to excrement.

Some claims have been made that pathogens are killed by the enzymatic activity in worm's digestive tracts. Alas, it's not so. In fact, worms are potential carriers of certain pathogens, although this is uncommon in healthy populations.

The manufacturer of the AlasCan composting toilet system likes to use two or three types of worms found in compost piles: red worms (*Lumbricus rubellus*), brandling worms, or "red wigglers" (*Eisenia foetida*), and white worms (*Enchytraeids*). It's the red wigglers—sometimes called sewage worms—that are commonly used in larger-scale composting of organic wastes.

Culturally, advocating adding worms to your processor may turn off many people. You might think twice before mentioning the worms in your system to guests.

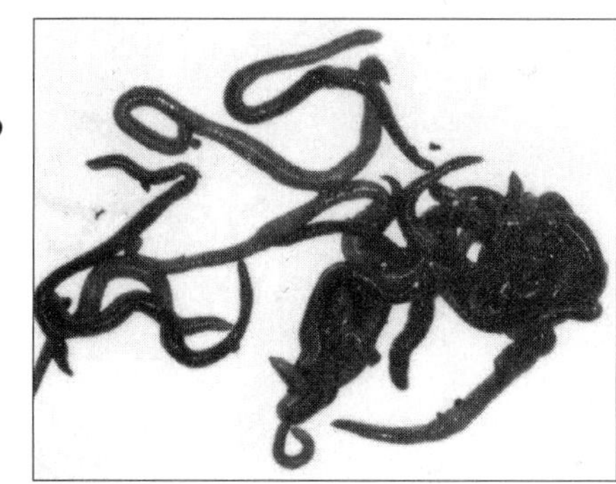

Worms from a composter.

And, no, worms will not crawl up through your toilet stool and surprise you. Worms don't like light, and will not leave moist organic matter for the cold, dry world of your bathroom.

Bottom line: Consider worms a compost helper but not a key player in your composting toilet system.

Worms can be purchased through many mail order garden suppliers as well as from worm growers that specialize in fast-acting worms. They usually run about $20 for two pounds. You may also find some regular earthworms in your yard waste composting pile.

Resources

Worm Digest is a newsletter that features all kinds of information on composting with worms, including suppliers: Worm Digest, Box 544, Eugene, OR 97440-9998; www.wormdigest.org.

Worms Eat My Garbage, a book by Mary Appelhof, discusses worm composters for kitchen wastes. (Flower Press: 1982)

Making Your Own Compost Starter

Just like some people make their own sourdough starter for bread, you can make an inoculum for your composter to make sure you have the right microorganisms at work. This is also a good idea for a planted aerobic graywater system.

You could purchase prepared compost activators that contain a few different organisms, but why do that when the best and lowest-cost additive is all around us!

The main players in an aerobic system are a healthy population of living organisms that transform the pollutants into safe and usable byproducts. In nature, the soil and roots of plants provide the environment for a diverse population of interdependent microorganisms and small animals that are busy at the work of decomposition and transformation. The enzymes and other chemicals produced by these organisms are critical to both the rate and completeness of the transformation. The more of these microbes present and the healthier they are, the better composting you get.

Because bacteria, fungi and other beneficial organisms are ubiquitous on this planet, providing a the right growth environment and nutrients is sufficient enough to attract the organisms. If you build it, they will come is one effective, albeit slow, way of bringing the players to the game. But a better way is to collect the player populations from an existing community.

Note that in activated sludge municipal wastewater treatment systems, a certain amount of sludge is recycled back to the first processing chamber to "activate" the process by adding the right microorganisms.

Remember that a particular diverse community develops in response to particular kinds and qualities of nutrients and environmental factors, so you must find a community where the nutrients are similar to the constituents you are planning to treat and transform. Essentially, go to where the microbes can be "seen" in action.

For example, a shovelful of:

- ☐ Warm compost or end-product from a neighbor's working composting toilet
- ☐ Warm, composted animal manure
- ☐ Composted plant materials—they are rich in aerobic organisms geared to carbohydrates, such as sugars, cellulose, lignin and plant proteins.

To transform urea and ammonia in urine to nitrates for plant use, you need aerobic *Nitrobacter* and *Nitrosomonas* bacteria, which can be found in composting litter (urine-soaked sawdust or straw) from outdoor animal pens and barnyards.

* Note that adding composted material from outdoors may introduce fly larvae to the system.

Note that when compost is warm, it means it is at its peak of microbial activity.

Millions, possibly billions, of living organisms are in a mere gram (0.03527 ounces) of healthy soil or compost. There are a greater number of organisms in the root zone of healthy plants, because the roots secrete "exudates," which are microbe nutrients such as sugars and the sloughed-off root cells. Dig around root zones first.

Kept warm, moist and oxygenated in your composter (or graywater system), they will multiply rapidly. Those that find the nutrients they need will thrive and multiply. Those that do not will die or adapt to the new situation.

Do-It-Yourself Inoculum

For a home composter, collect a gallon of sifted (take out big stones and sticks) soil or compost. Moisten it thoroughly with a pint or two of warm (100°F/38°C) non-chlorinated water to which you have added a half-cup (.12 liter) of sugar or corn syrup and a half teaspoon (2.5 ml) of liquid soap or dishwashing detergent. The former provides quick-start nutrients for the organisms, and the latter breaks the surface tension (makes the water wetter). Mix this with your additive or sprinkle directly on the composting material.

For a graywater garden, use the same recipe as for the composter, but use four gallons (15 liters) of warm water. Stir for two minutes, and then allow the solids to settle for five minutes. Pour the liquid down the drain, so it will be dispersed into the pipes and growing media or soil.

How often?

Add inoculum during system start-up, after removing material from your composter, and once or twice a year. Spring and fall are best.

If you use bleach regularly, you might want to reinoculate your graywater system more frequently. However, remember that growth of these organisms occurs whether you want it to or not—don't spend too much time or effort on this process after start-up if your system is working. An extra boost, however, it can only benefit the system.

Adding a few tablespoons of compost microbes available in powder form from the garden supply store.

Ongoing Maintenance Issues

Long-Term Periods of No Use

If you plan on going away for two or more weeks, and shutting off your fan and heater, remember that insects or odor could become a problem (see "Troubleshooting" for ways to remedy this). To let your system passively compost in peace during this period of inactivity, tape a sheet of plastic film under the toilet seat and lid (of the dry toilet stool). Unplug the electrical power. Remove the in-line fan and replace it with a length of pipe to keep air flowing. Cover the top of the active compost pile with four to six inches of moist additive.

Removing Composted Material

Different systems have different retention or throughput rates as specified by the manufacturer or designer. Some require emptying after a few months, some a few years. For larger systems, especially continuous composters, remove the end-product at least every two years to prevent compaction; even better is to remove a shovelful every six months.

When it is time to remove the end-product, it should not look or smell at all like the original waste material that went in. No feces should be identifiable. Ideally, it will be dry, and look and smell like very rich garden soil or leaf humus from a garden composter or the forest floor. This humuslike material is an excellent soil conditioner that can be safely buried under six inches of healthy soil next

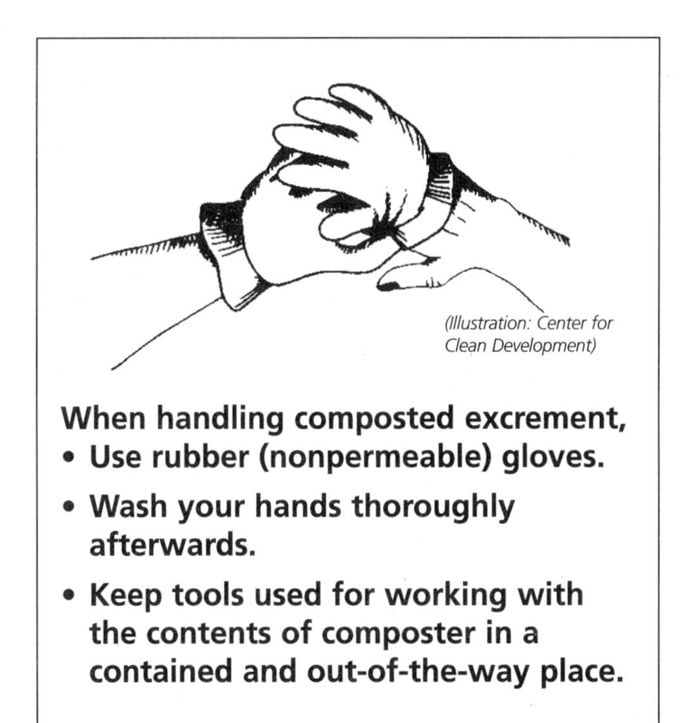

When handling composted excrement,
- **Use rubber (nonpermeable) gloves.**
- **Wash your hands thoroughly afterwards.**
- **Keep tools used for working with the contents of composter in a contained and out-of-the-way place.**

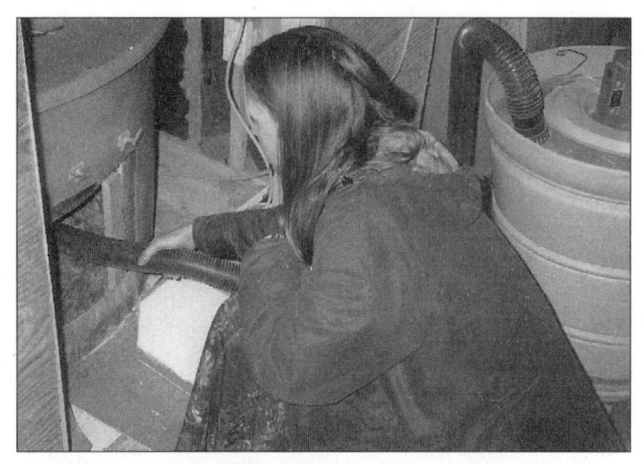

One of the authors empties a chamber in a Carousel system with a wet-dry vacuum.

Many owner-operators use a spade or a shovel (a larger one would be better than this one—and wear gloves!).

Joyce Evans empties her Phoenix system. (Photo: Mike Evans)

Removing Material from a Self-Contained Composting Toilet

Here is a way to remove the end-product from composting toilet systems that have removal trays (Sun-Mar, Envirolet, BioLet) to minimize spilling the contents onto the floor:

(1) Reach into a large empty plastic trash bag as if you were putting on a large glove. (2) Grasp the tray through the plastic bag and (3) pull the tray into the trash bag, rolling the trash bag over and around the tray as you drag it out of the composter. Now the tray and its contents are in the bag, and both can be transported through the house to the place where you want to store it. Or trench it in around nonedible trees and shrubs (in some states, it must be buried under 6 to 12 inches of soil). If you use a biodegradeable, environmentally friendly bag, you can compost or bury the bag and its contents.

Taking an end-product removal tray outside. (Photo: David Del Porto)

to the roots of any nonedible tree or shrub, where it will do the most good. Or, have it removed by a licensed septage hauler. For more about this, see "What Do You Do with It," later in this chapter.

Most of the large systems can be shoveled out with a spade; you can also use a wet-dry vacuum cleaner. Remember, though, that any odors you vacuum up will come out into the room. Some vacuums have an exhaust port that can be connected to a long hose so that odors can be directed to where you want them, not in the toilet room.

Emptying Uncomposted Excreta

Most composters have been designed to fully contain excreta (when not overloaded) and prevent contact with it until it has been transformed into harmless humus. However, if you have to remove ETPA before it is composted, perhaps because you are moving the toilet or because it is overloaded, take great care to avoid direct contact with the material.

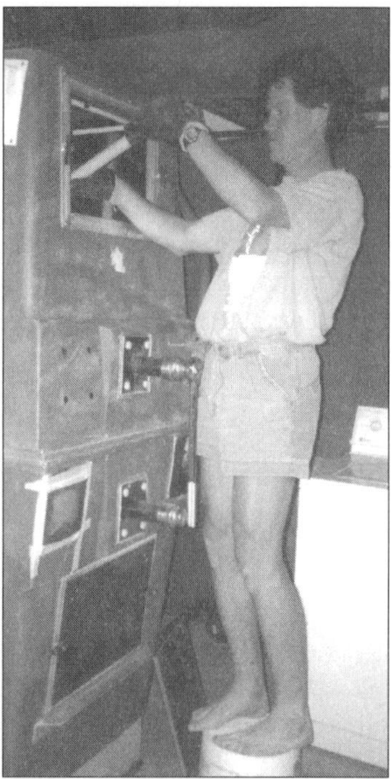

Leave plenty of room to maintain the composter. (Photo: Carol Steinfeld)

One solution is to call a licensed septage hauler and have the hauler pump out the unit with a vacuum hose and transport the unprocessed ETPA to a treatment plant.

Absent that, the excreta can continue composting in another composter, but it should be mixed with wood chips: four parts dime-sized wood chips or additive mix to one part unprocessed excreta. Place in a covered well-aerated container (a recycled plastic 20- or 55-gallon (76- or 208-liter) drum with a screened air intake is good) and store in a warm area where people and animals will not come in contact with it until composting is complete. Two years at 65°F/18°C should do it.

Wear gloves and, if possible, eye protection while handling excreta, and be sure to wash yourself, your tools, the surrounding work area and your clothes with soap and water afterward. Disinfect with a disinfectant such as a 3 percent hydrogen peroxide solution (available in all drug stores), especially if you suspect that some ill people have used the toilet.

Mixing It

Within certain limits, the more you mix the processing material, the better it composts. However, if the volume of the composter is right, it shouldn't be necessary—the material is getting enough oxygen thanks to an adequate surface-to-volume ratio. For this reason, movable container batch systems such as the 55-gallon (208-liter) drum types or Carousel-style composters, should not require mixing.

Mixing the composting material—just as you would a pile of composting yard waste—gets more oxygen into the material, releases carbon dioxide and more evenly distributes water, nutrients, microbes and carbon. In some larger systems, such as the Clivus Multrum and Minimus and large double-bin systems, you might give the top of the system a mix with a rake. In others, a mixer is built in: The AlasCan features an augur, and the Phoenix has rotating tines operated externally. Sun-Mar composters are rotating drums operated by hand or motor, and the electric BioLet models feature manual and electric leveling arms, which mix things a bit.

Raking the Top, Raking Forward

In the larger systems, particularly the single-chamber systems with which dry toilet stools are being used, material right under the toilet chute may start to form a cone. This is only a problem if the hill starts to ascend into the toilet chute. If that happens, knock the pile down with a stick or level it with a rake. In a batch movable container system, such as the 55-gallon (208-liter) drum system or the Carousel, you may be able to move the active container from side to side to control this pile. In Clivuses, you have to open the removal door and rake the material forward, to help keep things moving along.

Controlling Insects

Controlling insects is a particular concern for those using waterless toilets or toilets without traps, which provide a barrier between the composter and the toilet room. Flying insects, which can spread disease-causing organisms if they come into contact with fresh waste, are your major concern (see "Vectors and Insects" in Chapter 3). Non-flying organisms in the composter should not be killed unless they become a nuisance. In fact, they help the composting process by increasing the flow of air through the ETPA. They will disappear before the finished humus is finally removed in a batch composter, simply because they run out of nutrients. Composting toilet owner/operators' experiences with insects appear to vary widely: Some rarely or never see a fly, others do commonly. If there is an outbreak, it is usually seasonal.

Many owners find that simply placing fly screen over air intakes and vents and not adding kitchen scraps takes care of the problem. Although some live with the occasional fly, it is important to take precautions to keep pests out of the composter.

1. Deter Them.

This is your first line of defense. Keep them out and eliminate conditions favorable for breeding.

Dust with diatomaceous earth—it is like broken glass to insects. Ashes also work.

- Inspect and maintain insect screening. The air intake and the exhaust stack must be screened to prevent the entry of flies. Seal faulty gaskets, spaces between doors and hatches, etc. Placing a smoking incense stick inside the composter can help you identify these.
- Glue gasketing around the inside periphery of the toilet seat lid, if your composter has an air intake elsewhere. If not, consider making a fly-screened toilet bowl opening cover by simply cutting some screen and tucking it into the opening. This way, the air can get in, but flies cannot (just remember to remove it before use!).
- Properly maintain the composter. Material that is too dry or too wet appeals to flies.
- Avoid putting fruit and vegetable scraps in the composter. In addition to being highly appealing to flies, they can also introduce fly eggs, which may hatch.

This fly screen over an air-intake opening keeps flies out and serves as a trap for any flies inside the composter. Flies are drawn to light. Right: Always cover your exhaust chimney terminus with flyscreen.

- Keep the toilet room door closed to prevent insects from entering.
- Keep the toilet seat closed when not in use to help ensure that any insects that do get into the toilet will not then fly out of the composting chamber.
- Cover fresh deposits with compost additive, which keeps flies away from the feces, the primary attractant.
- Periodically dust the material and connecting pipe with diatomaceous earth, which is essentially microscopic jagged particles that kill insects when they breathe in or ingest them.
- Deter insects with incense, insect strips and aerosol sprays containing natural plant-derived compounds such as citronella (derived from the extracted oil of leaves and stems of *Cymbopogon winteratus* or *Cymbopogon nardus*). Or try tickweed, neem, pennyroyal (*Mentha pulegium*), lavender, geranium, melissa quassia, lemon grass, black walnut or tea tree oil.
- Introduce some predators. You can safely introduce fly predator insects such as spiders, parasitic wasps and beneficial nematodes, which kill the fly larvae before they hatch. Fly parasites (*Nasonia vitripennis, Muscidifurax zaraptor*) are gnat-sized, nocturnal, burrowing insects that do not bite, sting or harm humans or animals. The insects are shipped in the form of parasitized pupae in a sawdust medium and are dispersed by releasing a small handful at "hot spots" throughout the fly season. Natural predators are available through many garden supply catalogs, and from The Source Biological Fly Control, 1368 Grouse Drive, Redding, CA 96003. Phone: 800-380-3560 or 530/221-7391

2. Trap and Kill Them.

- Cover your exhaust pipe terminus with fly screen. The flies will be attracted to the light outside, fly to the screen, ultimately exaust themselves and fall back into the composter and be composted.
- Sticky traps are paper or plastic covered with a very sticky substance that has an attractant in it. If you can, hang them inside the composter. The flies land on the sticky paper and can't fly away.
- Try a "lightbulb-lure trap." Attaching a simple plastic bottle with a light source to the composter will attract the flies, which then cannot find their way out, and eventually die (consider putting fly paper inside it). Other traps include bottles or bags that use a lure to attract the flies. Baits range from exotic pheromones to homegrown recipes such as mixing a packet of baker's yeast granules with one beaten egg, a teaspoon (5 ml) of baking powder and a 12-ounce (.35 liter) can of beer or soda, and setting it in a warm area for seven days or until it is "ripe."

3. Organic Sprays

Several organic or biodegradeable sprays are effective. Spray a solution of water and biodegradable soap, an ultra-fine oil, or the organic pesticide pyrethrum inside of the composter with a solution of water and biodegradable soap, biodegradable ultra-fine oils. Pyrethrum, a short-lived organic pesticide prepared from the flowers of certain chrysanthemums, paralyzes insects' digestion processes. Concern® makes an insect-killing spray containing pyrethin, a pyrethrum extract. The soap and oil will smother the insects and their larvae. Also, a juvenile growth hormone marketed under the name Precor inhibits flies' proper development. These will not harm the composting organisms.

4. Toxic Sprays and Vapors

WARNING: **All pesticides are dangerous and should be used with extreme care in order to protect you and your family.**

A five-second burst of any airborne insect-control spray in the toilet room and in the air intake can be repeated each day until the problem goes away. This will not harm the composting process—although some of these insecticides are not very healthy for humans!

According to Ohio State University, College of Food, Agriculural and Environmental Sciences some pesticides are very effective in fly control. A popular product in Europe consists of strips of Vapona-impregnated plastic that can be placed in the composter. These strips can be used at the rate of one strip per 1,000 cubic feet of enclosed area. They will need to be replaced, as they lose their effectiveness after about three months. One popular brand is the No-Pest Strip by Shell Oil Company.

Methomyl (Golden Malrin) fly belts by Novartis can be attached to surfaces out of reach of food-producing animals. The belt may be cut to any desired length and attached to surfaces such as walls and ceilings. An insect growth regulator known as cyromazine (Larvadex) will control manure-breeding flies. Premix, produced by Ciba-Geigy Corporation, kills fly larvae before they reach adulthood, and does not adversely affect natural predators and parasites.

Reference
Ohio State University, College of Food, Agricultural and Environmental Sciences' website:
http://ohioline.ag.ohio-state.edu/b853/b853_5.html

What Do You Do with It?

Check your local and state regulations for what you are legally required to do with the end-product that you removed from your composting toilet. In Massachusetts, for example, state regulations call for either (1) removal and on-site burial under six inches of soil (preferably around nonedible trees and shrubs, where the oxidized nutrients will do some good –Ed.); or, (2) removal by a licensed septage hauler to a treatment plant.

If you claim this is "fertilizer" in any documents you publish, you could run afoul of the USEPA Part 503 (Biosolids) of Section 405 of the Clean Water Act. In this case, you must test the material for bacteria, viruses and heavy metals according to Part 503 (any sewage that is biologically processed to be used for fertilizer is considered a "biosolid"). This was originally written to regulate municipal wastewater facilities that process sewage sludge to be used as plant fertilizer and now applies to homeowners, too. Contaminants must be tested for, as combined wastewaters from communities will inevitably contain some. This is a classic case of "one rotten apple spoiling the whole bushel."

Burying composted human excrement six to 12 inches (152-304 mm), near plants that can use it.

Many composting toilet users/operators trench material from their systems right into their flower gardens or soil in their lawn or forest. If composting is managed properly and *ascaris* (roundworm) is not present in toilet users, then this is probably safe.

Remember that composted human excreta from composters from which leachate was not removed could have very high concentrations of table salt from urine. You would have to significantly dilute the material with other soils before using it on sodium-sensitive crops.

Using Composted Human Excrement on Food Crops

Healthy humans excrete healthy *E. coli* (only some strains of this bacteria can make you ill) and other healthy bacteria. If you are healthy, and you compost your excrement, then you use this on food crops, which you eat, you will not infect yourself with any diseases *that you did not have before.* **Because one can never be sure what's going on in this material unless it is tested, we are not recommending using this for food crops.** However, if you do, follow these guidelines.

- Ensure that *all* of the material has reached temperatures of at least 160°F/71°C, so pathogens have been killed.
- Place it only on crops that do not touch the earth, such as orchard crops, vines and berry bushes. Do not to apply it to root or tuber crops.
- Trench it in or bury it at least six inches (152 mm) into the soil, no matter how safe you think it is (because you never know), preferably within the root zones of plants.

Urine is usually sterile, and contains 90% of the nutrients in excrement. Most of the pathogens of concern are in feces. Pathogens are inadvertently swallowed and can multiply in the intestinal tract and be excreted in feces.

Urine is too strong to use alone. Dilute it with eight parts water. It is best to irrigate with this under the ground (subsurface).

Further Composting and Containment

If you are not confident with how processed the end-product is, consider further composting it in a well-aerated contained system. An inexpensive method is to place it in a 55-gallon (208-liter) drum composter as described in the section on site-built systems. In a rural environment, you could choose the co-composting approach outlined by Joe Jenkins in his *Humanure Handbook* (see the description of this method in Chapter 7) or a typical multi-bin yard waste composting system. As noted in the Pathogens section in Chapter 3, proper long-term composting will reduce most disease-causing organisms to safe levels. However, to kill roundworm, a

Putting the end-product of an alternating-container composting toilet into an outdoor composter with kitchen wastes. *(Graphic: Vera Miljø)*

Left: End-product from a composting toilet is added to an outdoor contained composter. Right: Here, it is added to an outdoor yard waste composter.

Below: Several types of yard waste composters in which end-product from composting toilet systems can be further composted. (Graphics: New Hampshire Governor's Recycling Program)

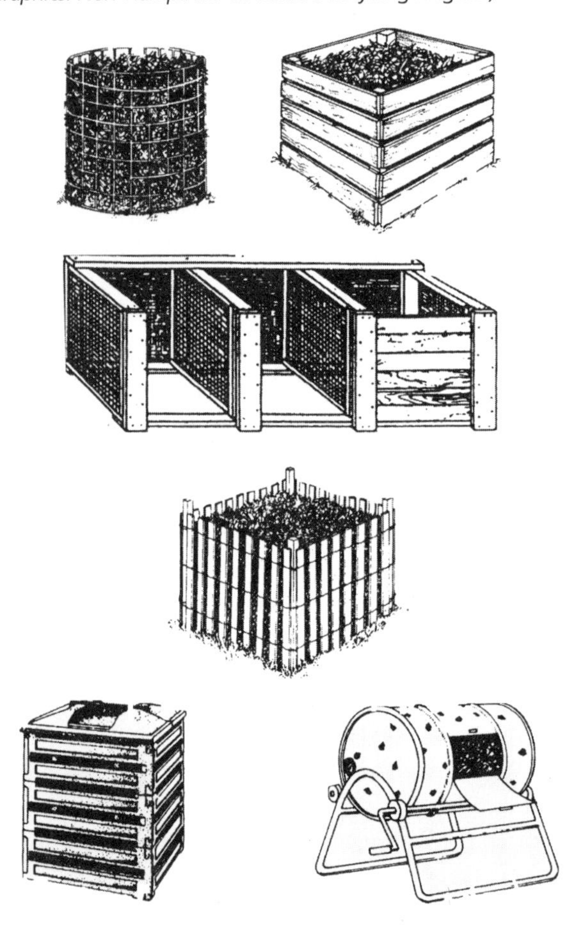

long-lived tenaciously surviving parasite, will require either longer containment (not less than two years), pasteurization, drying or disinfection.

Pasteurization

For those who want to go the extra mile to assure pathogen destruction, a commonly accepted way to kill or reduce the number of pathogens is through pasteurization, which destroys pathogens by heating contaminated matter for one hour at 143.6°F (62°C), then cooling it rapidly to prevent re-inoculation. The most common method is applying sufficient heat from external sources such as electric or liquid fuel heaters. In some cases, microwave or solar energy is used.

Solar heating is one way, although not always easy: An infinite number of variables determine whether or not pasteurization temperatures are reached reliably day in and day out every year. The issue of human disease transmission is too important to leave to a homegrown approach. That said, if you want to pasteurize your compost, some do-it-yourself means (aside from putting it in a microwave oven) include a solar pasteurizer, which you can make by building a solar cooker-style shallow metal box with a solar window or a system with a parabolic reflector. Heat transfer, however, is a complicated

A solar pasteurizer built by Paul Lachapelle to process insufficiently processed material.

thing; it is hard to reach sufficient temperatures through casual solar treatment, especially in places that are not routinely sunny, warm and dry. Place a thermometer in different parts of the material as it heats to ensure that it is reaching at least 143.6°F (62°C) throughout. Thin layers are better, as in large masses, the heat can dry out the outer mass, creating an insulation layer that prevents heat from going to the central core. If you can assure that it has been pasteurized, then the pathogens have been killed.

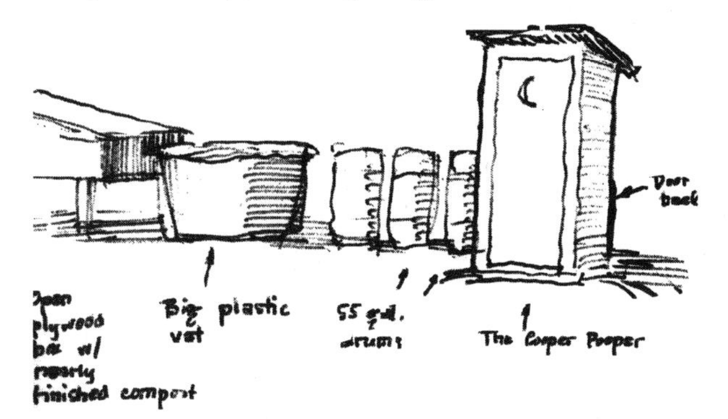

In this composting toilet system along Vermont's Long Trail, material undergoes three stages of composting and containment in various containers before being applied to the soil or packed out.

Maintaining It

The following are periodic preventive maintenance steps for maintaining your system. Again, read the manual that comes with your composter for specifics!

Inspecting the Composter

Initially, inspect the composter at least once a month by looking down through the inspection hatch or through the dry toilet stool with a flashlight. If the composting material appears to be very dry, sprinkle a quart (or liter) of fresh warm water down through toilet seat periodically, so that the pile looks very moist but not soaking wet (it should have the consistency of a well-wrung sponge).

Inspecting Insect Screens

Periodically check the insect screens on the air intake and on top of the exhaust pipe and make repairs as necessary to keep flying insects out of the composter. Also check to make sure that the screens have not become clogged with dirt or other materials that could block air flow. Use a dry brush to clean the screens at least once a year to keep them from getting clogged. Note: There is a screen on the Sun-Mar Bio-drum that may need to be unblocked if liquid accumulates in the drum. This can best be accomplished by rotating the drum so the screen is on top, and brushing the screen with a long-handled stiff-bristle brush.

Inspecting the Exhaust Pipe and Air Intake

Periodically check the air intake and exhaust vent pipe to make sure that there are no obstructions blocking the air flow. If either of these pipes is made up of several connected pieces, check the connections to make sure that there are no leaks. Repair with silicone or other caulking compound.

Inspecting the Heater

Periodically (and whenever liquid is accumulating too fast), check to see that the heater is working. Check that the fuse or circuit breaker associated with the power outlet is not blown and that the plug has not been pulled from the outlet.

Maintaining the Fan

If a fan is installed in the exhaust pipe, do not turn it off and on regularly—this will shorten the fan's life by condensing moisture on the motor, leading to corrosion of the fan's components. Also, if the fan is not running, the natural draft up the vent pipe will be prevented because it is an obstruction in the pipe. However, if you leave for a month or so, consider taking the fan unit out of the exhaust pipe and replacing it with two flexible couplings and a short piece of pipe without the fan in it. Brush the fan and motor occasionally to remove dust and particles; doing so will increase the lifetime of the fan motor.

Cleaning Accumulated Salt and Sludge

Over time, table salt left from evaporated urine and other solids will build up in the bottom of the composter where the leachate is collected. Remove this every other year by pouring some hot water into the bottom of the container. After soaking a few hours, let the water drain out through the leachate drain or pump it out into a container. In the case of small self-contained composters, it might be necessary to shim up the front of the composter to encourage drainage.

If the drain opening becomes blocked with solids (and this is a familiar complaint), flush it out thoroughly to dilute the solids. Consider widening the diameter of the drain outlet and pipe, and installing a permanent cleanout, like that of all residential and commercial plumbing drains, to make the job easier next time.

Inspecting Gaskets

Occasionally check the gaskets to be sure they are air- and watertight. Replace them with closed-cell elastomeric foam gasket or silicone or other watertight material.

Troubleshooting (Problems & Solutions)

Odor:

There is an odor of fresh excreta in or around the composter.

Some odor may be present in or around the top of the roof vent for the first week or two after a composter has been started up, before the composting process begins. After this, there may be a slight musty, sweet odor occasionally, if the wind blows air from the exhaust pipe down toward the ground. There should not generally be any objectionable odors in the building if the exhaust pipe was installed properly.

Improper Ventilation Issues: A foul odor in the bathroom may mean that air from the composter is venting into the room through the toilet seat or perhaps a leak in the exhaust system. Changes in wind and weather may sometimes bring odors down from the top of exhaust pipe above the roof, but it may be difficult to tell where the odor is coming from. Whole-house fans in the summer and fireplaces in the winter can draw air down from the composter's exhaust into the room. Provide separate warm make-up air for the composter if possible.

Fans and fireplaces can draw air from the composting toilet into the room. Use a "smoke test" to check air-flow direction.

One source of the problem could be that the dry toilet stool is acting as a chimney, drawing air from the chamber up into the room. Normally, air should flow down through the toilet seat into the chamber. You can check this by holding a smoke generator such as incense, a candle, or a pipe or cigarette over the toilet and seeing which way the smoke blows. If air is coming up from the composting chamber into the room from the toilet seat, do the following:

- Check to see that the fan is running!
- Make sure that the toilet seat has been closed when not in use.
- Make sure that the bathroom door remains shut and that there are no windows or large openings in the leeward walls or ceiling. The only openings along the wall should be small and near the bottom. Make repairs as necessary.
- Make sure the top of the exhaust pipe has not become obstructed. Remember, an installed in-line fan that has been shut off is simply an obstruction in the exhaust pipe. Perhaps the rainhat has slipped down to cover the opening or the insect screen has become clogged. The insect screen should be around the outside of the rain cover, so that it will not restrict air flow and it will be washed by rain.
- Make sure the air intake pipe is not obstructed and has no leaks.
- Make sure that trees and vegetation are not blocking the exhaust pipe or influencing the draft.
- Tall, dense foliage near the top outlet of the vent pipe may interfere with the draft, as well as create odors. Heavy, tall vegetation (trees) by the outlet of the exhaust pipe may cause problems with the draft, resulting in odors.
- If the smoke test shows that the air is flowing properly down into the toilet, the odor may be coming from a improperly sealed pipe or fitting. Check for leaks and cracks, and repair as necessary.
- Are all connections tight?
- Does the toilet room have sufficient ventilation?
- Is the exhaust pipe open, unblocked and as straight as possible?
- Is the top of the exhaust pipe higher than the peak of the roof? If the roof peak is nine feet or less horizontally from the top of the exhaust pipe, the pipe should be at least two feet higher than the peak or ridge top.
- Are the emptying hatches and the inspection door closed and tight?
- Are the make-up air vents open and unblocked?
- Is there sufficient ventilation where the composter is situated? Vents in the toilet room close to the floor are often sufficient to prevent odors.
- Is the bathroom or whole-house fan still connected? The fan will create a negative pressure or vacuum and suck odor from the composter into the room.

If you have checked all of the above points and still experience odors, try covering the air intake openings on the composter. This will allow the system to draw more air from the room via the toilet stool.

There is a strong odor of sewage or rotten eggs.

These strong odors indicate that the composting chamber may have become anaerobic or that the microorganisms do not have the right balance of nutrients.

- Increase aeration, add more bulking material and, for good measure, add some more compost microbes.

What else can you do to control odors?

• *Filters*

There are filters available from a variety of suppliers that can scrub odors out of the exhaust. Most contain granular activated carbon (charcoals of coconut husks are the best) or Zeolite, a naturally occurring mineral. These are porous materials with a high surface-area-to-volume ratio that can adsorb odorants (odor-causing chemicals, particles or gasses) and other chemicals so they are removed from the air stream. Zeolites are highly regular structures of pores and chambers that allow some molecules to pass through and cause others to be either excluded or broken down. It is in many ways the inorganic equivalent of organic enzymes, many of which also have specifically sized chambers that trap chemicals, holding them where they either break down or react with specific chemicals.

Potassium permanganate can be added to the filter media to reduce or destroy certain odorants, such as ammonia, sulfides and mercaptans, as well as organic-based odors. This comes in the form of dark purple odorless crystals with a blue metallic sheen that dissolve in water to produce a purple-red color. Potassium permanganate is a powerful oxidizing agent that is also used in water treatment as both an oxidizer and a disinfectant.

Finished compost also is an excellent filter material, and is often used in large municipal sewage sludge composters. In this case, the pressurized exhaust is pushed through the finished compost, where adsorption, filtration and biochemical processes operate on the odorants to prevent them from becoming a nuisance.

In every case, a fan or blower will most probably be required to overcome the exhaust pressure drop that occurs when filters are used. Ask your composter supplier about this (Sun-Mar and Orenco offer odor filters).

• *Odor Control Additives*

Throughout the world, cold wood ashes are sprinkled on compost piles and human manure pits to control odor. In addition to the stable ash content, the active chemical in wood ash is potassium carbonate, or potash. This highly alkaline and aggressive chemical is also used to make soap by leaching the wood ash with water to make potassium hydroxide, or caustic potash. This is boiled with vegetable or animal fats to make liquid soap. Potassium is an essential nutrient for plants and animals, so it is good for the final end-product. With proper use, it will not interfere with the composting process. Use it sparingly, so as not to make an impenetrable layer of ash in the compost.

Another option is using livestock odor control additives, such as neutralizing agents, oxidizing agents, absorbents, digestive deodorants and chemical additives. These were evaluated for swine manure odor reduction by North Carolina State University, which found that the neutralizing agent and the oxidizing agent reduced odor intensity and improved odor quality. Absorbent, digestive agents and chemical deodorants did not significantly improve odor parameters. Contact your agricultural agent for more information.

Moisture Issues

How do I know when the moisture is within optimum range for composting?

Ideally, the contents appear moist—again, like a well-wrung sponge—and are not sitting in leachate. Moisture can also be monitored with an inexpensive moisture meter. Stick the long probe deep into the compost mass, and the attached meter tells you if it is too dry, too wet or ideal. Check different depths to be sure. No batteries are required to operate this meter. They are sold for use with houseplants and are available from garden supply outfits.

• *The material is too dry.*

Symptom: ETPA is building up and not decomposing. Toilet paper is visible and, in models with mixing systems, winding around the mixer arms and tines.

Cause: This is very common in underutilized systems and self-contained systems with powerful fans and heaters. It often means that the evaporation rate is too high for the volume of water and excrement put into the composter, and the material is drying out. Spritz the pile regularly with a spray bottle (or the spritzer supplied with the composter), and turn down the heat.

This moisture meter indicates that the contents of the bottom of this BioLet XL is quite moist. The rest of the material should be somewhere in middle to high end of the moisture range.

• ***The material is too wet.***

Symptom: Saturated ETPA. Visible liquid in the composter. Smell of rotten eggs, indicating that anaerobic processes have taken over. Process rate slows and soupy solids accumulate.

Cause: Overloaded system—too much urine and/or the evaporation rate is too low and/or the leachate is not draining away from the composting mass.

Solution: Reduce the urine load. Determine if the exhaust pipe is blocked, the fan has stopped or the leachate separation mechanism—whether it is a grate, screen, hole or drain line—is blocked. Also, try adding more absorbing additive or install a urine-diverting toilet. Check the heater to make sure it is operating properly.

Special Moisture Issues

• ***Large, Single-Chamber Continuous Composters***

With the large vault or bin composters, urine and flushing water drains to the bottom and often leaves the top of the pile too dry. Consider adding a gallon (3.8 liters) or so of fresh water every week or two, as needed. However, it is best not to pour leachate back over the composting mass, as this rich liquid contains concentrated salts, ammonia, etc., which are detrimental to the health of composting microbes.

What's important is to provide sufficient irrigation over the entire surface of the pile, without drenching it. Some commercial models provide internal sprinkler heads with automatic sensing and timing systems by the leading names in landscape and garden irrigation, such as Rain Bird or Netafim.

• ***Small, Self-Contained Appliance Composters***

In small, self-contained appliance composters, the front section of the mass in the composter (under the seat opening) gets all the urine, and is often way too wet, while the back of the unit, where no urine can reach, is bone dry. This is exacerbated by high-volume internal fans and heaters intended primarily for leachate evaporation. Both tend to dry out the higher ETPA, hindering composting and hardening the ETPA. Composters with tumbling drums (Sun-Mar) tend to provide a better balance than those without (BioLet, Envirolet, etc.). In these systems, your options are to manually mix the wet front material with the dry rear material. You can also manually irrigate the rear of the unit with a long-necked watering can or a hand sprayer connected to a pressurized water supply (sink, faucet).

If it is too dry, you can also turn off the heater for a while and/or add a few liters of water to the system to give the material a chance to absorb and process.

Other Issues

The composter does not seem to be filling up, even after a year of use.

This is usually not a problem at all. It means that the ETPA is decomposing faster than it is being added. That means that the process is working very well. As long as there are no other problems and the composting chamber does not get full, you should keep using it—unless you wish to use the compost sooner than later.

• ***The material isn't completely composted.***

This is one of the biggest concerns of composting toilet owners. Usually the problem is improper management of the composting process, such as not providing enough heat, air or moisture and/or removing the humus too soon or too little capacity. The problem can be solved by better management, including taking care to not overload the system (too many users in too short a time period). It could be that the temperature is too low, and the composter was designed to work at a specific minimum temperature for the rated capacity. Most manufacturers specify 65°F as the minimum ambient temperature for their performance ratings.

• ***Organic concrete forms in the unit.***

The formation of rock-hard matter in the composter is a sign of improper management and possibly design flaws in the composter. The formation of this matter can be caused by:

- Compaction of the mass due to infrequent removal of finished product.
- Densifying of the mass by flush water from flush toilets. These can settle the large particles and rinse away small particles as the water drains through the ETPA.
- Drying of the mixture of the salts in urine and ETPA. This is usually due to inadequate moisture control or overheating.
- Bacteria and fungi naturally producing a material called glomulin that acts like glue to hold together particles. This is very important in producing soil structure. These work to bind smaller aggregates to each other to form larger aggregates. Protozoa and beneficial nematodes and arthropods bore through and make holes, increasing porosity.

- Concrete-forming bacteria at work in the presence of certain minerals excreted in excrement and in some additives. Many common types of bacteria use urea, the main component of urine, as their source of nitrogen. They break down urea, creating carbon dioxide and ammonia. The ammonia reacts with water to form ammonium hydroxide, which makes any nearby calcium precipitate out as calcium carbonate crystals or limestone.

What to Do About Hardened Material

Once it has formed, the only solution is to break it up with a probe and/or remove it from the system as unprocessed ETPA. Here are some preventive measures:

- First and foremost, keep the composting material uniformly moist and porous. A healthy and moist compost will encourage protozoa and beneficial nematodes and arthropods to colonize the system, helping process material and adding porosity.
- Remove the end-product at least once every two years if you have a large, single-chamber composter, such as Clivus Multrum or Minimus, CTS and Phoenix.
- Manage the heat input so it evaporates leachate yet keeps the ETPA moist. This is sometimes difficult with the small, self-contained appliance composters, such as BioLet and Envirolet.
- Mixing will certainly help, but not in a large, single-chamber system where you might not be able to access the lower levels of the pile. That makes regular removals of end-product and the addition of fluffy additive that much more important.

One of the authors takes a well-aged sample from a Carousel system.

General Composting Toilet Care

Cleaning the Composter and Toilet

Biodegradable soap and water will not harm the composting process. In fact, the soaps can actually help improve the availability of organic molecules to the enzymes of the microbes by reducing water's surface tension, thus making it wetter. However, strong chemicals, disinfectants, bleach and other poisons should never be dumped into the composting chamber. This will kill the beneficial organisms that carry out the composting process. When using household disinfectants (we like 3 percent hydrogen peroxide, available in all drug stores), apply them with a sponge and only wipe the surfaces of the commode. Don't pour it down the toilet. A small amount of any disinfectant shouldn't hurt the process much. In fact, diluted hydrogen peroxide (not more than a 3 percent solution) can benefit a composter by adding more oxygen, so that would be the only exception to the "don't pour it down the toilet" warning.

Other Tips for Improving the Composting Process

- If you add vegetable scraps, be sure to cut them into small pieces.
- Try to control the moisture—be sure the compost is not too wet, not too dry.
- Get air into the compost with frequent mixing, or add bulky additives (see "Additives for Composting Toilets" for a complete explanation and list). These help make the compost less dense.
- Add warm garden compost from time to time—it adds beneficial microorganisms to the compost.

Do not use antibacterial cleaners on your composting toilet (a little on the outside surfaces is fine). They will kill the active microbes in the system. (Illustration: Center for Clean Development)

Do's and Don'ts

Do:

- Keep the toilet seat cover closed and/or keep the waste valve to the toilet when not being used, to prevent odors
- Put toilet paper down the toilet (paper used for urine only could be excluded)
- Put a handful of organic bulking agent down the toilet periodically
- Use a mild soap when cleaning the toilet seat area
- Bury finished compost in a shallow hole or trench around the roots of nonedible plants

Don't:

- Throw trash, cigarettes, matches or burning material into the toilet
- Use harsh chemicals, chlorine bleach or toxic chemicals in the washbasin, shower or the toilet
- Pour lots of water down the toilet
- Empty composter until it is ready
- Remove compost from a filled external composter unless it has been composting for 6 to 12 months or longer

Toilet Paper, Tampons And Diapers

Yes, toilet paper should break down in a composting toilet system. If it doesn't, you may need to wet it. (In the BioLet XL, it tends to wrap around the turning bar. If this happens, spray it with water.) Or, use cheaper one-ply toilet paper or any tissue that has a reputation for being not too tough.

Tampons—the cotton and paper parts, not plastic applicators—will break down in composting toilets, but not very fast. They are best left out of composting toilets with grates, such as some BioLets and Envirolets.

With tampons, feminine pads and biodegradeable diapers, the issue is bulk: These will fill up a composter because they will take a while to decompose. Deposit them with care. Some composter owners dispose of plastic diaper liners in their systems, and pick them out when they remove end-product from the composter.

In case you were wondering...

When discussing composting toilets, people often ask what the end-product looks like when it is removed from the system. Here are some photos of removal openings of three larger-capacity composting toilet systems. In these, the end-product looks more like soil than excrement, and is not at all offensive.

A Clivus Multrum

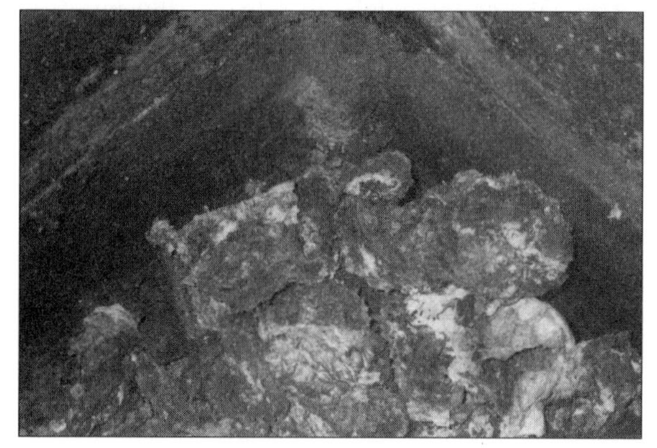

A chamber in a Carousel composter used with two SeaLand microflush toilets. Compost from systems used with flush toilets tends to have denser end-product.

In this Rota-Loo on the Pacific island of Yap, you can see seed pods added to the system still intact.

CHAPTER SEVEN

The Systems

Some Manufactured and Site-Built Designs And Toilet Stools

This section features both manufactured systems that are available "off the shelf" and several site-built systems you can construct yourself or have a contractor build. A section on wet and dry toilet stool options follows.

Prices, sizes and design features often change; for the most up-to-date information about these systems, including local dealers and installations, call their sources. Nearly all of the manufacturers sell plastic or fiberglass dry toilet stools and some kind of micro-flush toilet. Remember that you can mix and match: buy one company's toilet and another's composter, or two different types of composter for different locations in the house.

The authors do not endorse or recommend any particular system in this section but instead describe basic design and operation and some maintenance and application considerations. Nor is every model and configuration described—contact the manufacturers and designers for the rest of the story.

Most drawings and photos of the systems in this section were supplied by their manufacturers and designers.

Manufactured Composting Toilet Systems

Buying a manufactured composting toilet system has obvious and not-so-obvious advantages:

- They are ready to install.
- They come with operating instructions.
- They are guaranteed and backed with customer service.
- Often their performance is certified by a third-party lab such as NSF International or the CSA International (the Canadian Standards Association), and thus are easier to get permitted than a site-built system.

To some, the prices of manufactured composting toilet systems seem high, especially when compared to a simple septic tank or to build-it-yourself systems. The factors to consider are the high costs of marketing, business overhead, support and testing by NSF International or the Canadian Standards Association (which costs literally thousands of dollars annually) and state and county regulator education and approvals. If your regulator or health agent is amenable to composting toilets, it is likely due to the dogged education efforts of a manufacturer or a dealer.

As in many industries, some manufacturers in this one can be a bit "this and no other" about the systems they sell; it will be up to you to compare them. Make some calls, shop around and visit installations when possible. Several systems have been around for more than 20 years, and most have improved in that time.

Some other points:

- Be careful about buying on the basis of price alone: Consider a wide variety of other issues.
- Every manufacturer offers a photovoltaic-powered fan and/or heating system and usually a dry toilet stool and a water-flush toilet option.

If you need a system with state approval, contact your state (see Appendix for contacts), for a list of approved systems. For an updated list of NSF International- and CSA International-listed systems, contact:

NSF International
P.O. Box 130140
Ann Arbor, MI 481130-0140
1-800-NSF-MARK
E-mail: info@nsf.org Web: www.nsf.org.

CSA International
Etobicoke
178 Rexdale Boulevard
Etobicoke, Ontario M9W IR3
1-800-463-6727; Web: www.csa.ca

Note! A sample power requirement is listed for each system in this section. However, this can vary widely and depends on the configuration of a particular system and its components.

Other Sources

The following are catalog retailers (listed alphabetically) that sell composting toilet systems. The advantage they offer is that they sell more than one brand, as well as accessories and other innovative and resource-efficient systems for your home.

Ecos: Tools for Low-Water Living
P.O. Box 1313
Concord, MA 01742
978/369-3951
www.ecological-engineering.com/ecos.html
Catalog $2
Several brands of composting toilets and other water conservation tools.

Jade Mountain Catalog
P.O. Box 4616
Boulder, CO 80306-4616
800-442-1972
www.jademountain.com
Catalog $5
Lots of solar and other off-the-grid tools, systems and information.

Lehman's Non-Electric Catalog
One Lehman Circle, P.O. Box 41
Kidron, OH 44636 USA
330/857-5757
www.lehmans.com
Catalog $3
Traditionally aimed at the Amish market, Lehman's offers all kinds of basic hard-to-find tools and goods for simple living.

Real Goods Renewables Catalog
555 Leslie Street
Ukiah, CA 95482
800-919-2400
www.realgoods.com
Free catalog
Features all kinds of solar and off-the-grid items (as well as another catalog of appliances and decorative items for healthy homes).

AlasCan Organic Waste and Wastewater Treatment System

The AlasCan Organic Waste and Wastewater Treatment System represents one of the most high-tech forms of composting toilet system technology: a highly mechanized system for those who want (and don't mind paying for) a wastewater solution that is as automated as possible. An optional computerized monitor that reports system status to a central maintenance provider is certainly the future direction for many building technologies, not just wastewater systems.

This system was originally developed by Clint Elston for use in frigid Alaska, and so it is super-insulated. There are several installations in Alaska and some in Minnesota.

The AlasCan composter tank features two motor-driven mixing harrows that can be programmed to mix the composting material at set intervals. AlasCan supplies worms to help the process along. They work to digest wastes and aerate the mass.

When leachate accumulates at the bottom of the tank, an automatic self-priming pump and sprinkler system spritzes it evenly over the composting mass. Either micro-flush, foam-flush or vacuum-flushing toilets are used to transport excrement to the tank.

An optional companion graywater system is essentially a mini-packaged aerobic treatment plant. It uses extended aeration with suspended plastic media, with optional systems for filtration and ozone purification of the end-product. A hand-operated sludge pump moves sludge from this system into the composting tank for further treatment.

Maintenance: According to company literature, the tank must be emptied about four years after start-up, then quarterly to annually afterward. The company offers service contracts, as well as financing.

The system features three inputs—toilet, kitchen wastes, and sludge from the graywater system—and two stacks, one for air input, one for air output. An internal bulkhead divider separates fresh ETPA on top from the more processed ETPA at the bottom.

The large size of the system may require special installation considerations, and so it may be best suited to new construction and parks and recreation areas.

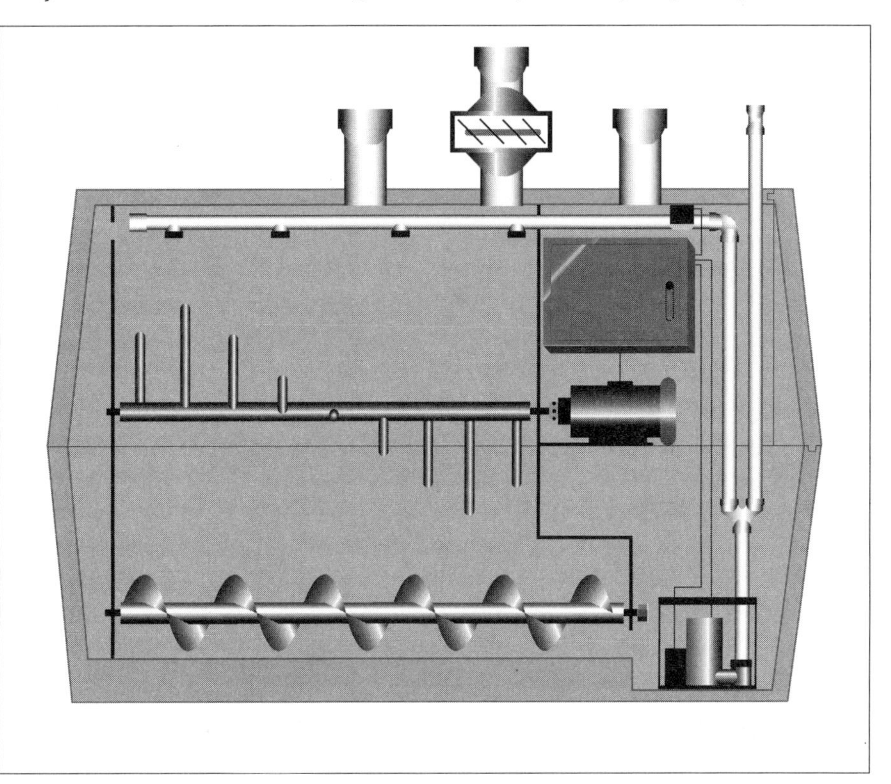

A cutaway shows the internal process controls of the AlasCan Separation Tank (composter).

Price range: About $5,000 for the composter, $2,500 for the wastewater treatment tank. Nepon foam-flush toilets cost $1,500 plus, a SeaLand micro-flush

AlasCan

MODEL	CAPACITY	DIMENSIONS
AlasCan	12 to 24 people	56" high, 48" wide, 92" long Working volume is 77 cu. ft.

ETPA = Excrement, Toilet Paper and Additive

toilet about $200, and the SeaLand VacuFlush system can run about $1,400 per toilet.

Power usage: Fan uses 27 watts, gear motors 1.4 amps, sump pump 1 amp (five minutes a day), continuously operating air compressor takes 2.4 amps, 57 watts for about 600 watts per unit time. Manufacturer estimates it costs $7 to $10 month to operate, without supplementary heat.

Materials: Welded polypropylene sheet material plastic.

Also available: An optional telemetric computer monitor checks liquid levels, pH, electrical usage and temperature, and will transmit this information to AlasCan or an AlasCan service provider, if the user has a service contract. Optional back-up 12-volt battery.

Advantages: A highly mechanized composter. Automated turning of the ETPA. Manual transport via ratchet-driven augers of the finished compost out of the system to the clean-out area.

Disadvantages: This is a big, highly mechanized system, featuring a lot of moving parts. However, those who balk at the maintenance aspects of a simpler system may find this more suited to their lifestyles.

AlasCan of Minnesota, Inc.
8271 90th Lane
P.O. Box 88
Clear Lake, MN 55319
320-743-2909, 800-485-4354 toll free
mail@alascanofmn.com, http://www.alascanofmn.com

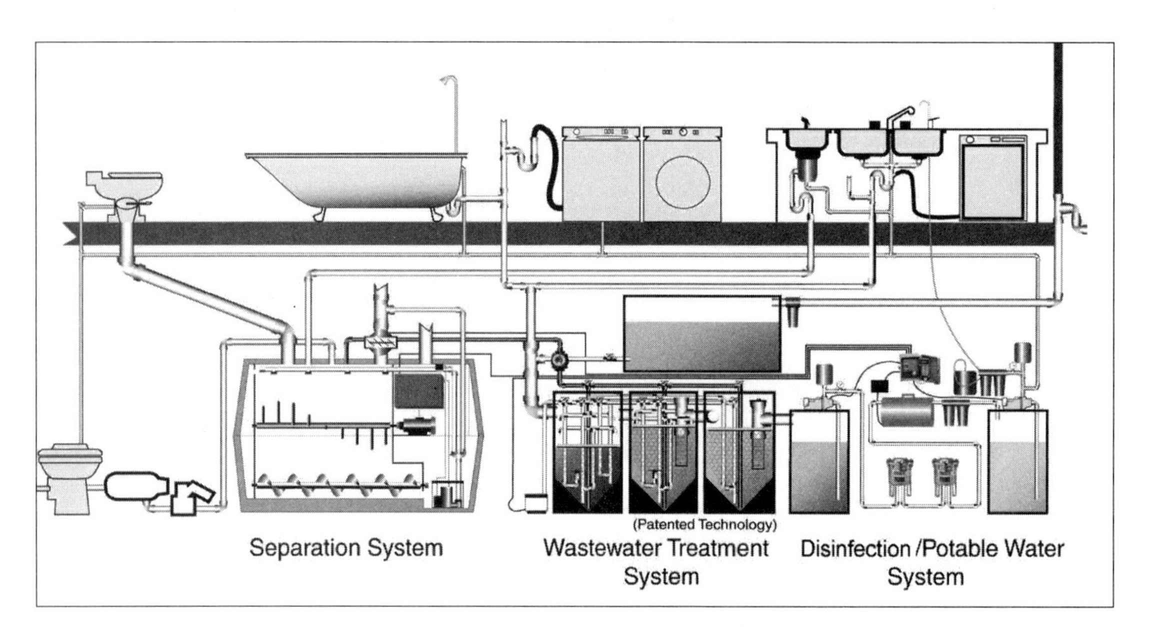

ETPA = *Excrement, Toilet Paper and Additive*

BioLet

BioLet, called "Mulltoa" in Sweden, was introduced to North America in the late 1970s as the "Humus" toilet. Numerous changes later, it was renamed the "BioLet." BioLet principally manufactures self-contained cottage toilets, but recently introduced the BioLet UFA, which is advertised as a small under-floor composter.

The BioLet Automatic, Manual and XL are continuous composters. The BioLet NE (non-electric) is a batch system.

What's unique about the BioLet is its pressurized air recirculation system, by which it constantly recirculates reheated air through the composter and exhausts it simultaneously. The pressure chamber forces air in one direction and pulls it from another.

The BioLet XL, BioLet's NSF-listed model, features a primary chamber in which two motorized arms slice through the composting waste, aerating it, leveling it and pushing it through a grate. Finished compost falls into a removable tray. It features two heaters: a thermostat-controlled radiant heater in the floor of the unit and a convection heater that circulates heated air around the composter to warm the compost and evaporate leachate. When the seat is lifted and then returned, the mixer is automatically activated and rotates one revolution. The BioLet Manual mixes with a hand-operated turning arm.

The BioLet NE consists of two interchangeable small containers, with perforated pipe for aeration. After the first container fills up, the user removes the top of the composter, and moves the first container to the back and brings the rear container to the front. When that fills, the older container is supposed to be ready for emptying and then interchanged with the full container. A clever compost pile

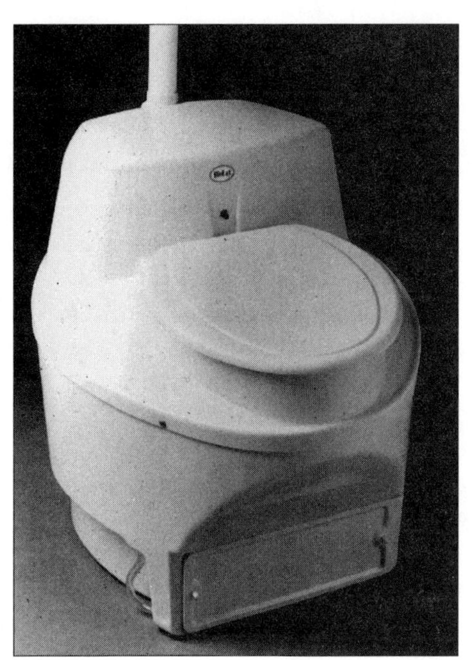

Left: The BioLet XL. Below: Diagram of the BioLet Automatic shows the mixing mechanism and the air flow paths.

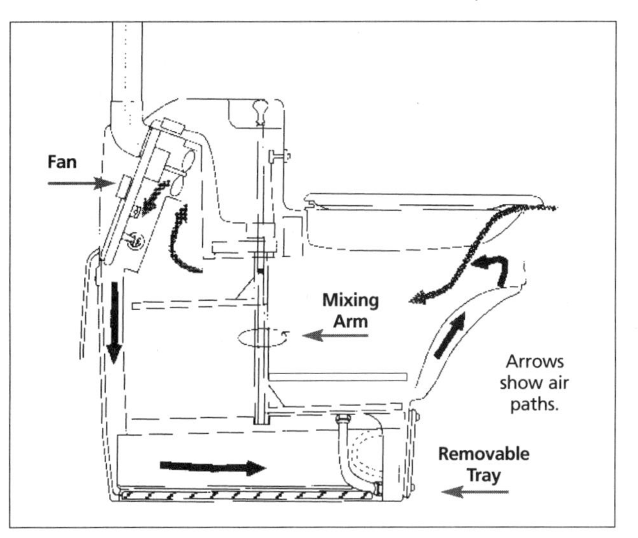

A sampling of some BioLet models

MODEL	CAPACITY	DIMENSIONS	COMMENTS
BioLet XL BioLet Automatic	6 people weekend/vacation use 4 year-round	26.4" high 33" depth 25.6" base width	Leave space for drawer
Biolet Manual	4 people weekend/vacation use	27" high 29" depth, 22" base width	Leave space for lifting top
BioLet NE	4 people weekend/vacation use	26" high, 30" depth, 22" base width	Nonelectric
BioLet UFA	6 people weekend/vacation, 4 people year-round	47" high, 26" depth, 24" base width	Placed under the floor

ETPA = Excrement, Toilet Paper and Additive

mixing lever is operated manually from the top of the toilet.

All models feature a white view guard—essentially a compost cover that scissors apart when the user sits on the seat (thus pushing a button).

Leachate rarely evaporates completely in BioLets, and should be drained before taking out the removal tray. Two half-inch clear tubes outside the unit show if the unit is filling up with leachate and must be drained. Unfortunately, as with all the small appliance composters, the drain lines are so low to the ground that only a very shallow container, such as a sheet cake pan, can be used to collect the drained leachate.

The shear pins in the motorized mechanism prevent the stripping of gears and the burning out of motors. However, in the BioLet XL, this causes a problem in unweatherized cottages in the winter: If the compost is frozen, just plugging in the composter causes the mixer motor to turn, which breaks the shear pin. The shear pins are made of a different metal than the shaft, and corrosion often makes them hard to remove (consider replacing the shear pin with a stainless steel nail). Disassembly of the unit is required to reach the broken shear pin in the automatic models (as always with an active composter, not a fun job). For that reason, BioLet XLs should not be turned on when they are frozen.

BioLet's latest model, the BioLet UFA, is an under-floor version of the BioLet XL, with a turner and a grate that "sifts" ETPA to the removal tray. It was first introduced with a urine-separating toilet, the BioLet Separera, an idea for which the United States wasn't ready at the time. BioLet's urine-separating toilet comes with an optional fan to help prevent odors from the toilet. The fan motor and thermostat, mixer and heater are housed in a removable cassette, which makes for easier service. The mixer operates on a timer, rotating at factory preset intervals. Other features include a liquid level indicator, a 200-watt convection heater (with adjustable thermostat) and a 50-watt heater under the removal tray.

Maintenance: Empty every three weeks; longer if there is low to moderate use or long intervals between usages. Usually the end-product is not completely composted.

Sample power requirements: BioLet XL
Moving air: 25 watts
Heating: 350 watts
Process control: 45 watts mixer motor

Materials: ABS vacuum-molded plastic. Metals are zinc electroplated carbon steel and epoxy-painted carbon steel (which are prone to corrosion), with stainless steel grates and mixers. All the piping is metric, so ventilation piping may be hard to find in North America.

The BioLet UFA

Notable accessories: A toilet stool with a view guard (a great feature for any of the self-contained composters) for the BioLet UFA. Metric exhaust pipe, roof flashing, fans for BioLet NE.

Advantages: Lower priced and small in size, ensuring fewer installation issues. A nice view guard.

Disadvantages: All of the usual issues with small cottage composting toilets apply. Due to their small size, they are often overloaded and processing can be incomplete. Excess leachate is the bane of the smaller composters. Heaters, fans and mixing motors must be occasionally replaced (as with many small composters). Thin walls of the composter are succeptible to puncturing and cracking at screw holes. Exhaust pipes are metric. Toilet paper winds around turning arm and builds up (except in the NE). Shear pins often break when the contents become hard and dry.

BioLet USA
P.O. Box 548
150 East State Street
Newcomerstown, OH 43832-0548
800-5BIOLET
info@biolet.com
www.biolet.com

Bio-Sun® System

The Bio-Sun System, designed by Al White, is a forced aeration system, whereby large volumes of air move through the ETPA. The composter can be a freestanding large, high-density polyethylene cannister-shaped tank or it can be built into the basement of a building. Built-in systems may be the cost-saving wave of the future, and Bio-Sun is increasingly designing and installing systems in which the reactor is part of the building's concrete foundation.

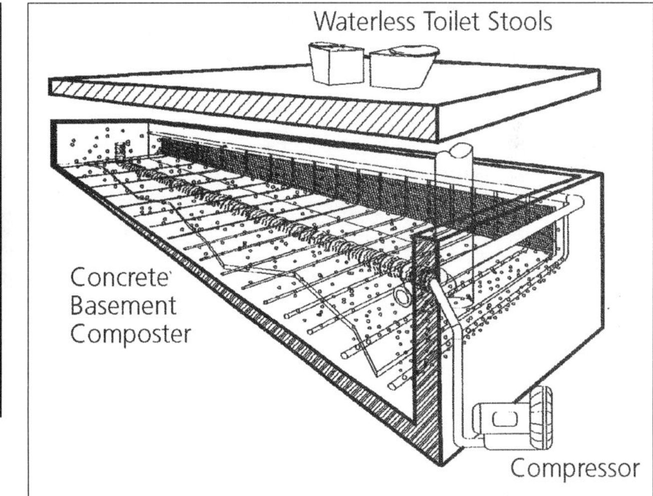

Top: The basement of a public facility acts as its composter, serving several toilets. Bio-Sun also offers tank composters.
Above: A diagram of a basement system shows Bio-Sun's extensive aeration system.

Bio-Sun's aeration process components, called the "Exel-Aerator," are installed into the HDPE tank or in a cast-in-place concrete reactor.

Bio-Sun recognizes that high-rate composting requires significant amounts of oxygen. To provide it, compressor-supplied air is piped from a manifold through perforated two-inch pipes (equally spaced along the length of the reactor) into the leachate at the bottom of the tank, and then bubbles through the composting mass. The manufacturer says that there is a particle-surface-to-mass-volume ratio of at least 1:1 from start-up through full capacity. If composting doesn't take place, dehydration should.

A bark-filled "denitrifying filter" for urine

Exhaust air is drawn down from the upper area of the tank (from the dry toilets above) and then through a large-diameter perforated pipe that is centered at the bottom of the tank and runs the full length of the reactor beneath the composting pile. At one end, it is connected to a large-diameter exhaust pipe with an exhaust fan that exits above the roof.

Maintenance: The pile must be raked occasionally to avoid mounding under the toilet piles. The temperature of the make-up air must be maintained so that the composting pile is not cooled when temperatures drop below minimum composting temperatures.

Sample power requirements: Eight-person Bio-Sun
Moving air: 109 watts (intermittent)
Heating: None
Process control: None

Advantages: Bio-Sun provides enhanced aeration, which promotes more thorough and faster composting and drying, and evaporation of liquids. Capacity of the in-foundation composters is determined by the size of the vault constructed. Larger composting vaults also potentially allow longer composting periods for smaller loads.

Disadvantages: Air blowers require significant power. The larger single-chamber tank units are more susceptible to compaction of the composting mass, and the one access door at one end of a very long tank could make removing end-product difficult. In the basement systems, one literally walks in and rakes the material, although, because the material is dry, this is not as unpleasant as it sounds.

Bio-Sun® Systems, Inc.
RR #2, Box 134A
Millerton, PA 16936
800-847-8840, 717/537-2200
Bio-Sun@ix.netcom.com, www.nota.com/bio-sun

Systems are sized for specific applications.

ETPA = Excrement, Toilet Paper and Additive

Clivus Multrum

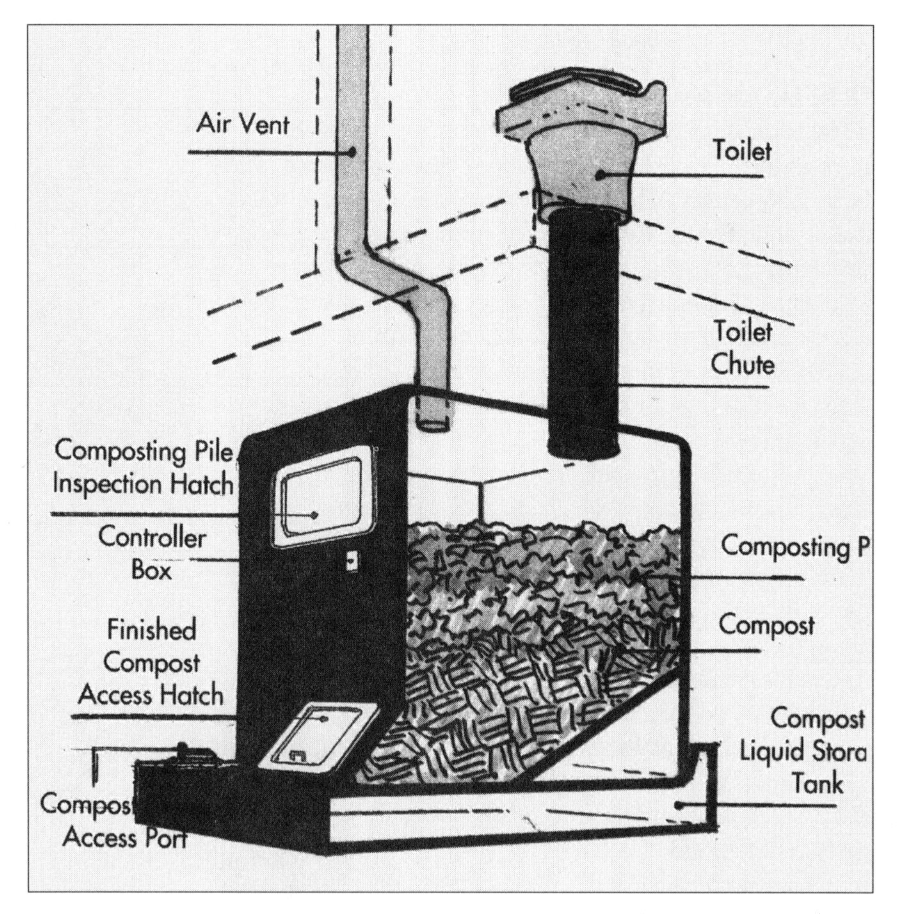

One of the best-known composters in the United States, the Clivus Multrum is a large, sloped-floor, single-chamber composter into which material is added to a starter bed of wood chips or sawdust. The incline aims to prevent compaction and promote aeration by slowing the passage of the ETPA to the lowest point of the composter. A baffle at the base of the unit keeps unprocessed ETPA from filling the finished compost compartment at the bottom of the composter. Material is removed through an access hatch at the lower front base of the tank.

In the larger models, an upside-down rigid, V-shaped channel serves to aerate the compost. Baffle panels on both ends of the compost chamber support the air channels at each end, allowing air to circulate in the composting mass and preventing compost from blocking the air circulation pathways. This feature has been eliminated in Clivus Multrum's smaller residential models, as material would occasionally "bridge" across the two pipes.

Leachate drains to the bottom of the unit to a collection container, where a pump moves it to a separate leachate tank to be emptied. In the large model M-35, leachate drains into a polyethylene tank underneath the unit.

Clivus Multrum is so well known that many folks consider its name synonymous with "composting toilet." Some models have better track records than others. Most are installed in public facilities. The design has been copied many times and constructed in many materials (fiberglass, plastic, wood, concrete, etc.) all over the world, with varying success. Not surprisingly, many report that the composter works best when a leachate drain is added.

Over the years, Clivus has maintained its sloped-bottom design, steepening the angle to promote the gravity movement of the finished compost collecting compartment at the base of the composter.

Clivus offers 11 models, all based on the same technology, but each model is sized for a different application, from a vacation home to a public facility. Most are NSF listed.

Materials: In recent years, Clivus has switched from fiberglass to high-density linear polyethylene, a less costly and more recyclable material.

Price range: Costs range from about $2,500 for a seasonal-use Clivus (M1) to more than $10,000 for an institutional unit.

Maintenance: Periodic leveling of the top of the pile and raking forward of material at the removal hatch is required, often weekly, using a rake provided. Some users also turn the material when they do this, to prevent compaction. As in nearly all continuous composters, urine and water leaches through the material to a removal compartment at the base of the composter. The screen between the end-product tank area and liquid storage tank may need to be cleared off occasionally. It is advisable to remove some material at least every two years to prevent compaction (a shovelful every six months is better), and and to remove leachate yearly or as needed.

Sample power requirements: M12

Moving air: 43 to 93 watts

Heating: None

Process controls: 10 to 50 watts for optional moistening system

Notable accessories: Clivus Multrum offers a graywater filter, a waterless urinal, and an optional 900-series automatic controller that monitors liquid levels, air flow, temperature, pump operation and composter usage. It also controls pump operation in response to liquid levels, automatic daily compost mass moistening, automatic filling of freshwater supply tank, automatic fire suppression and internal chamber light. The 905 displays information through an LCD panel and provides maintenance alerts through an audible alarm. In some locations, Clivus offers maintenance contracts.

On some models, an internal (or external and modular) freshwater storage tank supplies freshwater to the Automatic Moistening and Fire Suppression System. A built-in moistening system sprays fresh water to the compost mass at timed intervals. The fire-suppression feature engages if internal temperatures reach 165°F, spraying water on top of the contents.

The public-facility models feature a freshwater storage area under the sloped floor. This can be used to irrigate the compost when it is too dry. In the Model M-35, this tank is inside the unit; in all others, it is an optional outside container that also supports the unit.

Advantages: The big box systems allow long-term retention, which permitting officials love for its potential for thorough processing and infrequent handling of the end-product.

Disadvantages: It may need some aeration help occasionally—by you with a raking tool or pitchfork. The large mass in one place makes it more prone to compaction.

Clivus Multrum, Inc
15 Union Street
Lawrence, MA 01840
1-800-4CLIVUS
forinfo@clivus.com, www.clivus.com

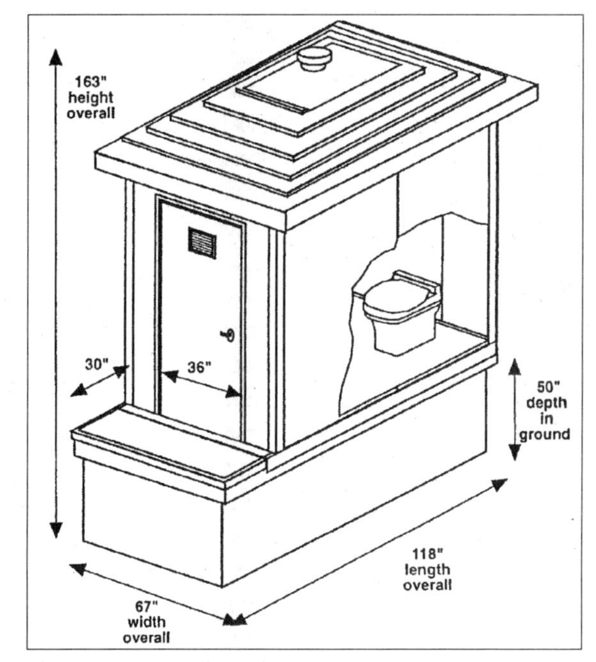

The M54 ADA Trail Head

A sampling of Clivus Multrum models

MODEL	CAPACITY	DIMENSIONS
M1-W	10 uses daily	56" high, 65" deep, 33" base width
M2-W	15 uses daily (2 bedroom house)	66" high, 65" deep, 33" base width
M10		60" high, 104" deep 47" wide
M12-W	80 uses daily (7-bedroom house)	58.5" high, 105" deep, 62" base width
M15	100 uses daily	99.5" high, 41" wide, 119.5" deep
M18	120 uses daily	83" high, 105" deep, 62" wide
M22	80 uses	64" high, 115" deep, 62" wide
M25	100 uses	98.5" high, 122" deep, 41" wide
M28	120 uses	89.5" high, 115" deep, 62" wide
M32	110 uses	67" high, 103" deep, 70.5" wide
M35	180 uses	89" high, 103" deep, 70.5" wide
M54	60 uses	Complete building: Floor length 118", roof width 84", total height 163"

CTS

CTS—Composting Toilet Systems—is essentially the Clivus Multrum (Clivus's patent expired in 1981). The key difference: It's fiberglass. It also features the double- or triple-pipe aeration system, which was dropped in some Clivus models. Price range: about $3,850 to $5,560. Most models are NSF listed. See Clivus Multrum for maintenance, advantages, disadvantages and power requirements.

Composting Toilet Systems, Inc.
P.O. Box 1928
Newport, WA 99156-1928
509/447-3708
cts@povn.com

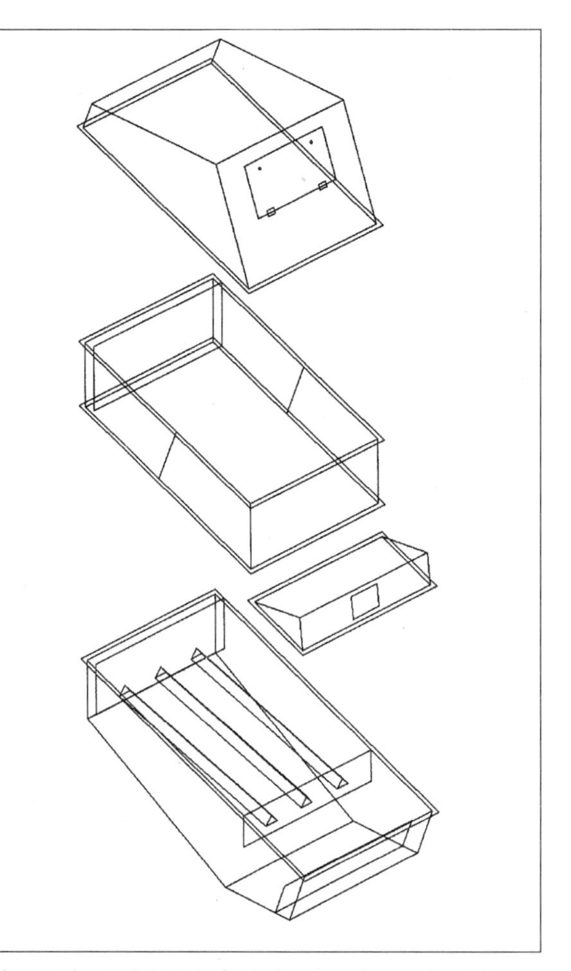

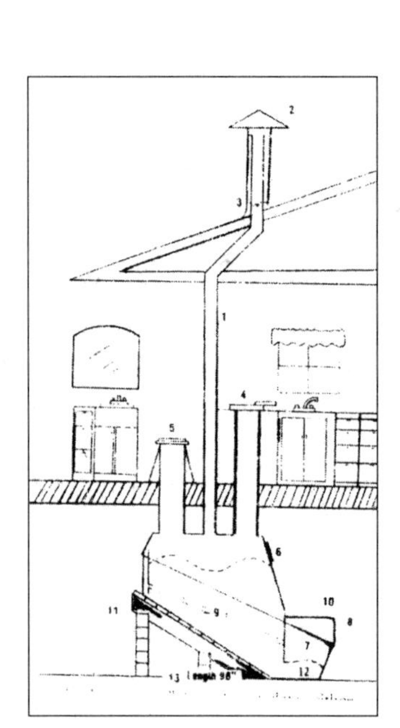

Above: The CTS-914. Left: A diagram shows the CTS with a dry toilet and a chute for kitchen scraps.

A sampling of CTS models, more available

MODEL	CAPACITY	DIMENSIONS
CTS-410	18 uses daily	6' high, 9' deep, 4' base width
CTS-710	40 uses daily	7' high, 9' deep, 4' base width
CTS-1010	75 uses daily	8' high, 9' deep, 4' base width
CTS-904	80 uses daily	5' high, 9' deep, 5' base width

EcoTech Carousel

Designed by a Norwegian inventor and first fabricated by Vera Miljø of Norway, the Carousel ("Snurredassen" in Norwegian) may be the best-selling central composter in Scandinavia, with more than 35,000 installed there alone. In North America, it was first distributed in California, and now is manufactured in the United States by EcoTech. The large model is NSF listed.

The Carousel is a fiberglass cylindrical container consisting of an outer case and an inner case. The inner case is divided into four composting containers with drain holes at the bottom. These rotate lazy-susan-style inside an outer case. ETPA is deposited into one chamber at a time. When one chamber is full, the next one is rotated into position. After about two years of accumulation, the chambers are emptied every three months to one year, depending on capacity and usage.

The level of the contents of the active chamber can be checked through an inspection hatch at the top. Leachate drains to the bottom of the outer case, where it evaporates or is piped away for utilization or disposal. This ensures liquid and composting solids are separated.

Like the Clivus Multrum, the design of the Carousel has been copied in other countries, where patent rights are not an issue. In Australia, the Rota-Loo is the Carousel, with the exception that the compost chambers are removable wedge-shaped containers. In Finland, the Ekolet is the same design available in a wheeled unit. In Sweden, it's one of Aquatron's three composter designs. A variation of this by Vera Miljø is the "Removable," which, instead of one reactor divided into four, simply features four removable containers inside the outer case.

A cutaway diagram of the EcoTech Carousel

EcoTech Carousel models

MODEL 80-A	CAPACITY	DIMENSIONS
Small Carousel	2 people for year-round use Weekend/vacation use: 12	20.87" high, 52" wide
Medium Carousel	3 people for year-round use Weekend/vacation use: 15	24.8" high, 52" wide
Large Carousel	NSF listed with heater and fan Capacity with heater: 5+ people for year-round use; Weekend/vacation use: 32 people per year	51.2" high, 52" wide

ETPA = Excrement, Toilet Paper and Additive

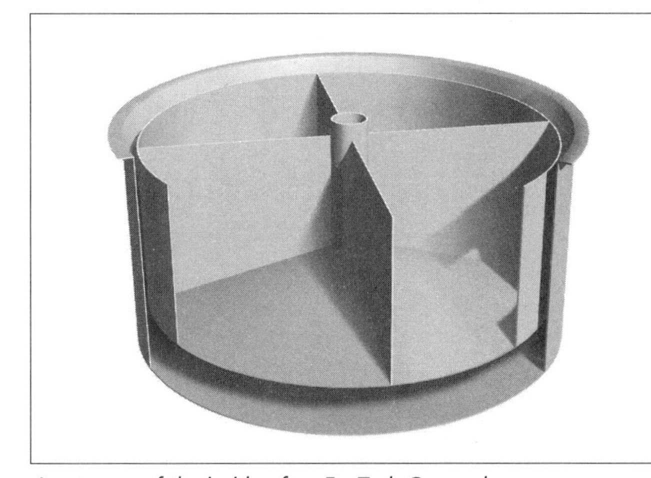

A cutaway of the inside of an EcoTech Carousel

Materials: Fiberglass
Sample power requirements:
Moving air: 43 watts
Heating: Optional heater for inside or outside the composter: to 350 watts
Process controls: None

Advantages: The Carousel offers long, uninterrupted processing time in four separate composting chambers, uncontaminated by fresh excrement, so the end-product is usually more finished. This can allow the Carousel to provide more capacity in a smaller space than other large central composters. Also, a batch system offers more opportunities for pathogen reduction—an advantage where pathogens are a problem. Manual turning of the composting material is not necessary.

Disadvantages: One must periodically check the level of the composting material, and rotate new composting chambers into place as they fill up. Rotating the containers could be difficult if the unit is not properly balanced. As in many composters, the removal hatch size is smaller than ideal for easy removal of the end-product.

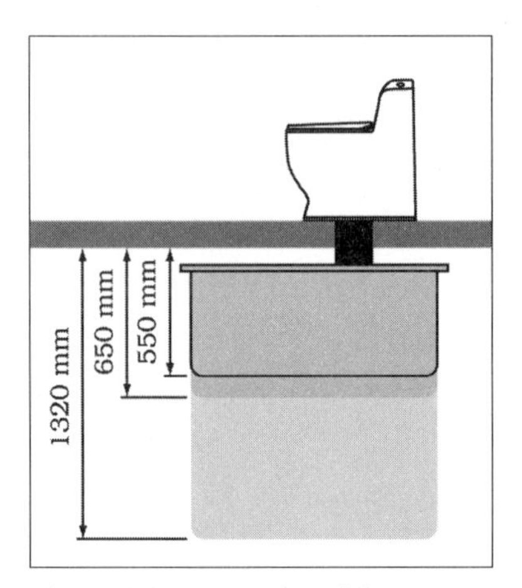

The various Carousel models.

EcoTech
(Manufacturer of the EcoTech Carousel and
North American distributor of Vera Miljø products)
P.O. Box 1313
Concord, MA 01742-2968
978/369-3951
watercon@igc.org
www.ecological-engineering.com/ecotech.htm

Envirolet

Made in Canada, Envirolets are much like BioLets, but they all feature a cage-like inner basket that holds waste, and a movable shaker grate—called a "mulcherator"—that is manually pulled and pushed to break up and aerate the composting material.

Dual muffin fans circulate air around the basket, which holds the processing ETPA. In addition to the mulcherator, a built-in raking bar helps sift material from the basket to the collecting tray below.

The composter is available as a self-contained unit as well as a remote composter that can be coupled with the company's plastic dry toilet stool or a SeaLand micro-flush toilet. Switches on the top of the unit allow the user to operate the heater separately from the fan, so if the composter sees little use, the fan-only option will save electricity. In the electric model, there is a 40-watt fan and a 540-watt heater.

One nice aspect is that the fans, heater and thermostat are in a removable power module to make servicing easier.

Envirolet models are essentially the same type of composter with different tops to couple with different toilet seat/stool options. The Multi-System 10 is CSA International listed.

Maintenance is the same as the BioLet self-contained models.

Materials: All models are constructed of high-density polyethylene plastic outer cases.

Sample power requirements:

Moving air: 40 watts

Heating: 540 watts

Process control: 45 watts mixer motor

The Envirolet

Price range: $1,149 to $1,700, including toilet stool

Advantages: Small and inexpensive. When the mulcherator works, it's a nice system.

Disadvantages: The units with heaters primarily dry the material, because high air flow and heat dry faster than composting can occur. It also makes it very hard for the material to sift through the mixing grates.

Sancor Industries Ltd.
140-30 Milner Avenue
Scarborough, Ontario M1S 3R3
Canada

A sampling of Envirolet models, more available

MODEL	CAPACITY	DIMENSIONS
Envirolet self-contained toilets and centralized toilets with dry toilet stool	Non-electric: 4 people weekend/vacation use; 2 people continuous With 12-volt fan: 6 weekend/vacation and 4 continuous 8 people vacation; 6 people continuous	25" high (height to seat: 17.75"), 33" deep, 25" base width
Multi-Unit System 10 540-watt heater and 40-watt fan (Envirolet's Remote System)	3-5 people weekend/vacation use 8 people vacation; 6 people continuous	28.5" high, 33" deep, 25" base width

Remember to add one foot to the depth for room to pull out the drawer.

Phoenix

The Phoenix Composting System was developed by former Clivus Multrum dealer Glenn Nelson, who sought to design a system that more actively aerated and mixed the composting material. The Phoenix is a vertical, single-chamber composting system in which the composter is encased in an outer container. Air vents on the sides of the composter help aerate the sides of the composting material.

The Phoenix features a horizonal mixer with tines that work like a rototiller to loosen, mix and aerate the material as it moves downward. The number and position of the mixers vary depending on the model. These are rotated via a removable handle on the front of the composter. These two key features—the mixing tines and air circulation around the composter—are the distinguishing characteristics of the Phoenix. The compost should be turned regularly to prevent compaction of the compost.

Leachate drains through a sloped screened floor, which separates the solids from the liquids, to an evaporating area with a peat moss filter. The Phoenix features a leachate pump which can spray this leachate on top of the material. The end-product is removed from a drawer at the base of the system.

Some Phoenix models are CSA-listed. Models cost $2,500 to $4,500 and up for public facility models.

Materials: Cross-linked and linear polyethylene

Notable accessories: Advanced Composting Systems offers an evaporator for managing leachate. The system is a 50- or 60-gallon storage tank and a counter-flow evaporation tower containing plastic curly-Q shapes, which provide added surface area for the evaporation of leachate, which is sprayed on the top. Air is forced through the media by a fan, evaporating leachate. Now and then, the unit should be cleaned to remove the dehydrated organic matter and salts from the leachate.

Sample power requirements: R-200

Moving air: 5 to 25 watts

Heater: None

Process control: By hand

Advantages: This is a large-volume single-chamber composter that has overcome some of the problems of other such composters by providing mechanical mixing, which promotes aeration, minimizes compaction and facilitates through-put of ETPA. Also, the side air vents allow for better aeration.

Disadvantages: Opportunities for rotating tines to get stuck and for the grate to clog.

A cutaway diagram of the Phoenix

Advanced Composting Systems
195 Meadows Road
Whitefish, MT 59937
406/862-3854
phoenix@compostingtoilet.com
www.compostingtoilet.com

A sampling of Phoenix models

MODEL	CAPACITY*	DIMENSIONS
R-199	2 people year-round	55" high, 39" wide
R-200	4 people year-round (30 uses daily)	73" high, 39" base width
R-201	8 people year-round (50 uses daily)	91" high, 39" base width

Several models for public-facility use. *At 65°F

ETPA = Excrement, Toilet Paper and Additive

Sun-Mar

Sun-Mar's composting toilet systems all feature the company's patented Bio-Drum™, a crank-turned tumbling cannister composter mounted horizontally in a case.

Founded by the Swedish inventor of the first self-contained composting toilet for cottages, Sun-Mar offers four self-contained models and three central composting systems.

In all models, ETPA (and in some models, water from micro-flush toilets) enters the system from the top and passes into the Bio-Drum. A hand-crank allows users to periodically rotate the Bio-Drum, which mixes and aerates the material (a motorized crank is also available).

Clockwise from left: A Centrex with a dry toilet stool; a Sun-Mar Excel; and a cutaway diagram shows the four processing chambers in a Centrex Plus.

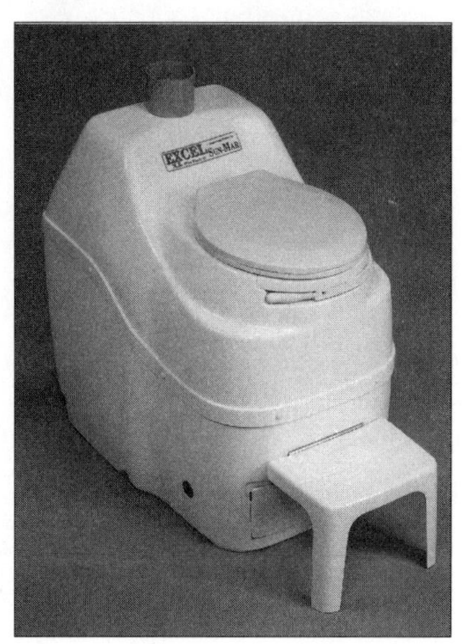

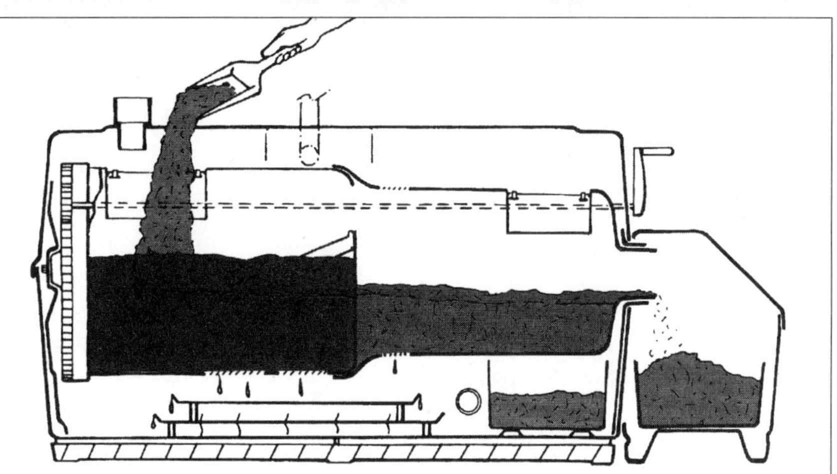

For routine mixing, one rotates the Bio-Drum every three or four days. When the Bio-Drum is about two-thirds full, one turns about a third of its contents into a finishing tray beneath the drum by unlatching a drum lock and revolving the drum counter-clockwise. Processed and unprocessed material then drops down into the pull-out compost finishing drawer, where it composts further. It is emptied when the drum fills up again.

Leachate drains to an evaporating chamber. A heating element under the case evaporates some leachate and gently warms the Bio-Drum. Any unevaporated leachate drains to the base of the unit and must be removed or drained away. In the Bio-Drum, a screen separates solids from leachate.

Materials: All Sun-Mar's major process components and outer cases are made of injection-formed fiberglass, with stainless steel drum-turning shafts.

As of this writing, the Sun-Mar Excel (electric) is this company's NSF-listed model; the Centrex Plus may soon follow.

The Excel is the company's highest-capacity self-contained toilet system. The vent and drain are at the rear of the unit, and the finished end-product is removed from the front via an unsealed drawer.

The WCM (Water Closet Multrum) is an Excel configured to be used as a central composter with a one-pint flush toilet or a dry toilet. The micro-flush toilet drain line runs to the top of the unit, which also features an access door for adding additive.

The Centrex differs in that the Bio-Drum is mounted tranversely. The vent pipe, electrical panel and additive door are accessed from the front. Owners of these systems often place them outside, under their cottages. As with any outdoor electrical system, remember to safeguard against potential electric shock: weather-proof the outlet box and consider installing a ground fault circuit interrupter (GFCI) receptacle.

Sun-Mar's latest model, the Centrex Plus, features a

ETPA = Excrement, Toilet Paper and Additive

longer Bio-Drum divided into two sections. An internal bar improves the mixing. When the primary chamber in the drum is full, it automatically spills into the secondary chamber. As the secondary chamber fills, it automatically spills into a finishing container. When that is full, the user places that in back and replaces it with an empty container. Twin 250-watt heating elements can be turned off, on or at 50 percent. A powerful air blower runs continuously.

Prices range from about $950 to $1,649.

The evaporating mechanism is two stacked cascading evaporating trays, which provide more surface area and trap suspended particles, increasing evaporation. The trays, combined with the surface area of the floor of the unit, give the Centrex Plus a total of 10 square feet of evaporating area.

Interesting option: The Centrex NE is available with a space underneath, where hot water can be circulated in pipes, in the style of radiant floor heating, to speed up composting.

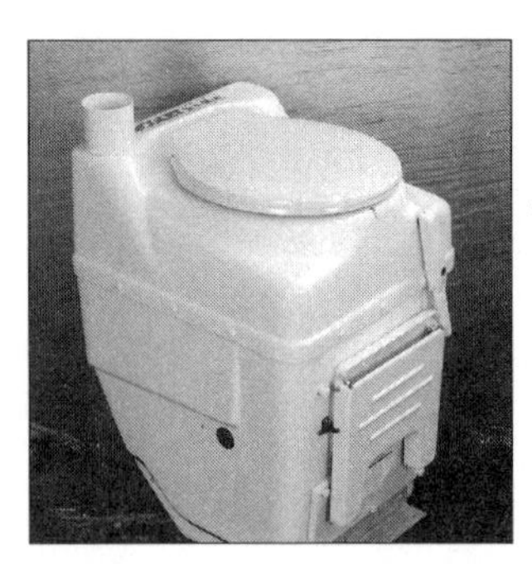
The Ecolet, for RVs and boats

The Centrex Plus is particularly interesting because it features some degree of batch composting with some tumble-mixing prior to static curing.

Performance reports or confirmations of the stated capacity are not yet available. Due to the amount of water flowing through the system, it is possible that ETPA could block the drain lines, and it might be difficult to service that from the access hatches. Sun-Mar's systems may be considered batch-continuous hybrids as they do have a three-chamber design, but are not as "batch" as the alternating container designs because unprocessed matter from one chamber is moved to another along with processed matter.

Sample power requirements: Excel

Moving air: 25 watts

Heating: 250 watts (but on only half the time)

Process controls: Optional motor

Advantages: Sun-Mar's drum composter provides mixing and aerating better than most composters. A good track record among the cottage composters.

Disadvantages: The drains for the drum and the leachate can clog. Occasionally, tumbling can "ball up" the ETPA.

Sun-Mar Corp.
5035 North Service Road, C9-C10
Burlington, Ontario
Canada L7L 5V2
905/332-1314
www.sun-mar.com, compost@sun-mar.com

A sampling of Sun-Mar models

MODEL	CAPACITY	DIMENSIONS
Compact	3-5 people weekend/vacation use	27.5" high (seat: 21.5"), 33" deep, 22" base width
Ecolet 110 Ecolet 110 Mobile	3-5 people weekend/vacation use	27" high, 21" deep, 19" base width RV is made to fit boat hulls
Excel	3-5 residential (NSF tested for four adults)	29" deep (51" required to pull out drawer), 22.5" base width (26" for handle)
Excel NE	An Excel, with dual vent stacks (a 2" and a 4" top-mounted)	
CENTRAL COMPOSTING SYSTEMS		
WCM (Water Closet Multrum) Used with micro-flush toilet	7-9 people weekend/vacation; 4-6 residential	32.5" high, 29" deep (51" required to pull out drawer), 22.5" base width
Centrex NE and AC/DC available	7-9 people weekend/vacation; 4-6 residential	27.5" high, 27" deep (40" required to pull out drawer), 32" base width
Centrex Plus	7-9 people weekend/vacation; 4-6 residential	27.5" high, 27" deep (40" required to pull out drawer), 32" base width

Nonelectric, AC/DC models and a dry toilet stool are available.

ETPA = Excrement, Toilet Paper and Additive

Vera Miljø

Vera Miljø's toilet systems are all based on batch composting, with a variety of alternating compost containers.

Well established in Norway, the company offers a wide range of systems for cottages and off-the-grid homes, as well as tanks, filters and appliances for collecting, retaining and using rainwater. Vera developed the Carousel-style composter, now licensed in North America to EcoTech. The company's products are sold in North America by Ecos however, high freight costs and metric components mean Vera is not yet a major player outside of Europe as of this writing. Vera also offers a dry toilet stool, a urine-separating toilet and a waterless toilet stool with a mechanical trap.

The company offers three self-contained composting toilet systems featuring two alternating containers (more can be purchased). One Toga model may be embedded in the floor. The "Removable" is essentially a Carousel with removable containers. The Toga 2000 consists of alternating rollaway trash containers—and will undoubtably inspire many to build their own.

Maintenance involves alternating compost containers.

Sample power requirements: All

Moving air: 15 to 30 watts

Heating: 108 to 260 watts

Process controls: None

Materials: Most toilet systems are made of polyethylene.

Advantages: Alternating containers allow for more complete composting. Their capacity is as great as the numbers of containers used.

Disadvantages: But you've got to monitor and alternate those containers.

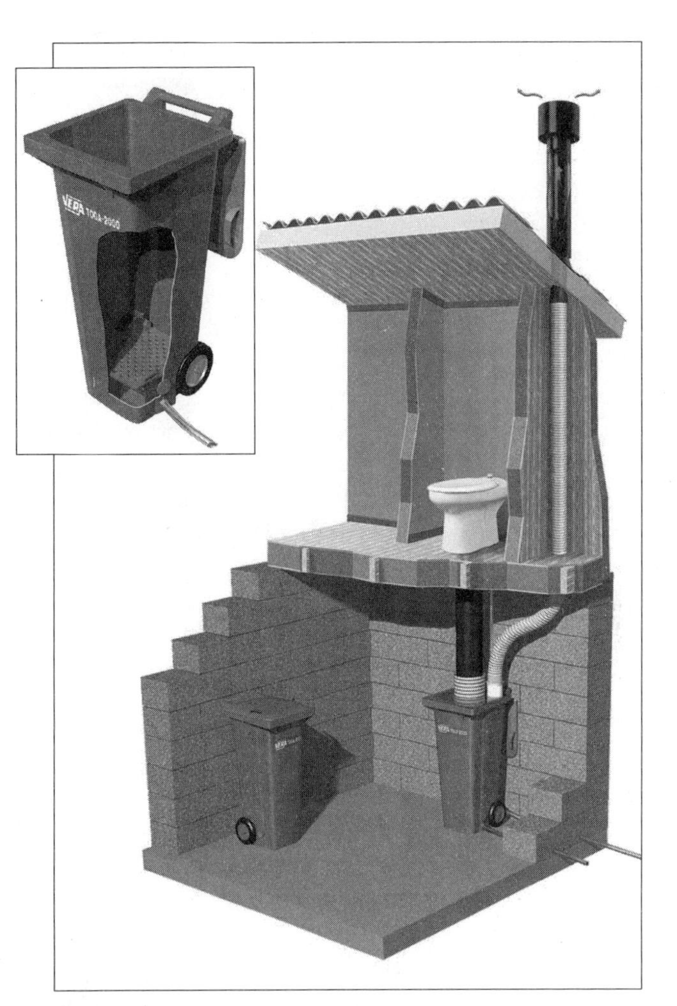

In North America:
Ecos
P.O. Box 1313
Concord, MA 01742
978/369-3951
watercon@igc.org
www.ecological-engineering.com/ecos.html

In Norway:
Vera Miljø A/S
Postboks 2036
N-3239
Sandefjord,
Norway
www.vera.no

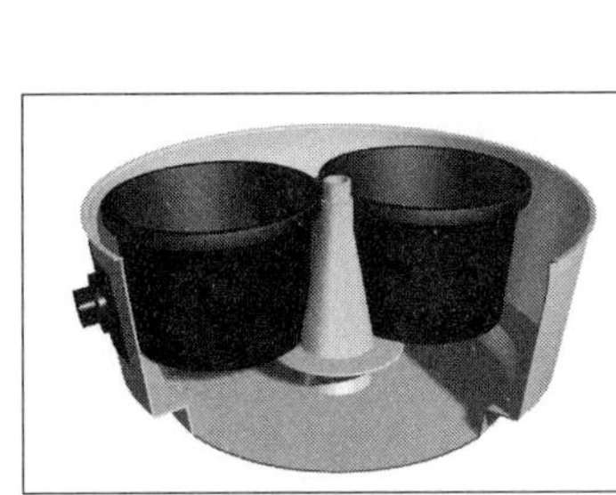

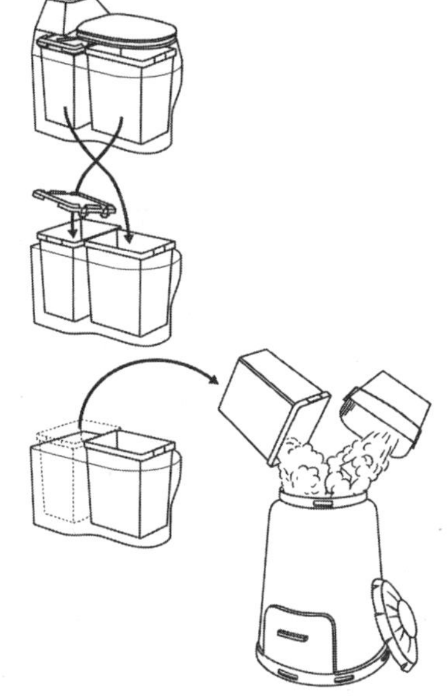

Clockwise from top: The Toga 2000; the composting bin of a Pioneer is emptied into another composter with kitchen scraps; the Pioneer and the inside of a Vera Carousel with removable containers.

Other Systems Worldwide

As of this writing, the following companies do not have strong distribution channels or approvals in North America. That may change; contact the companies for up-to-date information.

Aquatron International AB
Björnnasvagen 21
S-113 47 Stockholm, Sweden
+08-790- 98 95

Aquatron offers four models of continous composting reactors (in materials from coated steel to polyethylene) and one carousel-style model, serving cottages on up to 24-unit apartment buildings. One model, the Aquatron ALE system, is a solid waste and kitchen waste composter, using an electric motor-driven drum composter. The Aquatron is often used with an Ifö Cera adjustable 3- to 8-liter-flush toilet.

What's unique about Aquatron is its Aquatron Separator, a passive solids-and-liquids separator, which is described in "Toilet Stools for your Composting Toilet System," later in this chapter.

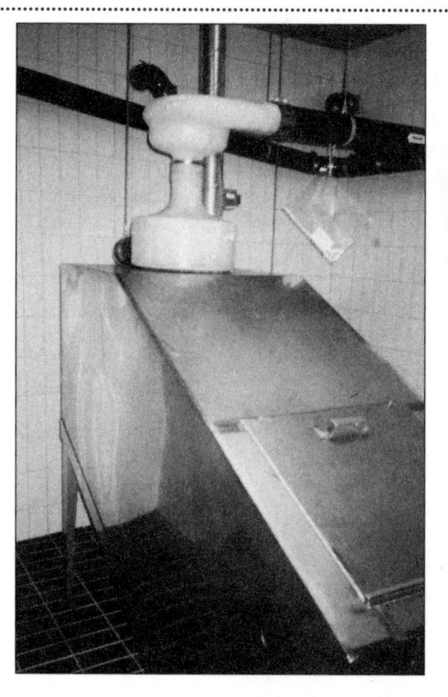

Left: The Aquatron separator and a stainless steel composter at a school.
Above: Ifö Cera adjustable-flush toilets are used with the system.

Bio-Recycler
5308 Emerald Drive
Sykesville, MD 21784
301/795-2607

Designed and patented by retired sanitary engineer Jeremy Criss, the Bio-Recycler is a three-part system that is not yet in mass production. Criss couples the system with a vacuum toilet of his own design that features a spray nozzle, much like some boat toilets.

In the cylindrical plastic composter, ETPA drops to a top area. After a while, a chute is opened, and the material drops to a finishing area, where worms help process it. Leachate is drained to a holding tank at the bottom, from which it is drained to an outdoor yard waste composter.

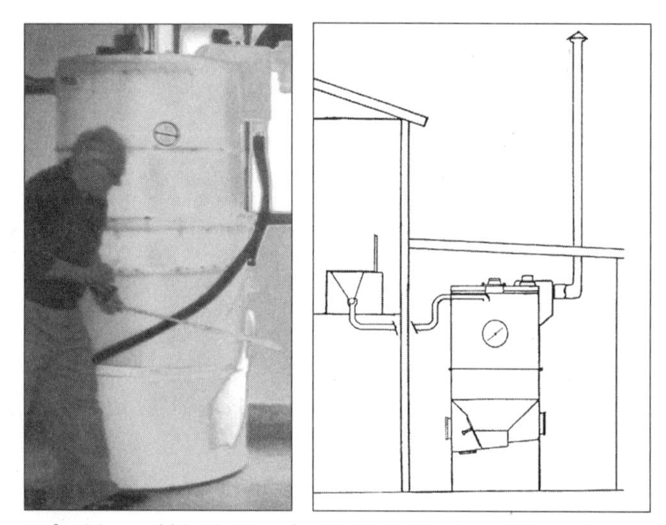

Left: Criss and his Bio-Recycler. Right: A diagram of the system.

Dryloo
P.O.Box 75619
Gardenview
2047 South Africa
27-11-6155328

The Dryloo™ is another carousel-style batch system, except that it has no walls. It is simply a rotating PVC plastic frame upon which one hangs six woven polyethylene bags, which serve as solids composters. Place the system in a watertight container. Add a toilet stool, a fan and a vent pipe.

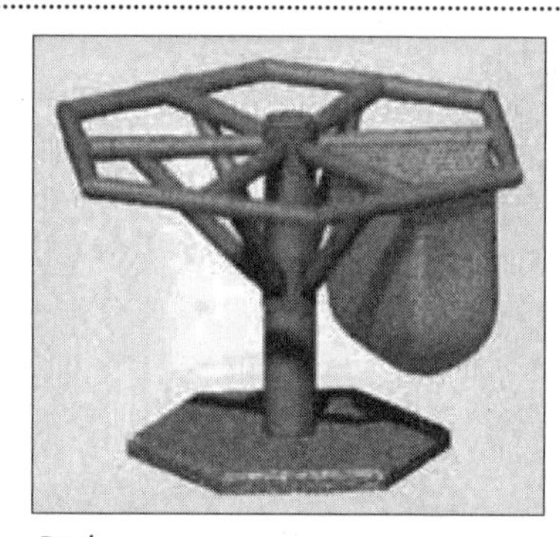

Dryloo

ETPA = *Excrement, Toilet Paper and Additive*

Cotuit Dry Toilet
P.O. Box 89
Cotuit, MA 02635
508/428-8442
www.cape.com/cdt

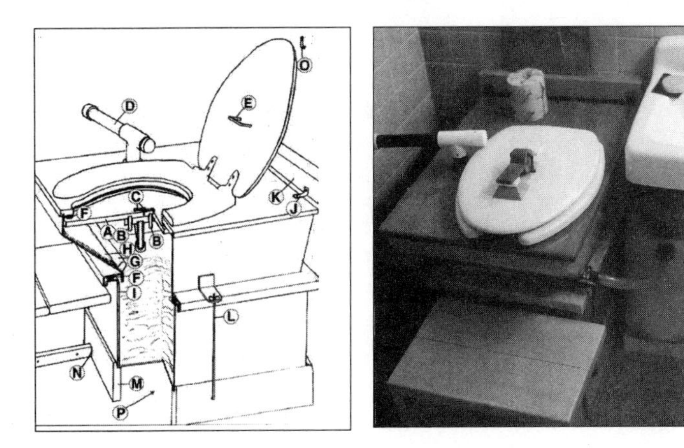

A self-contained system that is a plastic box within an attractive wooden box. Urine is diverted to a plastic tank. Also features a view plate that is moved away when toilet is to be used. The removable plastic bin allows for batching, but you still must have a place to put the processing bin and the urine.

Devap
BMS
P.O. Box 8248
Fort Collins, CO 80526
800-524-1097

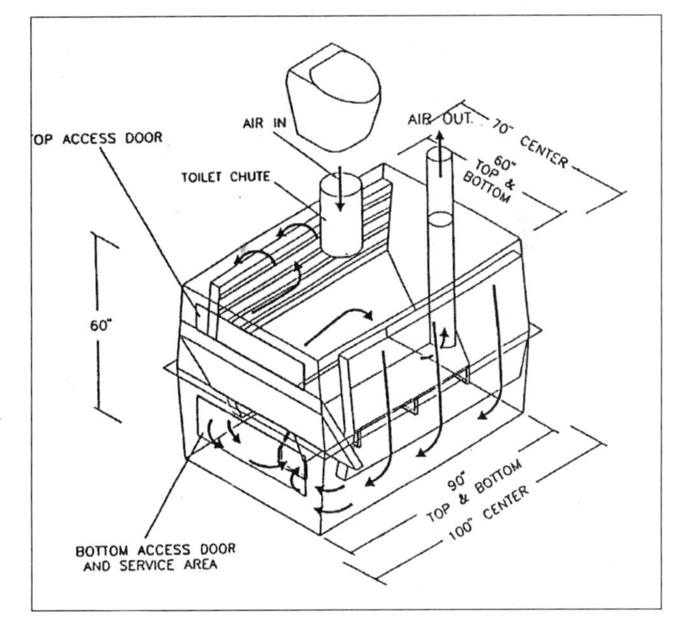

BMS calls its Devap system an evaporator; however, the process at work here is virtually identical to all composters especially those such as BioSun, that blow significant amounts of air through the processing material to dry it and evaporate leachate.

In the Devap, ETPA enters the system onto a bed of wood chips, where excess liquids are constantly evaporated by forcing ambient air through the material. Liquids from toilets pass through the compost/wood chips and are directed via liquid channels on the sides of the tanks. Like a Clivus Multrum, there is sloped-floor aeration system, a baffle and a freshwater spray system activated by timers.

The size of the air-moving blower system varies depending on the capacity of the models, ranging from 265 to 650 cubic feet per minute (CFM). The blowers are centrifugal in-line fans located above the composting/evaporating tank, and often enclosed in an electric-powered roof ventilator system. The system pulls a significant vacuum on the tank, allowing untempered outside air to be drawn through the system. If the air is too cool or moist to promote evaporation, then leachate accumulates and can be pumped to a collection tank.

This system is filled with significant amounts of wood chips which provide both carbon and surface area for evaporation.

One nice feature is the leveling tools that are permanently installed inside the tank and controlled from the outside. There are also view windows, so the operator can see the process.

However, material from the back must be moved manually from the middle to a drying area, where it remains to finish for several months. The manufacturer claims that fresh material cannot reach the more advanced material.

A baffle in back of the removal door keeps new material from spilling forward when emptying.

Essentially, if enough air and heat are added, drying and/or a biological process will result.

Price range: From $8,500 for a system and two toilets.

Advantages: High-volume fans help get a lot of air to the material, which either dries the material or assists composting. Works well where the air temperature is warm and humidity is low.

Disadvantages: High-volume fans require significant power. Also, since a lot of air is pulled into the system, evaporative cooling can take place. For that reason, the system requires supplemental heat in the winter. Without supplemental heat, the Devap can only be used when ambient temperature and moisture conditions are optimal for the process.

Dowmus
P.O. Box 400
Mapleton, Queensland 4552 Australia
dowmus@ozemail.com.au, www.dowmus.com

Imagine that all of your home's graywater, toilet flush water, excrement and kitchen wastes, as well as some paper and cardboard, are deposited into a large aerated tank. Leachate trickles through the solids, and is drained to a leach field. That's the Dowmus ("Domestic Organic Waste Management and Utilisation System"). It can also be used to manage just some of a home's wastewater and as a composting toilet system alone.

The reactor is a cylindrical single-chamber system with a coiled perforated pipe at bottom center and a fan for aeration. To remove end-product, a removable auger on the end of a long-handled shaft screws into the compost mass. When it is pulled out, end-product is caught on the auger's grooves, and is removed. This is a possibly slow but innovative removal mechanism. (Beware of creating a "cave" in the material, though.)

The manufacturer maintains that, when it is used as a whole-house wastewater system, the composting organic waste and carbon (paper, etc.) act as a trickling filter, just like top layers of soil, where aerobic bacteria help clean up the water. What may require more examination is how fast the water trickles through—biological treatment takes time, so a minimum residence time is critical. At the same time, too much water can flush out the beneficial microbes, so one must maintain the scouring velocity below the threshold that would remove the attached microorganisms and rinse them out. It will be interesting to see how post-filtered effluent quality compares to the performance of other trickling filters in terms of efficiency of pretreatment. Stay tuned.

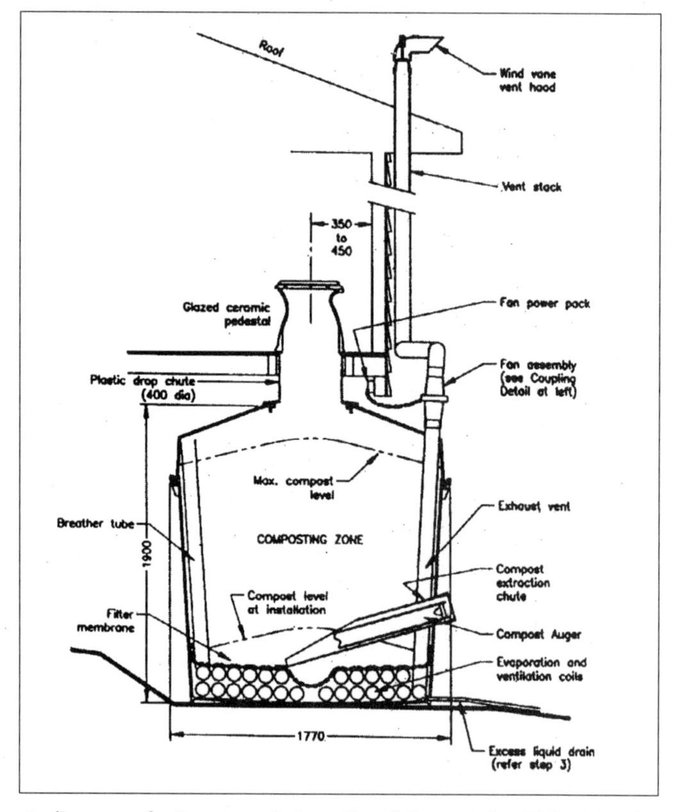

A diagram of a Dowmus (International Composting Toilet News)

Ekolet Oy
Estetie 3
00430 Helsinki, Finland
eko@dystopia.fi www.dystopia.fi/ekolet

This is a rotatable four-compartment carousel-style composter that turns on wheels on a floor-mounted frame. As one chamber fills, the toilet chute is disconnected, and the entire composter is rotated to position the next empty chamber underneath. The containers are banded together to form the full unit. For easy shipping, it comes in kit form. All components must be assembled with bolts, silicone sealant (beware of leaks), screws and special tensioning bands that hold the four bins together. Leachate pumps and heaters are available.

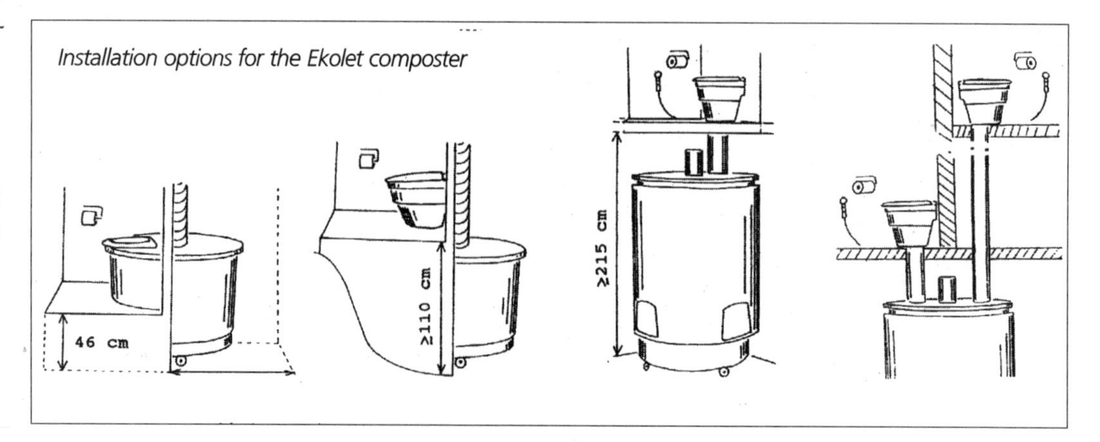

Installation options for the Ekolet composter

ETPA = *Excrement, Toilet Paper and Additive*

Enviroloo
Enviro Options Ltd.
P.O. Box 13 KYA SAND 2013
Johannesburg 2013 South Africa
2711 708 2245
eloo@mweb.co.za

Designed for hot African climates, this is a polyethylene continuous system whereby solids sit on top of a "drying plate," and liquids collect below to evaporate. The compost reactor is usually half buried in the ground. Includes a mixer and an interesting toilet stool with a mechanism that flips the solids into the composter after the user pulls a lever. The system requires a long-handled shovel to remove material.

Several Enviroloos installed in a school

Mullis
Finns M&C
Skogsvägen 16
S-89340 Köpmanholmen, Sweden

Mullis is a classic sloped-bottom continuous composter made of stainless steel and featuring four aeration channels ("air rails") and a baffle. It comes with a finishing composter, which has a perforated plastic pipe coiled in a container.

Nature-Loo
P.O. Box 1213
Milton, Queensland 4064 Australia

The Nature-Loo consists of alternating removable roll-away composting containers shaped like upside-down flower pots. It is also available in a do-it-yourself kit form.

Naturum
Luonto-Laite Oy
FIN-17740 Kasiniemi, Finland
luontola@sci.fi

This very interesting-looking fiberglass self-contained system features a clever internal polyethylene two-chambered composter with a removable polyethylene container with a handle for transporting the end-product. After use, one depresses a pedal. The solid wastes move out of sight, and a spring-loaded tension cable rotates a small compost container, which tumbles fresh and older material. After several rotations, the material gradually shakes out into the emptying container. Leachate is managed by adding a biodegradable absorbent package full of peat, which soaks up liquid. Urine is diverted to a separate collection area (with a very low drain!), keeping the composting material unsaturated. It uses no power and must be emptied monthly.

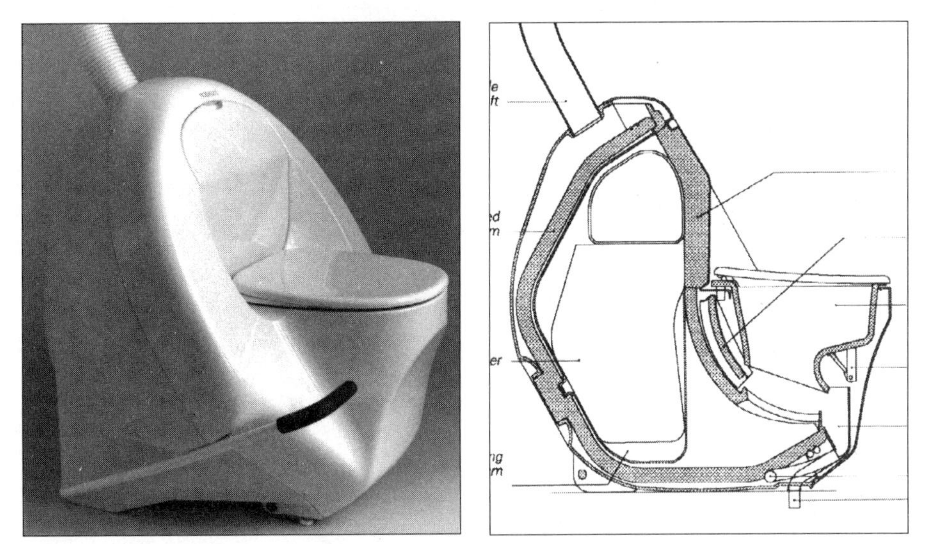

The Naturum from Finland takes the prize for most interesting-looking waterless toilet.

Romtec
18240 North Bank Road
Roseburg, OR 97470
503/496-3541

Romtec is a simple large-box composter with a toilet seat on it, an air intake and two solar windows. To remove material, you simply lift the top, suggesting that fresh matter will also be removed. Designed for 15 uses per day, it is made of double-walled insulated cross-linked polyethylene with Lexan passive solar heating windows, and stainless steel hardware. Primarily a single, partially aerated vault, this composter is mostly purchased for public areas that get occasional use. Romtec recommends stirring the contents every three days, and using two units as alternating batch composters. Buy two, and alternate them as they fill up—a good idea that may make this the least-expensive prefabricated double-vault system.

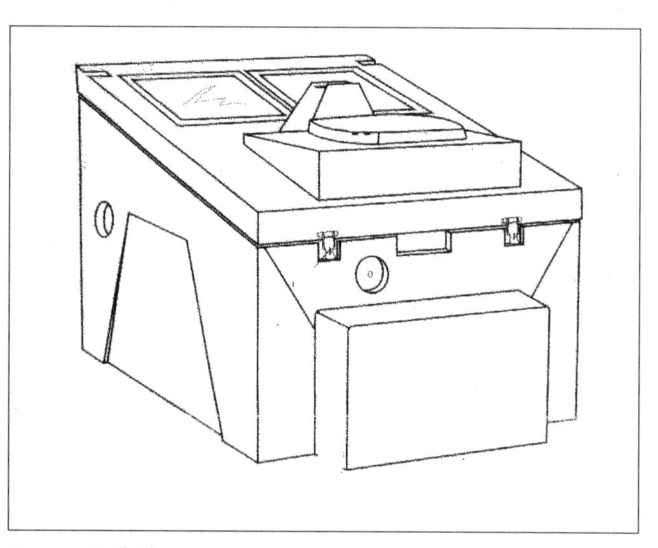

Romtec Trailside

Rota-Loo and (Aus.) Soltran
Environment Equipment Ltd.
2/32 Jarrah Drive
Braeside, Victoria 3195 Australia

Rota-Loo is a carousel-style system in which, depending on the model, either six or eight composting chambers are convenient removable bins.

Soltran is the passive solar assist described elsewhere in this book. The Australian version is a duplicate of the Ecos Soltran with some exclusions.

Right: Diagram of a Soltran (Aus.) system

Below: The components of a Rota-Loo. Removable containers allow easier emptying.

ETPA = Excrement, Toilet Paper and Additive

Separett® and Septum Torrdass
Servator Separett AB
Skinnebo SE-330
10 Bredaryd, Sweden

Manufactured since 1976, the Separett is essentially a self-contained urine-separating, heated and ventilated alternating-container system, whereby most of the composting must occur elsewhere. What is unique is that, inside the reactor of the Classic model, the solid waste is deposited onto a central plate that spins to disperse it evenly into the container by centrifugal action when a button is pressed. A fan and heating element encourage composting or drying and evaporation of the urine.

Separett Classic

When the reactor pail is full, the top of the toilet is lifted, the pail removed, and its contents emptied into an outside composter for further processing (the company sells a classic outdoor yard waste composter for this purpose).

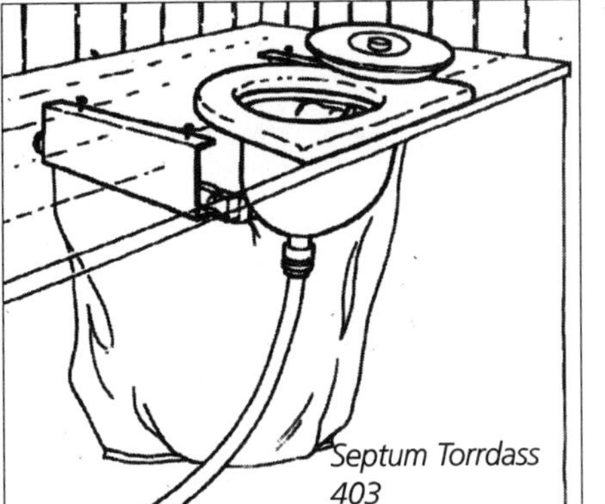
Septum Torrdass 403

Urine can be diverted to a tank. The company even offers an ejector system that, when used with an outdoor pressurized faucet, automatically dilutes the urine with eight parts water as you spray the solution with a garden hose nozzle for irrigation. Remember that this is fine now and then, but overuse of diluted and unoxidized urine can stress plants.

The case is made of polystyrene, a material that is prone to cracking.

For minimalists, the company offers the Septum Torrdass, simply a toilet bowl insert with a urine collection area and drain line in the front part of the bowl. You supply a bag or a container to collect feces, and a cabinet or supporting structure. It even comes with a fitted lid to close the toilet opening. This is a helpful device for those who want to make their own toilet system or a build-their-own toilet commode for a centralized composter.

Mökki Makki
Finn-Compos Oy
Vähäojantie 5
SF - 20360 Turku,
Finland

Simply a black metal pipe with a leachate tray. Might be fine for very occasional use in warm spots, but the open exposed leachate tray is not sanitary.

Tara Ekomatic Eko-Toilet
Ecora Oy Tara Finland Ltd.
00420 Helsinki, Finland

Two models: One is a simple ventilated bucket with a toilet seat and a mixing device, and the other is a remote unit that is a simple ventilated round container with wood chips.

Site-Built Composting Toilet Systems

Site-built composting toilet systems are constructed every day in communities where combining local materials and labor with proven designs produce affordable and effective sanitation.

They can be less expensive than a manufactured system, especially if the labor is free.

These systems are usually difficult to get approved by local permitting agencies, because precise performance data for these systems is not available or not guaranteed by a standards organization such as NSF International, and they are usually not built uniformly. However, there are occasional instances in which they are approved—usually when the local health agent is confident that the owner is well versed in their maintenance considerations, or a conventional system is already in place. For that reason, many site-builts supplement septic systems, and no permits are sought nor, perhaps, needed.

People have long made their own composting toilet systems out of a variety of new and reused materials, even old dishwashers and artillery shell cases! However, when building a composting toilet system, it is important not to scrimp on the quality of the construction—the results could be quite unpleasant.

As with all composting toilet systems, site-built systems will become more acceptable as more are constructed and their performance is evaluated. In the future, more of them will likely be built right into the foundations of new homes.

The systems described in this section represent several types that have been installed and successfully operated for at least two years, and usually many more.

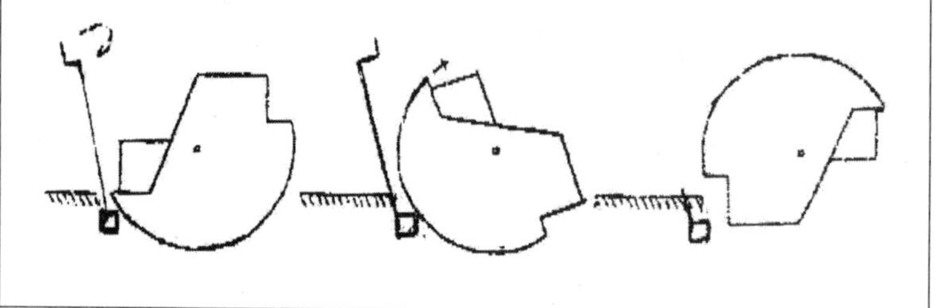

Clockwise from left: The entrance to an outdoor RAW toilet in England. (Photo: Kevin Ison); the access hatches for a ReSource composting toilet system in Mexico (Photo: ReSource Institute for Low-Entropy Systems); constructing a 55-gallon net batch system (Photo: David Del Porto), and a diagram of the De Twalf Continuous Composter in action (Illustration: De Twalf Ambachten).

Movable Batch Composting Toilet Systems

Sunny John, Fiji 55-Gallon Batch, Solviva Compostoilet, Wheelibatch, etc.

Movable batch composting toilets can be as simple as an alternating removable pail system or as sophisticated as powered roll-away containers with quick-disconnect couplings for toilet, exhaust and leachate drain connections. They first became popular when used 55-gallon steel drums were discovered by composter designers in the early 1970s. The problem was, the steel drums were heavy and would quickly corrode unless painted with expensive and toxic coatings.

Now these systems are enjoying a resurgence due to the availability of environment-friendly drums made of recycled polyethylene (a chlorine-free, thermoplastic polymer resin that can be melted and recycled infinitely). Food distributors buy "wet goods" in these polyethylene drums, which are ultimately given away or sold, sometimes rinsed clean, to the public for a low price ($10 at this writing).

Some systems employ roll-away trash bins, which come with built-on wheels. (Some manufacturers are making composters in this style—the Vera Toga 2000, for example.)

Some drums come with lockable water- and air-tight lids; others come with threaded fittings. The industrial drum industry offers a multitude of handling options, including special wheeled handtrucks for moving drums, wheeled dollies, liquid transfer pumps, add-on drain spigots, plastic bag liners, electric drum heaters, and even automatic tumbling equipment for mixing liquids in drums. Bulkhead fittings can quickly adapt standard plumbing pipe and fittings to provide quick connections for toilet, exhaust pipe and a leachate drain.

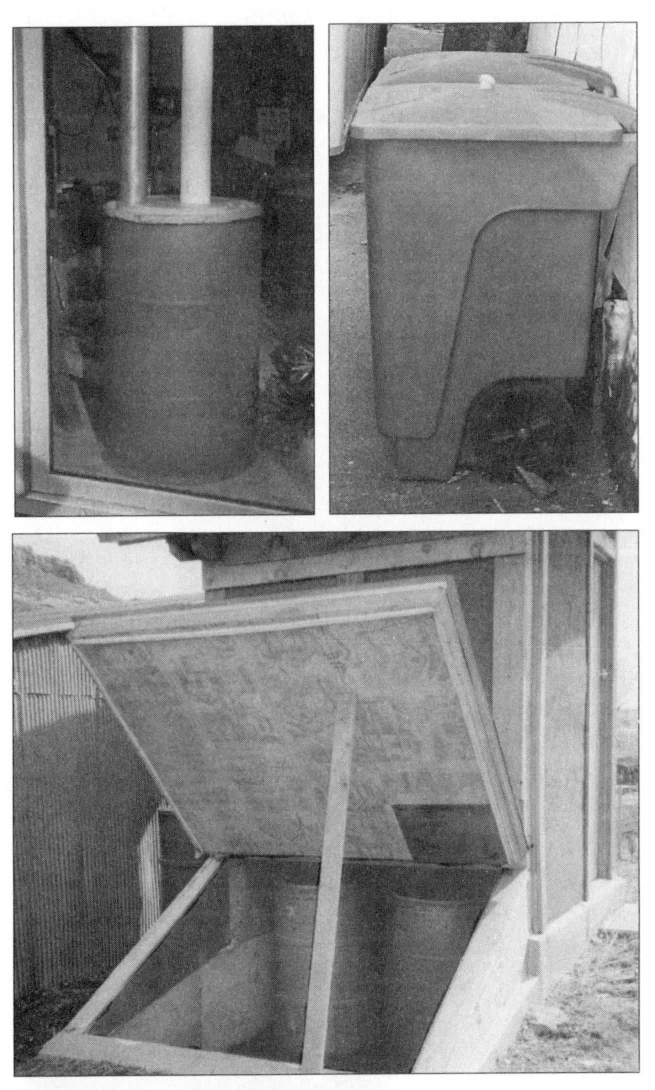

Top left and right: A reused 55-gallon drum and a trash bin are batch composters. Above: The back of a Sunny John. Left: A range of potential composter containers.

The Importance of Aeration Assists

As in any composter, the goal is to get air to as much of the material as possible, to assist composting and prevent anaerobic conditions (which lead to odors and slow processing). The key in aeration is the surface-to-volume ratio of the composting mass: The more exposed surface relative to the volume, the greater the opportunity for diffusion of air into the compost. A pipe drilled with little holes, an often-used option, does not make an ideal aerator; for one thing, the pipe can fill with carbon dioxide (produced by composting microbes) because that gas is heavier than air. The grates, stainless steel hardware cloth baskets and nets you see in the following batch system designs help expose more of the composting material to air, while suspending it up and out of the leachate as much as possible.

Advantages: Easily expanded capacity—just add more containers. Cheap. More opportunities for thorough processing. Containers can be taken away to be emptied.

Disadvantages: Must periodically check the level of the material in the active drum to determine when to replace it. Requires changing the containers.

Tips for Planning a Batch System

- ☐ Figure out how many containers you will need. A 55-gallon drum has 7.352 cubic feet of volume, 6.5 after you put in a net or grate to separate leachate and enhance aeration of the material. A conservative formula is: One person will fill one 55-gallon drum with a grate and a leachate drain in about 200 days at 65°F compost temperature (unless you are adding gallons of additive every week)—more days at higher temperatures.
- ☐ If you are not using wheeled containers or adding casters to your drums, figure out a method to move your containers. If you use a two-wheel hand truck, do not fill the containers high, to avoid spills when they are later moved. Remember that even well-drained composters get heavy—as much as 300 pounds, and 470 if the leachate was not drained!
- ☐ Establish a warm storage/processing area for the full composters if you are not keeping them next to the active composters and using the same exhaust pipe. Ideally, this is a warm, ventilated room, perhaps near your water heater or clothes dryer. A minimum of one year of processing at 65°F should do it—longer if it is cooler, less time if it is warmer. Install a flexible duct from each composter to the active composter's exhaust opening, or fit the drums with screened or perforated lids, and vent the room. Note, though, that they might not need as much active aeration during curing as they need during the initial active aerobic microbial digestion. Consider, too, placing them in a vented outbuilding with solar windows.
- ☐ Each drum requires a leachate drain with a valve that can be closed when you need to switch containers (unless you use a fixed leachate system). The leachate drain fitting on the drum must be higher than its destination, so it can drain by gravity. Otherwise you will have to pump it. Use at least a two-inch pipe for this if you are installing a micro-flush toilet.
- ☐ As with all composters, remember to inoculate each container with compost starter, such as warm sifted compost from your compost pile or purchased bio-activator. Or leave a little material, fluffed up, from the previous batch.
- ☐ Consider providing an inspection hatch or plexiglass window at the top of the container.

The Systems

Here are some representative systems for which plans are available—nearly all can be modified by adding different aerators or using different kinds of containers. Use any of these designs with a roll-away trash bins, available at nearly every home center and hardware store. The Consolidated catalog offers new plastic containers that make great composters, as well as tools for composting, such as long-handled mixers, scrapers, scoops, plastic gloves, scouring pads, etc.

• Wheelibatch Composter

Designed in Australia, this composter is a roll-away trash bin fitted with wire mesh bottom and two perforated pipes on the side for aeration.

Right: A Wheelie-Batch System in use (Ecological Sanitation, Sida, Uno Winblad, ed.).

The simplest form of 55-gallon net batch composting toilet system: When the active drum is close to full, place the toilet seat, lid and exhaust chimney connection on a new drum. Place the full one in a dry, warm and ventilated place to compost.

• CEPP Container Net Batch System

Following on the success of Sustainable Strategies' double-bin net batch composter in the Pacific islands, this system was developed for an eco-tourism resort in Fiji. Its owners wanted to use flush toilets with composting systems.

In this design, a piece of heavy-duty fishing net is suspended in each drum to optimize aeration of the composting material. The net stretches and expands with the material. It also features a quick disconnect leachate line, so changing containers is easy. Either dry or micro-flush toilets can be used, or it can be fitted with a urine-diverter.

Leachate can be drained to a Washwater Garden™, a planted evapotranspiration system. Both the leachate and the graywater are drained through a filter and into a subsurface 18-inch-deep trench, where perforated four-inch pipe has been set on 6 inches of gravel covered with 12 inches of sand. Fresh air is drawn first through the leachate drain line, up through the composter and out to the peak of the roof, through a four- or six-inch exhaust pipe (four-inch if power-vented).

Total materials cost for the double-drum composting system alone is about $60. The plans include an informative manual, plans for the integrated plant-based graywater system and a source list for the fittings.

Plans and maintenance manual are $30 from:
Center for Ecological Pollution Prevention (CEPP)
P.O. Box 1330
Concord, MA 01742-1330
978/318-7033
EcoP2@hotmail.com
www.cepp.cc

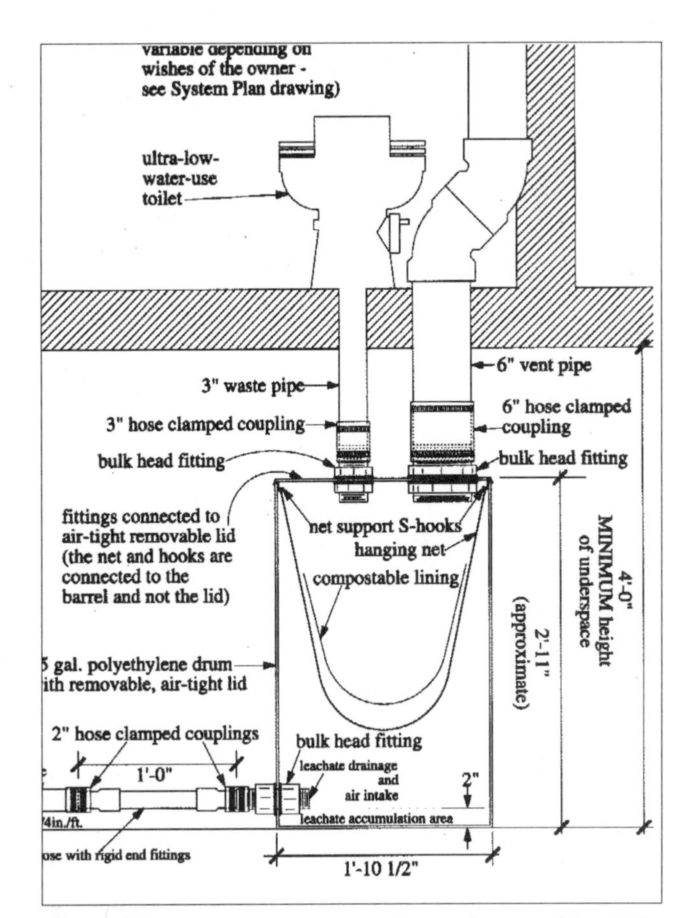

Top: An engineer's diagram of the CEPP Container Net Batch Composter. Left: This system can be used with a micro-flush toilet.

The system in use at an ecotourism resort in Fiji.

Above: The system uses a quick-disconnect leachate drain. Left: Here, a simple system (lid is also the toilet seat) is fitted with a urine-diverter. This photo also shows how the net can be kept stretched open with a hoop.

ETPA = Excrement, Toilet Paper and Additive

• Sol-Latrine

The Sol-Latrine design is a passive solar composting privy that houses three open composting containers. Their bottoms are perforated, so leachate drains through a false floor and into a shallow foundation designed to promote evaporation. Two of the three barrels are in a separate room with the leachate collection area. As the leachate evaporates, its vapor helps hydrate the contents of the two barrels.

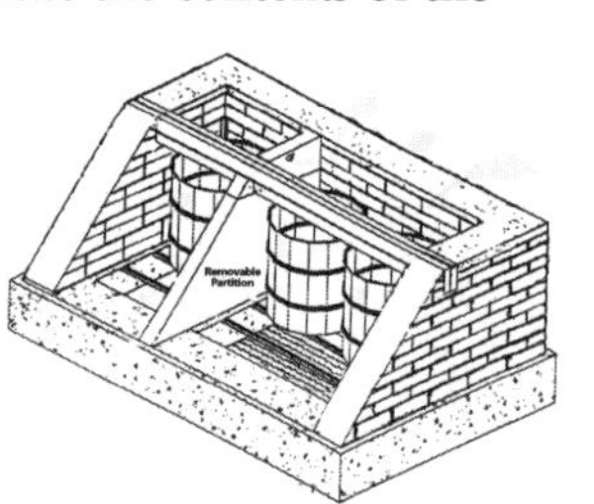

Top view diagram (Illustration: John Cobb)

Eventually, the oldest barrel is taken to a finishing area that is hot and dry. When sizing this system, remember to consider the significant weight of the barrels and provide the access and space needed to move them. Also, in areas that are not hot and dry, extra leachate may accumulate. One option is to drain away leachate—one of the designers drains it to a solar pasteurizer so that it can later be used on his vegetable garden. This is a simple system that makes sense, especially in hot, dry areas.

For more information:
Solar Systems Engineering
Rob Sutcliffe and Dr. John Cobb
Appropriate Rural Technology Association
HC 60 Box 137
Quemado, NM 87829

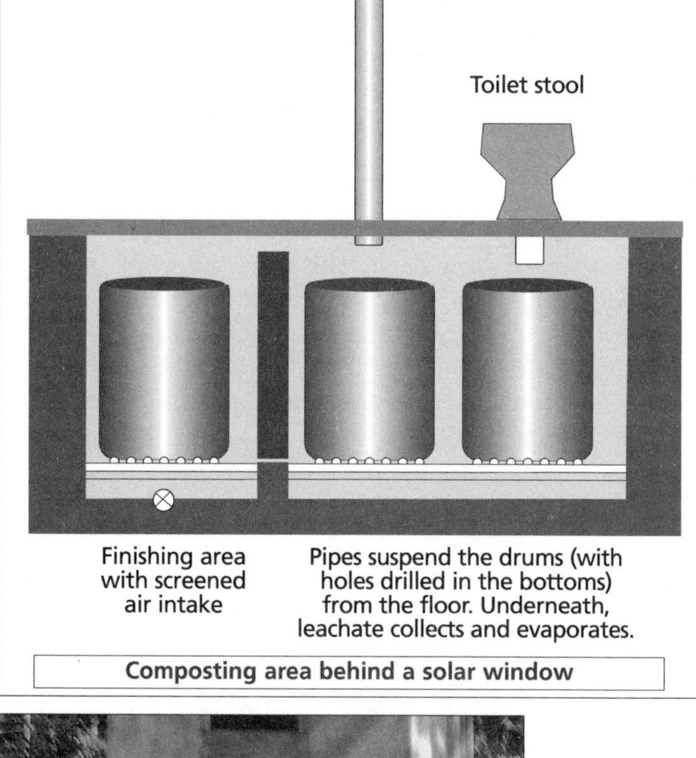

John Cobb's Sol-Latrine in New Mexico. (Photo: John Cobb)

• Solar Composting Advanced Toilet

This system is not a movable batch system, but it could be made to be. Instead of using one bin, and periodically raking fresh excrement back, use two and alternate them, so one can process uninterrupted by fresh deposits. Be sure to use a six-inch vent/exhaust chimney (or a fan with a smaller one). Plans are $50 from:
Larry Warnberg
P.O. Box 43
Nahcotta, WA 98637
www.solartoilet.com

Right: The inside of the system. Left: A unit to be delivered. Warnberg works hard to get Washington state laws to allow site-built systems.

• Carousel-Style System

Place four (or more) pails or drums on a rotating base with casters. Size it according to the capacity needed. Divert urine. Or, to avoid having to disconnect the leachate line every time a container is removed and added, provide a fixed leachate container: Place perforated containers *inside* a large wide polyethylene container, such as the type used for water gardens or water troughs, and put the whole thing on a lazy susan or rotating base on casters.

Just add internal aeration assist and some way to monitor the levels of material in the compost containers, as well as ventilation and warmth.

The Eco-San system, a proposed design by Uno Winblad for developing countries (from Ecological Sanitation, *Sida).*

• CEPP Removable-Bin Composter

Designed for a client who wanted a movable version of the CEPP Twin-Bin with Net system, this is a wooden box into which a plastic container fitted with a hanging net (to enhance aeration) collects feces. Urine is diverted to a urine composter or combined with graywater in an aerated planted bed. When the container is full, it is capped with a vented and screened lid and left to compost in a warm place.

Right: A newly made removable bin composter ready to be installed. Urine will drain to a urine composter, which drains to an aerated planted bed. This system features the TESEC urine-diverting toilet stool.

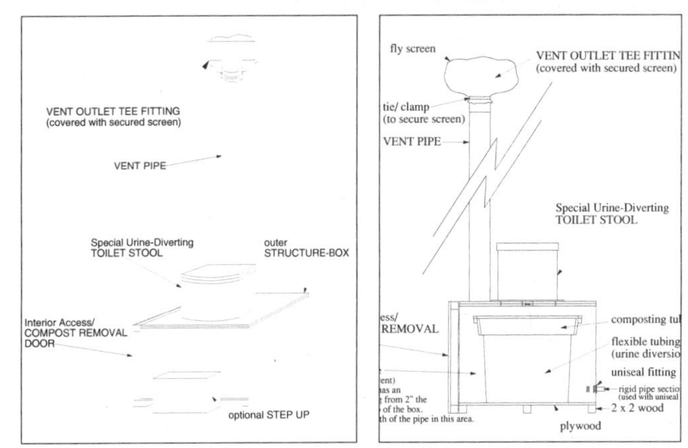

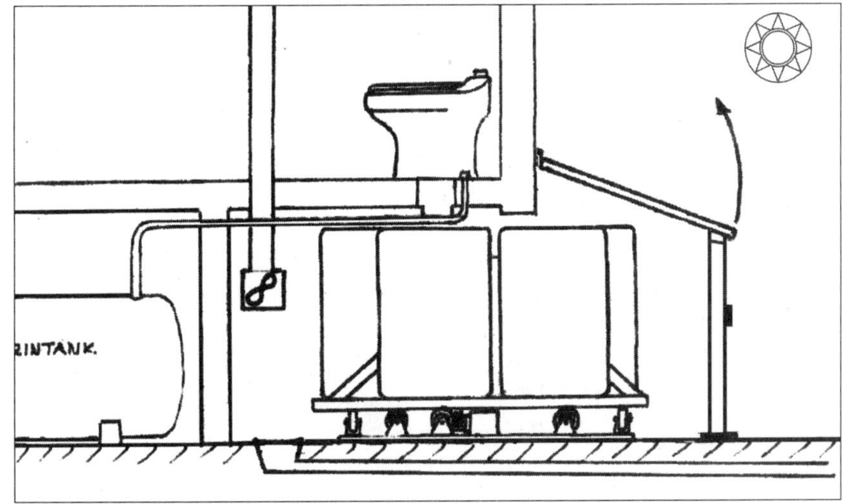

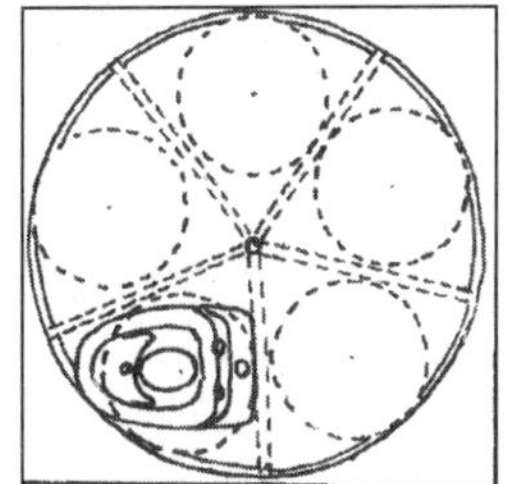

This diagram from Ekologen is a carousel-style composting system to use with the company's urine-diverting toilet at a botanical garden in Sweden. (This one serves the staff of a botanical center in Sweden.) Make sure you have a way of carting away containers that are ready to empty. Note that the room, not the composters, is vented, and includes a floor drain, a passive solar window, and an outdoor removal door. It is best to cover the containers.

• Sunny John Moldering Toilet

The Sunny John is a nonelectric integrated composter and passive solar building (privy) with a passive solar chimney. Its designer, John Cruickshank, calls it a "moldering system" to account for the slower processing that takes place in the system in the often cloudy and cold weather typical of his Colorado home.

The basic composter is a recycled polyethylene 55-gallon drum. The tops of the drums are cut off, perforated with holes and suspended from the bottom of drums with pieces of broken brick. The sides of the barrels are perforated with 5/16-inch holes for air exchange, and at the bottom are 3/4-inch holes for drainage. The open barrels are placed directly underneath the drop chute from the toilet, and the leachate drains out of the barrel composter to the floor, where, ideally, it evaporates.

Solar heat is passively collected and retained in an area below filled with river rock. The solar chimney—a black-painted vent pipe behind an enclosed solar glazing panel topped with a wind turbine—pulls a vacuum on the composting room, pulling air down into the barrel from the seat and out. It is augmented by a wind turbine on windy days. A ground-level screened opening provides air intake. Cool outside air is brought into a solar collector, heated on sunny days, drawn through the rock storage bed and into the room where the barrel composters are located.

The basic plans are for side-by-side toilet rooms with two identical toilet stools. Although it can be made of a wide variety of materials, the plans are for a concrete block foundation and a timber-framed building with 10-inch exhaust pipes. (An interesting twist: The designer likes to plant the roof with grasses and wildflowers.)

The estimated cost according to the manual (circa 1995) is $692 for materials. Solar-savvy folks may see opportunities for more direct use of solar heat, instead of routing it through the toilet room. Other drum composter designs can replace the perforated drums here, and leachate lines added. Also make sure you have a way of hauling up full drums from their below-grade location.

Plans and well-written manual are $20.
Going Concerns Unlimited
5569 North County Road 29
Loveland, CO 80538

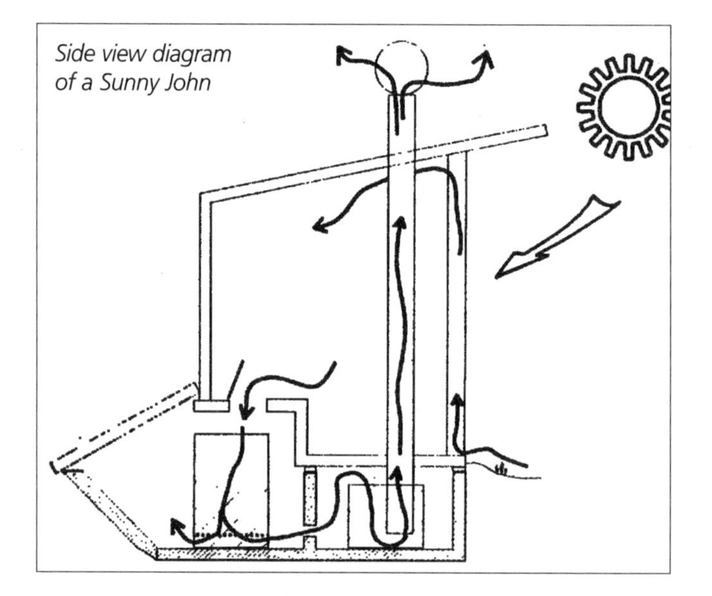
Side view diagram of a Sunny John

Top: This photo of the inside of a Big Batch system shows the perforated pipe aerator system. Right: A Big Batch installed.

• Big Batch Composting Toilet

This system, designed by Robert Fairchild at EKAT (Eastern Kentucky Appropriate Technologies), uses two tiltable roll-away Rubbermaid® one-cubic-yard tilt trucks with fiberglass resin-coated plywood covers. Inside is a "vent horse" (picture a stick drawn horse) made of 10 feet of four-inch PVC perforated pipe. EKAT recommends using a small fan. Twelve-inch sewer pipe is used for the chute and the toilet stool (if you wish to make one). In 1998, the Rubbermaid tilt truck cost about $550; a cube truck (not as easy to tilt and dump) cost $228. The manual describes some simple ways for optional urine diversion. Consider adding a leachate drain to the system.

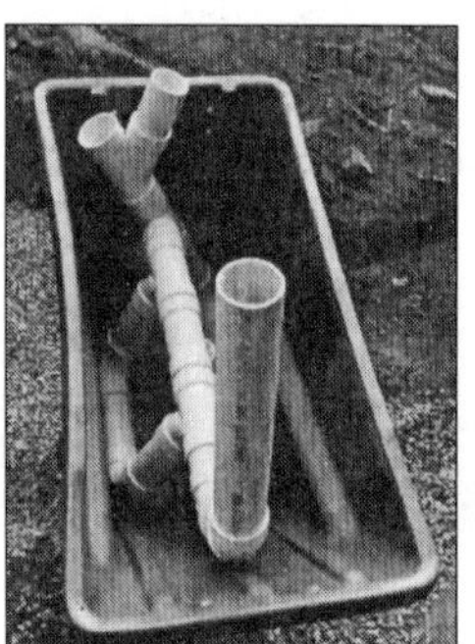

A very clear plan booklet is $7.
EKAT
150 Gravel Lick Branch Road
Dreyfus, KY 40426

• Other Barrel Systems

Other systems use perforated pipe for aeration. This provides relatively small surface area (think: small-diameter holes adding up to not a whole lot of surface area in contact with the compost).

Design plans for a 55-gallon drum system with perforated pipe aerator and urine diverter are available for $7.
Slaughter Energy Enterprises
3517 Virginia Avenue
Kansas City, MO 64109-2455

A system from Australia features both a false bottom and a perforated pipe aerator. Plans include a plant-based gray-water system and a substantial manual.
Plans and instructions are $75 AUS.
Eco Design Sustainable Housing
P.O. Box 2000
Fairfield Gardens 4103, Australia
+61-7-3848-0846
http://www.powerup.com.au/~edesign

The books *Septic Tank Practices* (by Peter Warshall, Mesa Press; out of print) and *The Toilet Papers* (by Sim Van der Ryn, Sierra Books) show diagrams of a 55-gallon drum system with a perforated pipe aerator, and the use of a scissors jack to raise the drum up to the toilet seat from underneath. Just remember that tightly gasketing the top of the open drum to the underside of the floor can be difficult, as can perching a heavy barrel on a jack.

A system design from EcoDesign in Australia. Consider making the vent/exhaust chimney connection more direct.

Keep the composting reactor in a warm location that also allows for convenient servicing.

Sources for Barrels, Accessories, Fittings and Industrial Containers

Cheap drums and other reusable containers:
Reconditioned 30- and 55-gallon open-topped polyethylene drums with lids and locking bands are available for $8 to $10 each. Check your telephone directory for reused containers dealers. Other sources include food production operations, such as juice vendors and distributors that repackage "wet" foods, such as juice concentrate and pickles. If that doesn't work, try:
Association of Container Reconditioners
800-533-3786
Their web site lists sources in most states:
www.reconditioner.com

Fittings, containers, tanks, casters, accessories (gloves, brushes, spatula mixers, etc.) and nearly every kind of plastic container suitable for composters:
Rubbermaid/Consolidated Plastics Company
8181 Darrow Road
Twinsburg, OH 44087
800-362-1000

Quick-disconnect flexible couplings:
Uniseal by AGS, Inc.
2530 County Road 775
Perrysville, OH 44864
419/938-7120

Fernco/Plumbqwik
300 S. Dayton Street
Davision, MI 48423
800-521-1283

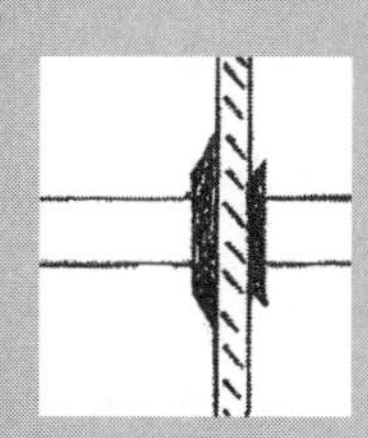

Uniseal for easy watertight seals

Bulkhead fittings:
Hayward
900 Fairmount Avenue
Elizabeth, NJ 07207
908/351-5400

Bulkhead fitting for leachate lines

Inspection hatches (sold as removable deck plates):
Viking Marine
1630 W. Cowles
Long Beach, CA 90813
888-268-3387

Fans:
800-225-5994
www.grainger.com

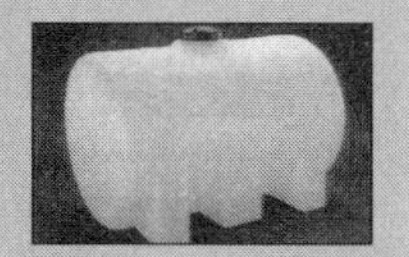

A wide variety of plastic containers, tanks, and fittings are available from the Consolidated Plastics catalog

Fixed-Batch Systems

Double-Vault, Twin-Bin, Farallones System, Gap Mountain Moldering, etc.

One of the world's first composting toilet systems, this design is now perhaps the most prevalent in the world.

This design consists of two side-by-side vaults or bins, ideally containing mechanisms for ventilation, aeration and leachate drainage. (We prefer to call this system a double "bin" composter, as "bin" better describes the dry nature of this kind of compost chamber. Public facilities often refer to their pumped-out systems as "vaults.") The bins are used alternately: When material in one bin reaches a certain level, it is closed and the other side is used until full. By then the first is usually ready to be emptied. Depending on loading rates and ambient temperatures, this can take a number of years to occur.

As far as we know, this system was first widely built in Vietnam in the late 1960s as unventilated bins that relied on time and dessication to kill pathogens. Improvements were made as different designs were tested in the field. The most significant was adding an exhaust chimney, which eliminated odors and enhanced the aerobic biological process. After this was added, the system was called the Ventilated Improved Privy, or "VIP Toilet." Thousands of VIPs were installed in Tanzania under the United Nations Development Programme sanitation program of the late 1980s.

These have been especially successful in hot climates where the biological rate of decomposition is high throughout the year. In dry African and South American climates, these are often more a drying system rather than a biological composting or moldering system. Many thousands of twin-bin dry and composting systems have been constructed in Central America, Africa and parts of Southeast Asia.

Over the years, these systems have been improved, and acceptance by users is high. Very likely, this system will become even more popular for its low cost, high efficiency and reasonable maintenance requirements. More and more, they may be built into homes rather than as stand-alone privies. More parks should also consider this lower-cost, lower-maintenance option.

The bins are often constructed of cement blocks, and coated inside with waterproofing in the interior to seal the vault and on the exterior if groundwater intrusion could be a problem. In most designs, each bin is about four feet by four feet by four feet for a total of 64 cubic feet each.

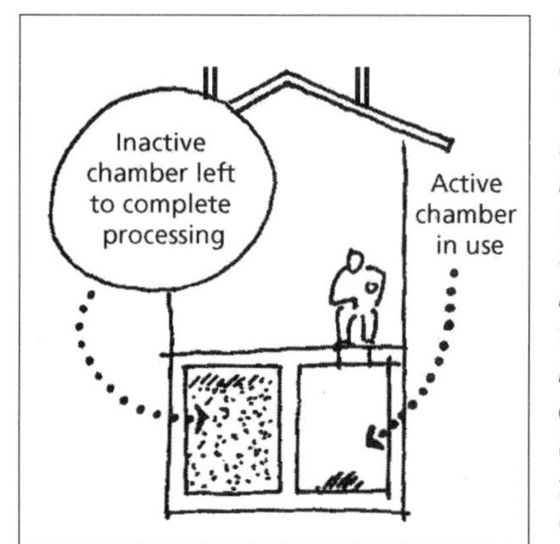

The twin-bin or double-vault system: When one side fills, ETPA is directed to the next, either by moving the toilet stool or using a micro-flush toilet and Y-pipe end. The vaults are typically made of cinder block or mud brick. (Illustrations adapted from graphics by César Añorvé)

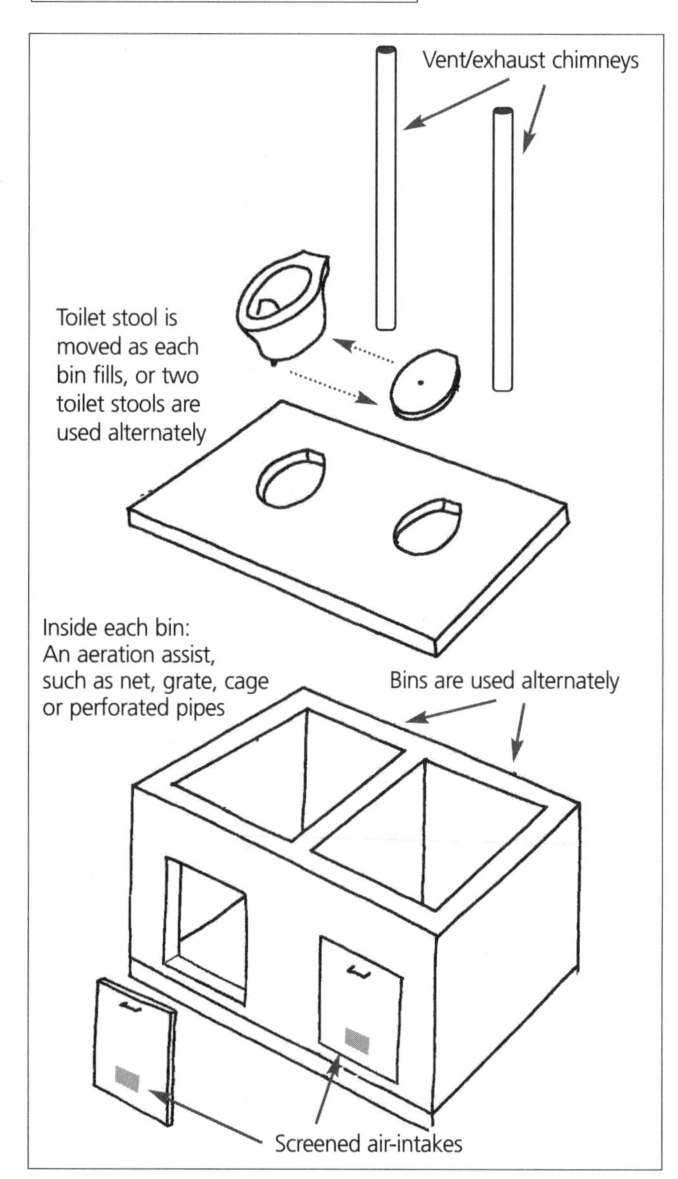

ETPA = Excrement, Toilet Paper and Additive

Keeping It Aerated

Since the earlier designs, internal cages, grates and nets have been added to these systems to provide greater aeration and leachate management, which expands the system's capacity for nearly no additional construction costs. Perhaps the most forgiving of the composters, they require little or no regular maintenance, except the addition of additive. Depending on the volume of the unit, finished end-product from each bin should be removed every two to four years. In fact, in some tropical climates, large extended families of 20 or more have yet to fill the first chamber even after four or five years of use!

Some of these divert urine to a leaching pit (a hole in the ground full of stones), a plant-based evapotranspiration system, or to storage tanks to be later used on crops.

Unless drained or evaporated, leachate will accumulate in the bottom of the bin, creating anaerobic conditions and possible leaks. For that reason, these systems should include leachate drainage, such as a pipe through the bin wall and above the sludge level.

Most twin-bin systems are located in outbuildings, or "privies," separate from homes. Sometimes this is due to local custom (in a large part of the world, it is considered barbaric to excrete in one's home). However, twin-bin systems can easily be incorporated in homes, as they are in Mexico, Australia, and even the United Kingdom. Mark Moodie reports from England that he has used an Aquatron water-waste separator toilet system with a twin-bin system for a few home installations. Anna Edey of Solviva uses a standard 1.6-gallon toilet with a small two-bin system, *but with an extensive leachate drainage and management system.* This shows that the toilet stool need not be moved—merely a toilet pipe redirected to the new bin.

In the Farallones system, when the active bin fills up, its contents are then shoveled into the second bin. It is thought that this aerates the material more. However, this can be an odious and unhygienic task, and we cannot recommend it.

Aeration, however, is important: Whatever system you choose, make sure it provides some kind of aeration grate, cage or net, and reliable leachate drainage.

Advantages: Relatively low cost and low maintenance for usually high effectiveness.

Disadvantages: Alternating the compost bins requires either moving the toilet or redirecting a flush toilet pipe. Size requires special siting issues.

Above: A twin-bin system in Fiji

Left: Espacio de Salud's two-vault drying toilet system in Mexico (from Ecological Sanitation, *Winblad, ed.)*

The Systems

The following are sources for plans for two-vault systems. Each features a different aeration assist, from a metal cage to a hanging net.

• CEPP Twin-Bin Net Composter

David Del Porto, one of the authors, designed a system for a Greenpeace initiative in the Pacific islands that uses a heavy-gauge fishing net suspended by hooks from the sides of each vault. The net is inexpensive, non-biodegradable and flexes as the mass changes shape, creating more surface area to improve diffusion of oxygen through the composting material. In projects sponsored by Greenpeace and island medical schools and government agencies, this system has been successfully introduced in the Micronesian islands and Fiji. In Pohnpei, the design has been duplicated by islanders, and the public utility assists in some installations. Some of these system have been coupled with evapotranspiration beds to handle leachate and graywater from washing facilities. Visited years after their installation, these systems were working well.

Plans and manual for the Twin-Bin Net Composter are $30.

Center for Ecological Pollution Prevention (CEPP)
P.O. Box 1330
Concord, MA 01742-1330
978/318-7033
EcoP2@hotmail.com

Clockwise from top: A strawbale structure is being constructed around a twin-bin system in Pennsylvania (Photo: David Cervenka); a side view diagram of a CEPP Twin-Bin with Net system; inside the vault before it is sealed for use, palm fronds or a degradeable mat are placed on the net; a front view. (illustrations: CEPP/Dan Harper)

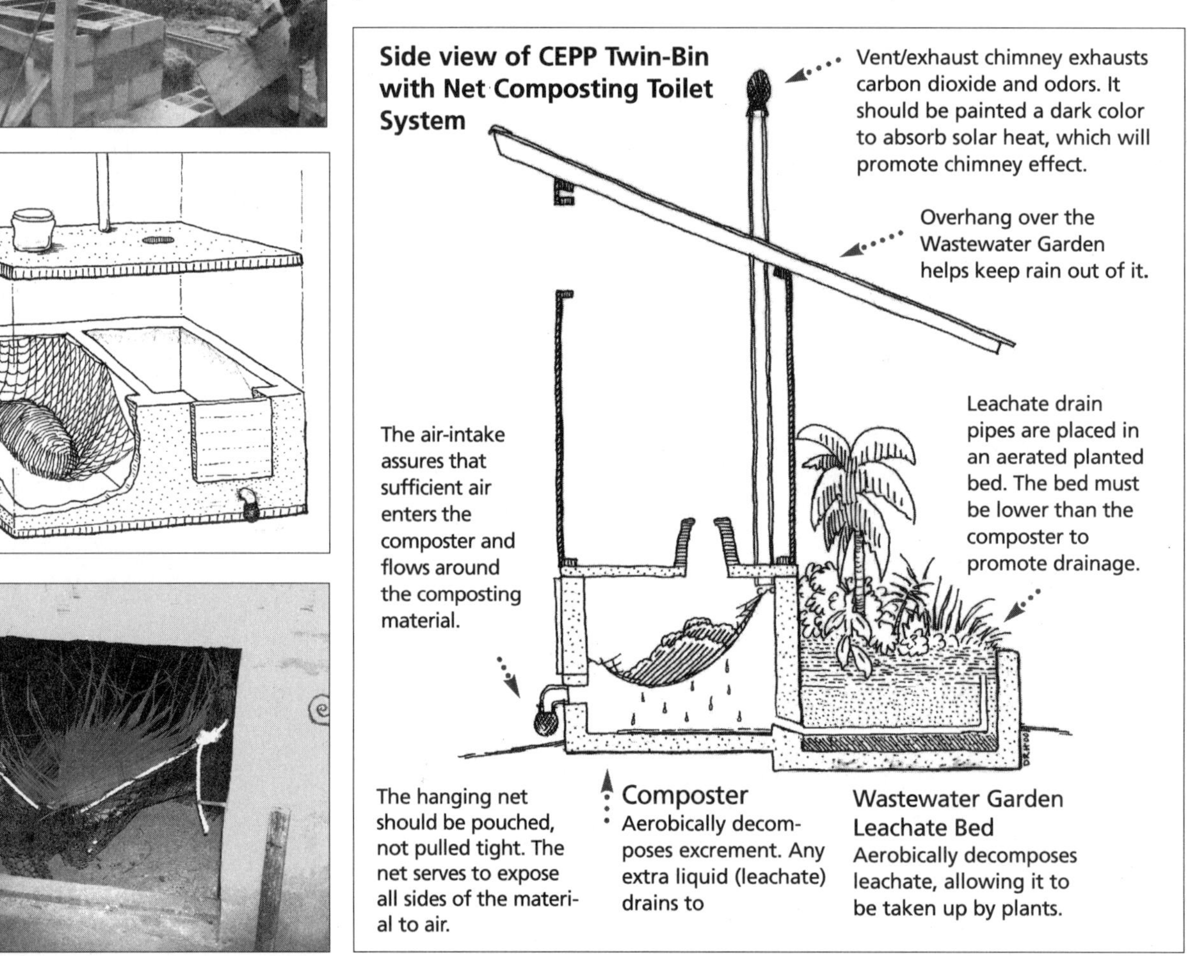

ETPA = Excrement, Toilet Paper and Additive

• Cage Batch Composter

Another aeration improvement is the cage, whereby three or all four vertical sides of the composter are a cage or wire fencing material, with a mesh false bottom. This, too, provides more surface area for aeration. Australia-based Dr. Leonie Crennan has piloted Cage Batch composters in the islands of Tonga and Kiribati, where, she reports, they have consistently produced a pathogen-free end-product. They can be built locally with local materials, which, she notes, "reduces long-term dependence on imported goods and aid."

Acceptance rates of double-bin systems are usually high, because they involve little contact with excrement, are usually odor free and are easy to maintain, often requiring removal of compost after a matter of years.

Also, Crennan reports, in some tropical places, locals prefer to use particular kinds of leaves rather than expensive toilet paper, which clogs water-flush toilets and must be imported.

Dr. Leonie Crennan
85 Dunbar Street
Stockton, 2295, Australia
Fax: 61-2-49284082

• CAT/COMPUS Twin-Vault System

Plans and construction and maintenance information for a twin-bin composting toilet system are found in the informative books about composting toilets from the United Kingdom, *Fertile Waste* and *Lifting the Lid* available from:

The Centre for Alternative Technology (CAT)
Machynlleth
Powys SY20 9AZ UK
www.cat.org.uk
Available in the United States from Jade Mountain.

Moving a floor slab onto a twin-bin with net system in Mexico. The sides of the upside-down buckets serve as forms for concrete toilet risers. Six-inch vent chimneys will be added.

• Gap Mountain Permaculture Moldering Toilet

Another common variation features a grate at the bottom. The Gap Mountain Permaculture Moldering Toilet is a double-bin system with a plastic-coated wire mesh (chain-link fencing) false bottom. Doug Clayton, who designed this system with fellow permaculturist David Jacke, recommends using a bucket of wood chips for a urinal. He uses this nitrogen-rich material on his fruit orchards. However, one can install a leachate line to drain or pump leachate to a management system. Clayton's system is in use in chilly New England.

Plans cost $19
Gap Mountain Permaculture Center
9 Old Jaffrey Road
Jaffrey, NH 03452

• Solviva Wooden Twin-Bin Composter with Flush Toilet and Green Filter Bed

Twin-bins do not have to be large concrete structures. Anna Edey of Solviva offers plans for her twin-bin system made of plywood and insulated with foam board. She couples it with a 1.6-gallon flush toilet and an extensive "Green Filter" leachate system. Consider adding an outdoor vent or exhaust chimney, using a dry or micro-flush toilet with it, adding some supplementary heating (or siting it for solar warmth) and perhaps providing a more convenient way to remove the end-product.

Plans for this system are $50 from:
Solviva Company
RFD 1, Box 582
Vineyard Haven, MA 02568
508/693-3341
solviva@vineyard.net

The Solviva insulated wooden twin-bin system outside a house (remember to keep it warm!)

The Farallones System

Not all variations of the multiple-bin toilet design have been successful. In California, this design was experimented with in the mid-1970s by the Farallones Institute, a non-profit organization that developed and demonstrated appropriate technologies. Its system's method of operation was different from the traditional one: They frequently added large volumes of bulking agent (such as wood chips), and mixed the material often, usually by hand. When one vault filled up—which occurred rapidly due to the large amounts of bulking agent—its contents were then shoveled to the next vault (and sometimes there was a third vault). The rationale was that this aerated the material and resulted in high-temperature, or *thermophilic,* composting. However, when some of these systems were inspected in the early 1980s by the California Department of Health, they (like many others at the time) failed to meet minimum requirements of pathogen destruction, odor management and controlling pathogen vectors, such as flies. The construction was such that groundwater sometimes entered the vaults and flooded the systems. The removal doors leaked. Also, operators often delayed the not-so-pleasant task of shoveling the material. Pooling leachate was a problem with many, as they had inadequate drainage or none. It had been assumed that heat generated by the composting process was going to evaporate the leachate, but it rarely does in any unheated composter. Some operators diverted the urine, which helped.

Although content Farallones system users *do* exist, we cannot recommend a system that involves shoveling fresh excrement. Use Farallones plans as blueprints for a twin-bin system for which you alternate your use of the chambers—don't move their unprocessed contents. Add leachate drainage and management, and an aeration assist (a steel mesh false bottom, a net, etc.).

Sim Van der Ryn's book, *The Toilet Papers,* contains plans for a Farallones system. Also, Earthways in subtropical Australia offers plans for a Farallones system that features a third bin for extra finishing. Its designer, Brian Woodward, built his of soil cement blocks and he uses a squat plate. The third chamber can still be used for finishing, if you think you will need this extra step due to high usage or risk of pathogens.

A 35-page manual is available for AUS$20 from:
Brian Woodward
Earthways
Wollombi 2325
N.S.W., Australia
www.users.bigpond.com/nsw999

The Toilet Papers by Sim Van der Ryn, $17.95 from:
Chelsea Green Publishing
205 Gates-Briggs Building
White River Junction, VT 05001
800-639-4099

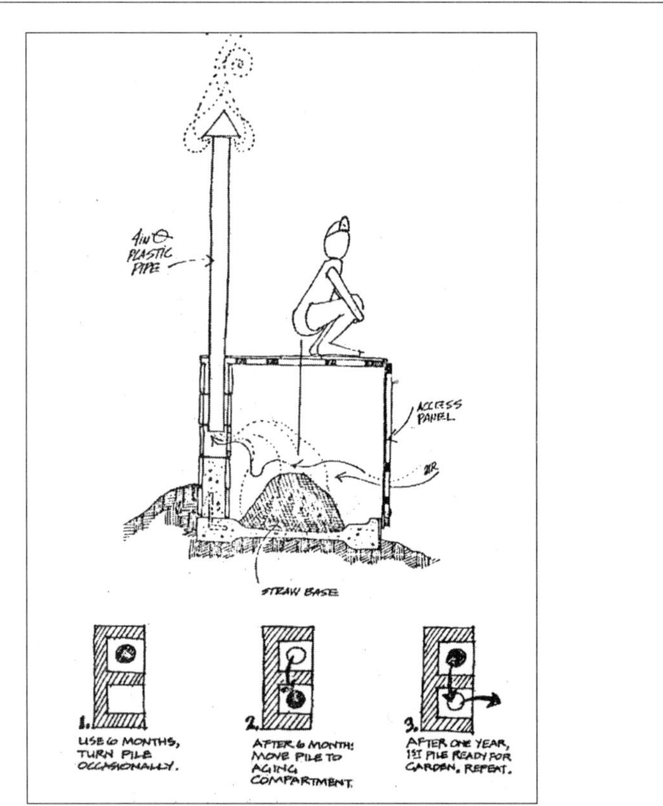

The Farallones system with a squat plate. The lower diagram shows the pattern of use: when the active bin fills up, its contents—much of it fresh excrement—are shoveled into another bin to process uninterrupted by fresh deposits. A better plan is to avoid the shoveling, and simply move the toilet stool. (From Septic Tank Practices, *Peter Warshall)*

The interior of an older Farallones system in which urine is diverted. Great amounts of sawdust are added to this one. (Photo: J.A. Creque)

Some Installation Options for Twin-Bin Systems

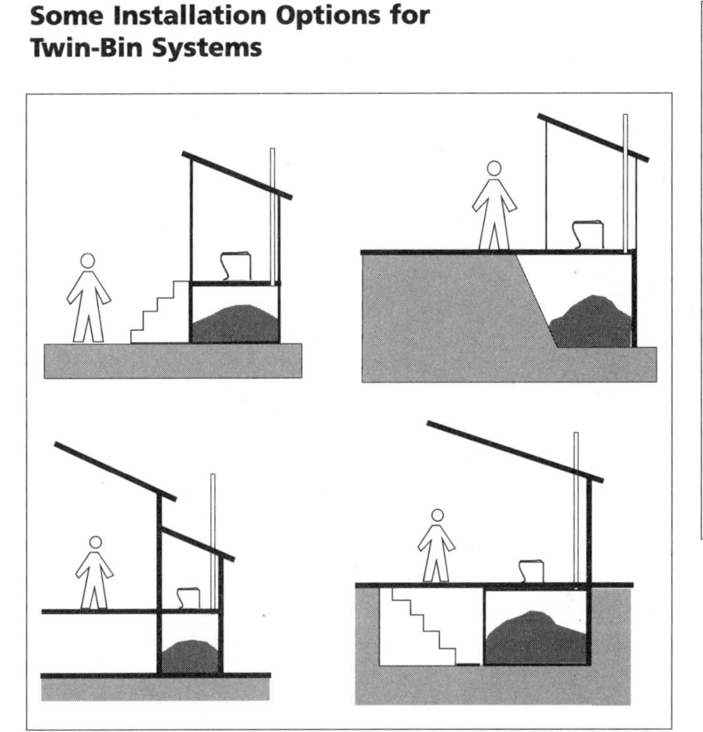

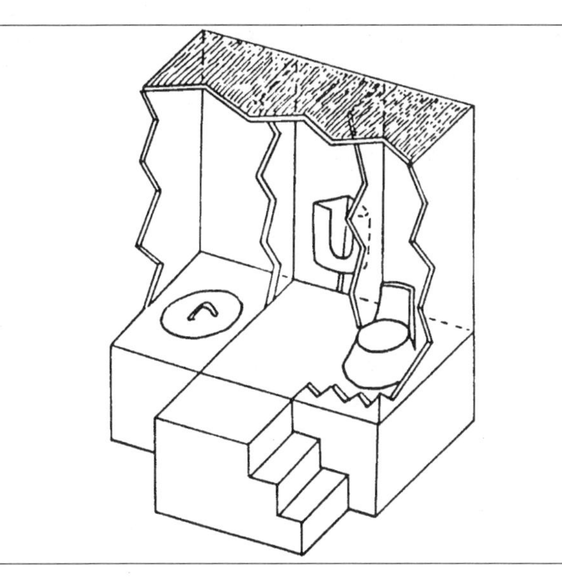

Left: Siting options for the twin-bin composting toilet system (adapted from an illustration by the Centre for Alternative Technology) Above: alternating toilets with a fixed urinal in between, which allows some urine-diversion (illustration: Grupo de Tecnologia Alternativa S.C.).

Far left: A CEPP Twin-Bin Net system coupled with a Washwater Garden to contain and use up leachate (Yap Non-Polluting Toilet manual, Center for Clean Development).
Left: Diverting urine helps control moisture and prevent anaerobic conditions in the bins. Here, it is diverted to a rock pit; it can also be combined with sink washwater and used for irrigation.

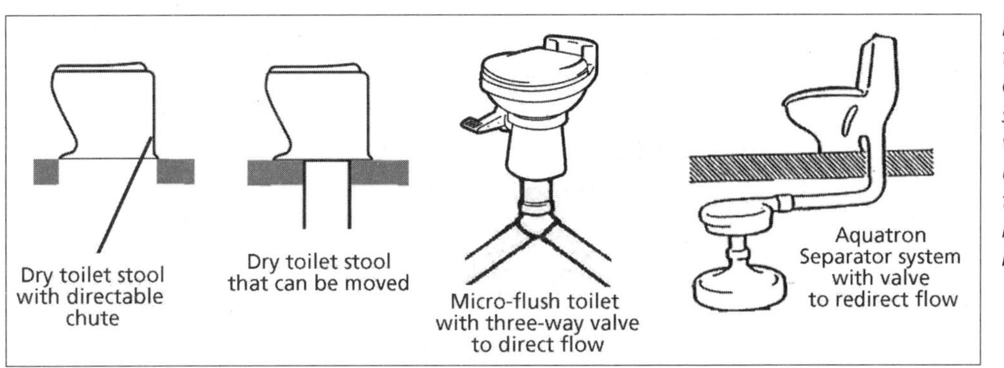

Left: Micro-flush toilets and toilet water separation systems can be used with two-vault systems—simply redirect a valve when one side has filled up. The toilet chute from a dry toilet stool can also be redirected—just beware of buildup.

Sloped-Bottom (Inclined) Single-Chamber Composters

Clivus Minimus, RAW

This system, simply a site-built version of the Clivus Multrum, has been built worldwide, usually of concrete block. It appears to be most successful in warm climates, although we have a report of a happy owner in Canada.

This design emerged in the 1970s, when the Clivus Multrum received much renown upon its introduction to the United States as the first large-capacity manufactured composting toilet system. Predictably, do-it-yourselfers picked up the design and ran with it. However, the premise--that the material would slowly glaciate through the composting process as it made its way down the slope--didn't always happen in reality. Some versions of this design today have steeper angles and no aeration channels (usually side-by-side perforated pipes), as designers found that solids could bridge on the channels.

In the late 1970s, McGill University's Center for Low-Cost Housing distributed plans for this design, which is called the "Clivus Minimus," perhaps taking advantage of the expiration of Clivus Multrum's U.S. patent. According to the 1977 book, *Goodbye to the Flush Toilet* (Rodale, out of print), several of these were built in the Philippines. More recently, the nonprofit ReSource Institute for Low-Entropy Systems, which is sponsored in part by the U.S. Clivus Multrum manufacturer, has built hundreds throughout Central America and China.

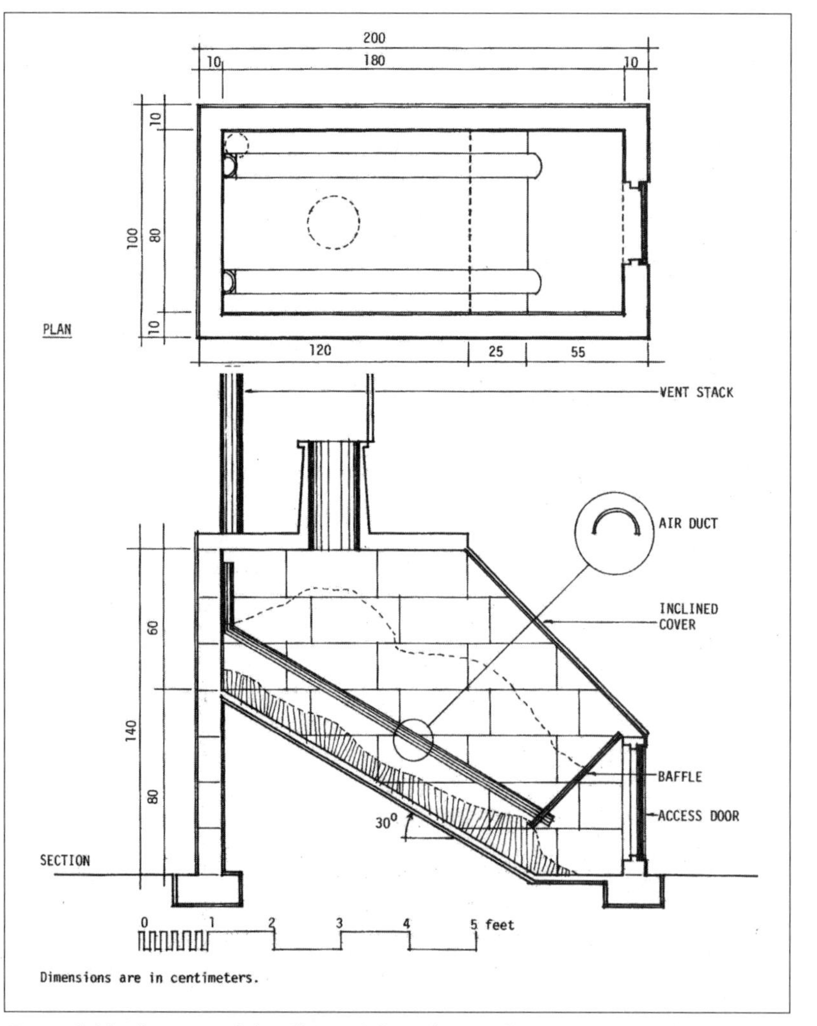

Top and side diagrams of the Clivus Minimus (Center for Low-Cost Housing, McGill University). Note that some subsequent designs feature much steeper slopes to improve throughput.

Several have also been constructed by homeowners in subtropical Queensland, Australia. There, Leigh Davison, a lecturer on water and wastewater topics at Southern Cross University, reports high success rates. Davison himself has had a Clivus Minimus in his home for 15 years (see "Composting Toilets at Work," Chapter 8).

Each variation of this design has sought to improve the aeration and the transit rate of the moldering mass as it moves down the sloped floor. A baffle, which forms the rear wall of the removal area, keeps fresh material from entering the removal area. As with all of the sloped-floor models, the weight of the large mass can cause the contents of the composter to compact, making the removal of the end-product difficult. For that reason, most sloping vault composters appear to function best when plenty of *lightweight* bulking agent (popped popcorn, sawdust or any airy, not-too-dense material) is added, and the end-product is removed at least every year or two years.

Remember that material builds up. Some thought that urine would act as a lubricant, but, because it is acidic and formed salt crystals, any benefit that the urine might have provided was lost. When building this system, make the interior slope as slick as possible. Concrete block has high friction resistance (is not slippery), because it is coarse. In addition, the humic acids created during the composting process, and urea and uric acid in urine, further pit and corrode the surfaces. Consider coating the cement with a skim coat of acid-resistant mortar or several coats of Water Glass (sodium silicate).

Capacity depends on the size and the usual composting success factors. Talk to the designer or compare sizes with Clivus and CTS models for estimates. Aeration systems are upside-down-V-shaped air channels (pipe cut in half lengthwise and placed open side down) or perforated pipes. However, beware of "bridging" of the material on these.

Costs are essentially for the pipe and concrete (or mud-brick) block. They can be built as part of the foundation of a building, sharing costs with foundation walls and floors.

Advantages: If one does not factor in labor costs, these can be less expensive than purchasing a similiarly sized manufactured composter.

The Recycle All Waste (RAW) Composting Toilet System

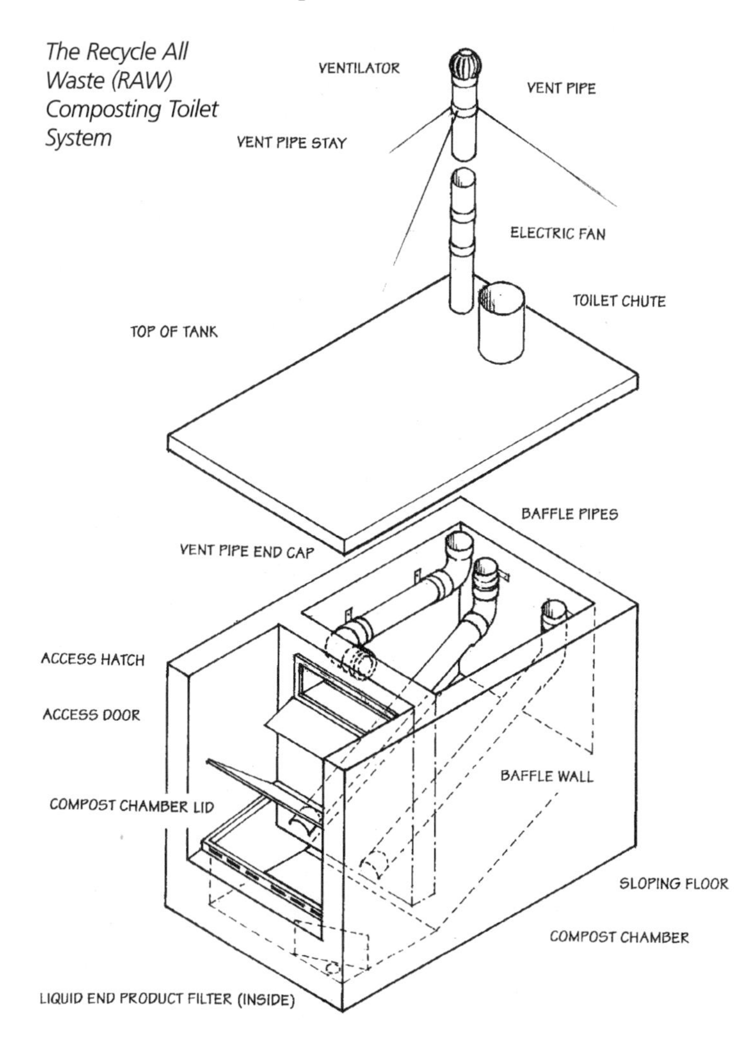

Disadvantages: Potential compaction problems. Hard, if not impossible, to access the interior for repairs. Hard to get permitted.

Some Sources for Plans:

Ron Davies has written a comprehensive 73-page book about building a Clivus Minimus system like the one he and his family has lived with for 20 years in Oregon. "How to Build a Compost Toilet...and Save a Piece of the Earth" is $19.95 from

Oregon Wells
814 Main Street
Cottage Grove, OR 97424
541/942-7174

The RAW (Recycle All Waste) System was adapted from the Clivus Minimus design by Kevin Ison and Ian Blandford. It is approved by health and building authorities in Queensland, Australia.

Plans and handbook £16:

Earthwise Publications
High Walk House
Kirkby Malzeard
Ripon HG4 3RY England UK

Plans for another Clivus Minimus for $10 (AUS):

Ray Flannagan
The Channon, 2480
New Lismore, Queensland
Australia

An outdoor Clivus Minimus in subtropical New South Wales, Australia. (Photo: Leigh Davison)

Bucket Collection for Direct Outdoor Composting

Nothing complicated to explain here: Use a bucket as a toilet stool/collector. Place a handful of sawdust or leaf mold onto each deposit. Every one to three days, take it outside and trench it into an active compost pile, adding plenty of carbonaceous material and leaf clippings for hot composting. Chances are you will have a significant amount of ammonia or nitrates from urine, so do not add anything that is highly nitrogenous, such as green grass clippings.

Because feces contains intestinal bacteria (which may or may not be disease causing), viruses and parasites, use this technique with extreme care to prevent contact with raw excrement. Use rubber gloves and perhaps a transparent face mask so you do not get anything splashed on you.

Each compost pile should be at least one cubic yard (3 x 3 x 3 or 0.76 m^3) for self-insulating purposes (bigger is better). Try to put the excrement in the middle—about 18 inches deep—to keep it warm and in the active thermophilic zone. Trench in the excrement and cover it over with organic matter.

Because of the risks of disease transmission, this composting pile system should be separated from others and constructed of sturdy hardware cloth (metal screening) to eliminate access by children and animals. Purchase a long-stemmed thermometer, and make sure the center of the pile is at least 143°F (61.7°C) for 36 hours; otherwise, you cannot be sure that all of the potentially pathogenic organisms have been killed. (In a composting toilet, time does the killing. In an outdoor pile, you kill it with heat.) If you are not sure that the pile is hot, put it in another compost pile until you achieve uniform thermophilic composting temperatures.

Consider covering or partly enclosing the compost pile(s) to keep out rain and minimize the formation of potentially dangerous leachate.

This method is very well detailed by J.C. Jenkins in his informative book, *The Humanure Handbook*. Jenkins notes that this method can allow you to achieve true thermophilic composting without supplemental energy. Hot composting helps assure pathogen kill (in an enclosed large composting toilet, you are relying on time).

Your health agent and your neighbors may not care for this method.

Consider lining the bucket with a biodegradable bag (see next page for source) to minimize handling and cleaning of the bucket. Also, like the other systems, this can be combined with urine diversion.

Advantages: Thermophilic (hot, fast) composting. Cheap. Simple.

Disadvantages: Dealing with excrement directly. Labor intensive. Requires a lot of carbon. Difficult, if not impossible, to get permitted.

The Humanure Handbook is available from:
Chelsea Green Publishing
802/295-6300
www.chelseagreen.com

Ole Ersson, a physician in Portland, Oregon, maintains a sawdust toilet for use by his family (a flush toilet is also available for squeamish guests). The contents of the toilet stool are emptied into the composting pile usually every two days and trenched in with kitchen scraps, wood chips and yard trimmings. (Photos: Ole Ersson)

Other Systems and Accessories

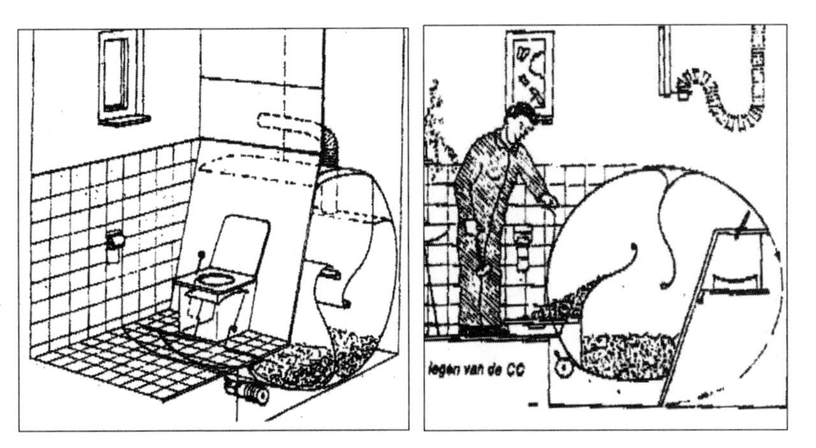

• De Twaalf Ambachten Tilting Continuous Composter

And now for something completely different. This is a unique rocking multi-chambered system mounted with a horizontal axel (think of a ferris wheel in the wall, with the toilet seat fastened to one side of the wheel). After a few uses, you rotate the wheel, which transfers the ETPA into a second chamber, where it composts. When the system is turned again, the material moves to a removal area. It requires a separate room behind the toilet system to house the system. Clever looking, it may be complicated to execute. Plans for this and other systems are $70 U.S. (postpaid; add $12 if by check).

Centre for Ecological Techniques
Mezenlaan 2
5282 HB Boxtel, The Netherlands
31/411-672621 Fax: 31/411-672854

• Solar Survival Solar Dessicating Toilet

Not designed to be a composter, this system is a high-temperature (200° to 400°F/93° to 204°C) solar cooker with a scraping mechanism. It is designed to turn excrement into, according to the literature, "fried ash." For $100, you can get construction plans for the system. The company also sells this system for $1,700 and a rotating drum model for $1,500.

Solar Survival Architecture Sales
P.O. Box 1041
Taos, NM 87571
505/751-0462

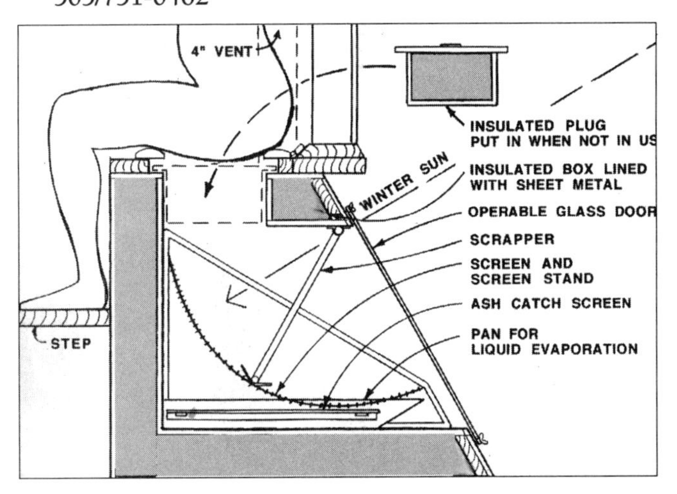

Accessories

Biodegradable Bags

Use these to bag your end-product to bury or to carry it to a composter. BioCorp's reSourceBag™ biodegradable bags are made primarily of cornstarch and a synthetic polymer, which is free of polyethylene. According to the company, this is "recognized as completely biodegradable and compostable under normal composting conditions and will leave no plastic residue in compost." They will disappear within 20 to 40 days in a composting pile. The bags are available in several sizes.

Biocorp
2619 Manhattan Beach Boulevard
Redondo Beach, CA 90278
310/643-1626
888-206-5658

Containers, Leachate Line Fittings and Interesting and Helpful Gizmos

Again, a great source for all kinds of containers and gizmos for building and maintaining composters (as well as rain barrels and worm composters) can be found in this catalog. Also look for gloves and face masks.

Rubbermaid/Consolidated Plastics Company
8181 Darrow Road
Twinsburg, OH 44087
800-362-1000

ETPA = Excrement, Toilet Paper and Additive

• Soltran™ Solar-Assisted Composter

Soltran™ is a solar-assisted composting toilet system, providing solar-derived heat, leachate management and aeration all within a modular building. Nearly any type of composting reactor can be used with it.

From the composter, leachate drains into a liquid storage system that is both a solar-heated evaporator and a heat-storage unit beneath the composter. It is connected to a solar-heated evaporation tank.

Outside air enters a solar collector that has a thermostatic damper (a heat-triggered air valve powered by the expansion of liquids inside a piston that opens and closes the damper door) to assure that no cold air enters the system. This heated air is directed first into the leachate evaporator. The evaporated moist warm vapor then passes into the composter, leaving the salts in the evaporator.

Designed by David Del Porto, this patented system also includes a solar chimney built into the roof, a solar pasteurization unit and process controls (such as temperature, moisture and pH detectors). The system was engineered to cascade urine and solar warmth through evaporators and composting systems to optimize the environment for fast, efficient composting.

The system has been configured for different applications, including for food residues and other organic wastes, and to use a variety of composters and solar collectors and evaporators in a specific sequence. All of the constructed sytems designed were for human excrement and built primarily in remote applications where no power was available.

Plan A has an integrated passive solar chimney roof. Plan B has a standard roof with a photovoltaic panel powering an in-line centrifugal DC fan. Some plans offer the option of exhaust pipes with photovoltaic-powered fans and/or passive Venturi/Extractor rainhats.

The plans are detailed in U.S. government bid document format, so every component and construction process is specified.

Price range: Depends on materials; a system for the South Pacific islands cost $1,000 for materials.

Advantages: No electricity is needed to speed up the composting process. Adds capacity.

Disadvantages: Not a mass-produced product, so approvals are dependent on the approval of the composter, not the entire building system. Prone to air leackage through the integrated roof/solar chimney. Can be expensive, because it is a complete integrated building system.

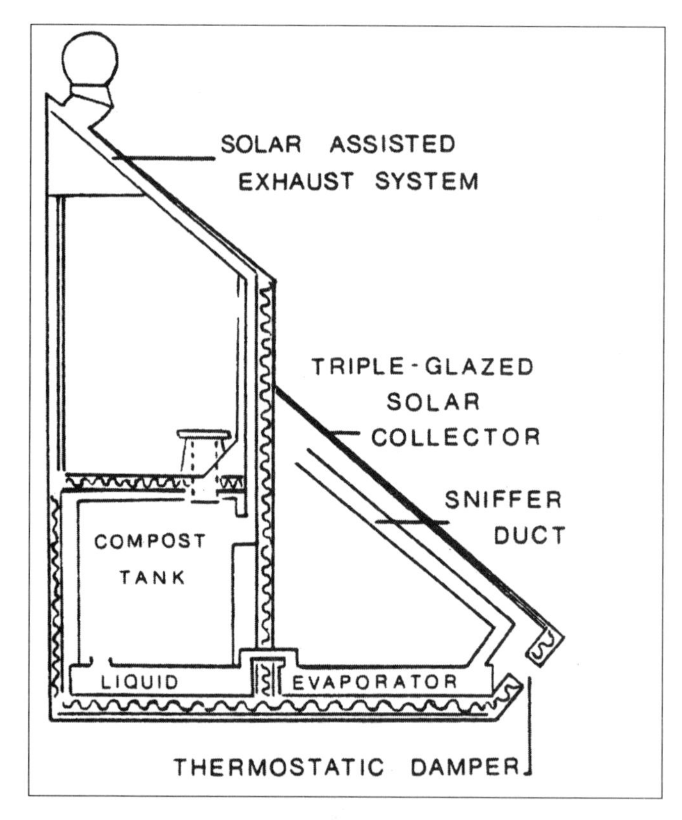

Top: A diagram of a Soltran system variation
Above: A Soltran in a park in Oregon

Plans, a construction manual and a maintenance manual are $50.

Ecos, Inc.
P.O. Box 1313
Concord, MA 01742
978/369-3951
watercon@igc.org

• CEPP Urine Composter and Wastewater Garden™ Planter Bed

So you decided to divert urine from your composter to expand its capacity. Now, what do you do with it? Most of excrement's nitrogen, phosphorus and potassium (NPK) is in urine, and if you want to use it, it must be composted, too. There are three general approaches:

- Direct application to soil, whereby you rely on soil microorganisms to transform the urine—but this is risky
- Dilution (graywater is a likely candidate for this) and irrigation
- Wet composting with the addition of sugar for carbon and fixed (nondegradeable media) to create air spaces
- Drier composting with carbon, such as cardboard, wood products, or other biodegradable carbohydrate fibers

Urine is typically sterile in healthy populations (see "Urine: Managing It Separately in Chapter 4 for a discussion of pathogen concerns). Its processing issues are that the nitrogen often immediately turns to ammonia and can be lost as a gas, produces acrid odors. So the goal is to immediately transform it and use it.

Top: A large family garden grown only with urine produced by the household (Photo: Conrad Geyser). Above left: A simple urine diverter made from a funnel. Right: Part of a urine composter with cardboard.

Examples of urine composting
Far left: A liquid composter uses sugar for carbon, perlite for air pockets and an air stone. Left: A large BioSun system uses a pipe full of bark for urinals. Below left: CEPP's design for a urine composter/planter used in Baja. Below right: A urine composter; a Wastewater Garden manages overflow and graywater.

For more information about urine processing and utilization systems, see "Using Your Urine," available for $9.95 from:

Center for Ecological Pollution Prevention
P.O. Box 1330
Concord, MA 01742-01330
978/318-7033
EcoP2@hotmail.com, www.cepp.cc

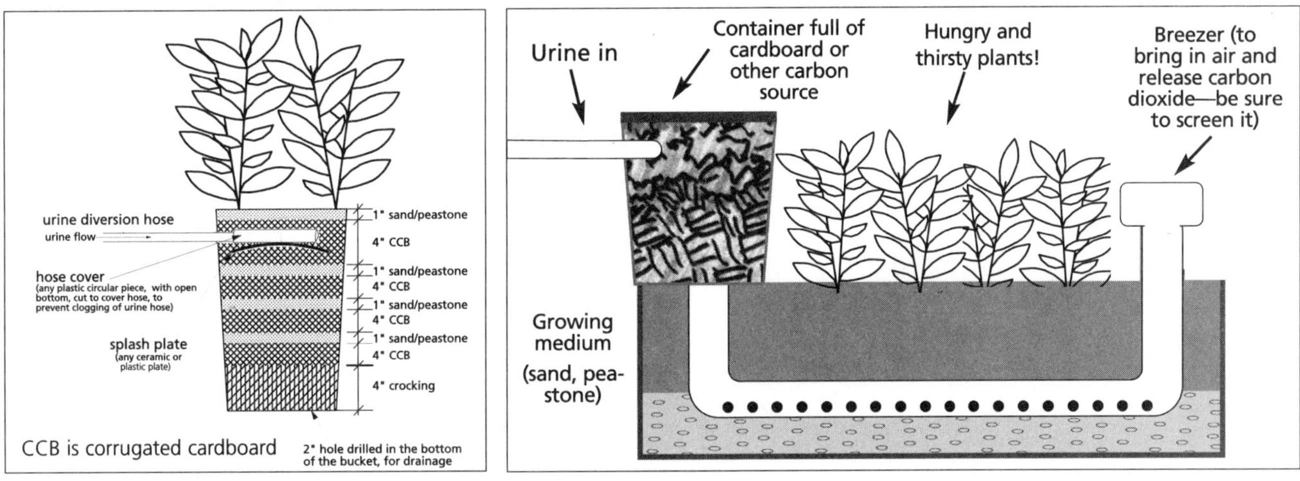

Incinerating Toilets

In an incinerating toilet, energy, derived from electricity or propane, is used to rapidly raise the temperature of a fireproof chamber, where excrement and toilet paper have been deposited. This high-temperature environment causes the excrement to combust as a fan pulls air into the combustion chamber. Depending on the fuel source, and volumes of excrement to be incinerated, the time ranges from a few minutes to a few hours to produce a dry ash. What remains is nonvolatile carbon compounds and residual salts. All human pathogens are destroyed.

Incinerating toilets rely on very expensive fuels (ultimately derived from nuclear power or petroleum). While they are effective, they are not sustainable as a solution for a world with increasingly scarce resources.

Consider your cost of power when looking at these systems.

• Incinolet

Because all of the valuable nutrients have been oxidized and volatilized, there is little nutrient value to the end-product, and less than 1 or 2 percent remaining solids. The best-known electric incinerator is the Incinolet. That has both a 220-volt model and a 110-volt model. The 110-volt model is typically used in cottages, but may require two hours or more between burning cycles while the combustion chamber cools down. Incinolet literature says the system burns up to 400°F (204°C) using two kilowatts of electricity per person per day. Uses a special bowl liner available only from the company. At 10 cents per kilowatt hour, this system costs about 20 to 40 cents per use.

Incinerating excrement produces extremely noxious odors, so the company provides an odor-reducing filter. These systems are all-metal construction. In practice, one places a special cellulose mat into the combustion chamber prior to each use. This prevents the fecal matter from sticking to the heating plate. Incinolet also supplies an incinerating urinal. These systems are typically used in industrial settings where there is limited sewage treatment access, but plenty of electricity. Price range: $1,379 to $1,700.

Incinolet
Research Products
2639 Andjon
Dallas, TX 75220
214/358-4238

• Storburn

Storburn developed its incinerator in response to the limited capacity of the Incinolet. The idea is that the system stores the excrement and toilet paper in a three-gallon tank during the day and incinerates it at night, using propane or natural gas as its fuel source. According to the company's literature, 110 pounds of propane incinerates 600 uses. It requires an anti-foaming agent for the urine. Prices start at $2,550 without a vent kit.

Storburn International, Inc.
48 Copernicus Boulevard, Unit #3
Brantford, Ontario Canada N3P 1N4
800-876-2286

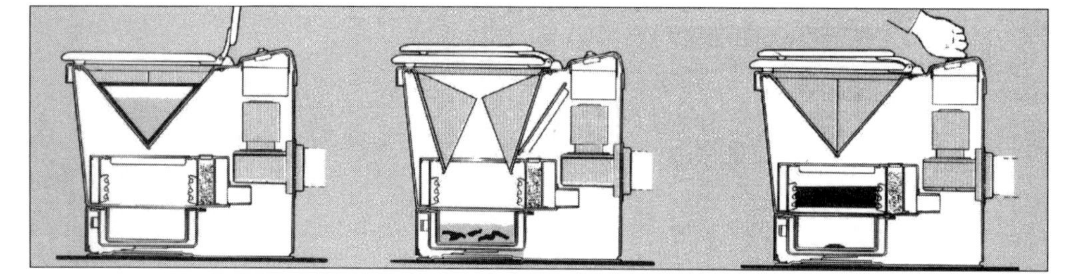

Operating an Incinolet: After a bowl liner is put in place, the toilet can be used. Step on a foot pedal, and the waste drops to a burning compartment. Finally, push a button to incinerate waste.

ETPA = Excrement, Toilet Paper and Additive

Toilet Stools for Your Composting Toilet System

A big question with composters that are remote from the toilet room is what to use as a toilet. In this use of the term "toilet," we mean the thing upon which one sits while "taking one's ease." Everything from a squat plate to a standard hole-in-a-board to a high-tech vacuum flush toilet can be used for this purpose. The criteria for your choice are aesthetics, cost, distance (from the composter) and how much leachate your system can handle.

What's particularly exciting about this field is that micro-flush toilets are now being used with composters, making this technology far more palatable to more people, especially those who prefer a barrier between the bathroom and the composter.

Whatever you choose, be sure you have considered its usage and future: Is it user-friendly to children and mobility-impaired visitors? Do you plan to sell the home someday?

Wet (Flush) Toilets

Increasingly, larger composters are installed with micro-flush toilets. We all can appreciate the significance of adding a flush toilet to a composting system, because in our culture, the ubiquitous flush toilet was incorporated into our earliest experience. Planners, designers and manufacturers recognize that to gain acceptance of a composting system, a water-flush toilet (preferably porcelain) will smooth the way.

Remember that composting is defined as an unsaturated aerobic process—that means that the spaces between the composting particles must be moist but able to carry free atmospheric oxygen to the aerobic organisms. The thin film of water that moistens the solids has a high dissolved oxygen content, which facilitates oxygen transport to the cell walls of the resident aerobic organisms digesting the solids. That means a little water is a good thing.

Saturated aerobic decomposition works, too, but it is not composting. Technically, if a liquid had sufficient oxygen dissolved into it (a minimum of three or four milligrams oxygen per liter of liquid) the aerobic bacteria would be able to survive in such a saturated environment. This method is used in aerobic wastewater treatment systems that pump air or pure oxygen into the effluent in small bubbles to force the gas to dissolve

The wide world of composter-compatible toilet stools (clockwise from far left): A wooden model (The Water Centre), an Ifö Cera, a Clivus Multrum toilet stool, and a urine-diverting toilet (EcoSchattweid Centre).

into the liquid effluent.

However, if solids are saturated under normal conditions (without bubbled aeration), the decomposition of the carbon quickly uses up any dissolved oxygen, and that condition is called "hypoxia" or an "anoxic" state. If the solids remain anoxic for too long, anaerobic organisms take over (anaerobiosis), and odors and inefficient processing result.

Composting works with micro-flush toilets because the small amount of flushing water—one pint or so—quickly drains through the compost so that the air-breathing microbes don't drown. In soil systems in nature, rainwater drains through porous soils and actually pulls air in with it by creating a temporary partial vacuum. In this way, the additional water can be helpful to the composting process. Temporary anoxic conditions do not necessarily lead to anaerobiosis. This is especially true at low temperatures. Below 41°F (5°C), microbial respiration ceases, and both aerobic and anaerobic bacteria will cease to digest nutrients, even if anoxic conditions exist, because they are not able to use oxygen anyway.

But if the toilet is flushed with only one pint for 5.2 times (the average number of uses per day per adult), then 5.2 pints (0.24 liter) of water is added to the compost in addition to the urine. That adds up to 237 gallons (897 liters) per year being used up and converted to wastewater. Add to that water used to wash the toilet stool and urine.

Using a flush toilet means more leachate to manage, either by draining it to an approved system such as a soil absorption system (leachfield or pit) or through the methods described in the "Leachate" section in Chapter Four. SeaLand one-pint-flush toilets have been coupled with nearly all of the larger remote composters. The extra water tends to remove finer particles and suspend them in the leachate. This can lead to increased compacting of the compost, but it does not seem to hinder the composting process. However, if the temperature of the water is extremely cold, flush water could cool the compost, slowing it down. If one lives in a cold climate, the toilet water might not have a chance to warm to room temperature in a tankless toilet like the SeaLand, requiring some supplemental heat for the composter.

This graphic from Sun-Mar shows a SeaLand one-pint flush toilet coupled with a Centrex composter. Note the 45-degree angle of the toilet pipe.

Enthalpy versus Entropy: The Physics of Flushing

Enthalpy is a thermodynamic property of a substance (in this case, water) defined as the sum of its internal energy plus the quantity of the substance (e.g., 1.6 gallons weighing 13.3525 pounds at 60°F—6 kg at 15.5°C). By elevating 1.6 gallons of water while filling the flush tank, gravitational energy is added and enthalpy is increased (by hydrostatic or head pressure). Entropy is the amount of energy unavailable for useful work in a system undergoing change.

By flushing water through the toilet and carrying away ETPA, the energy gained—enthalpy—is lost by the forces of friction (the irreversible transformation of energy into heat resulting from the momentum exchange between water molecules in laminar flow and between solid particles moving at different velocities in turbulent flow). Toilets that have more water than is necessary to flush the waste are high-entropy systems.

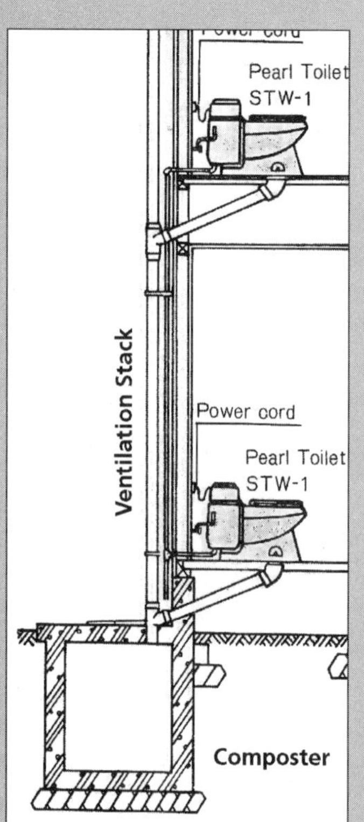

Flushing Coefficient of Performance (FCOP) is the ratio of the flush effect (moving wastewater out of the bowl and into the sewage pipes) to the enthalpy used in the process (output over input). Water closets are inefficient devices with a low FCOP. Remember, we often flush out one ounce (0.06857 pounds—319 kg)) of toilet tissue with five gallons (41.73 pounds—18.6 kg) of drinking water (high entropy) or a FCOP of .0016. By comparison, a good heat-pump has a COP of 3.0. Strictly speaking, the mean velocity pressure of the flush is calculated from the kinetic energy flux divided by the volume flow per unit time.

ETPA = Excrement, Toilet Paper and Additive

How Much Water Can Be Added to the Composter?

Apart from the poor eco-logic of using a lot of water, there is no upper limit to the amount of water that can be added to the composter, *so long as the process stays unsaturated* (not immersed in liquid). That may depend on the ability of your system to manage leachate, which is the combined liquid from the excreta and the flush water. Anna Edey of Solviva reports success with a 1.6-gallon (6 liter) flush toilet and a composter *with an extensive leachate drainage system* (part of her "composting biofilter" design). Keep in mind that if too much water is used, the enzymes and microbial populations might be rinsed from the composting particles, reducing the composting rate. It is best to not use more than one or two pints per flush. Many micro-flush toilets can be used with one composter for multiple-bathroom buildings. The key consideration here is drain-line carry—how far a little water can move the waste. The less water, the shorter the distance and the steeper the pitch of the pipe. Locate your composters as close to the toilets as possible.

Most micro-flush toilets are typically used for recreational vehicles and boats, where they flush to holding tanks underneath. However, we have seen composting systems in which one-pint flush toilets were used with a 20-foot (6 meters) horizontal distance and a 20- to 45-degree pitch to the composter. Because elevated water has so much embodied energy, it can overcome friction head and move solids horizontally with little pitch. That means the system should have a minimum of elbows and long runs of pipe. This seemed to be the maximum toilet-to-composter distance, as the users had to double flush occasionally. (SeaLand offers no guarantees for horizontal transport distances, as the company does not want to be responsible for pipe clogging.) It is important to use the correct diameter pipe—SeaLand recommends three inches. Larger is not better, because you want the flush water to fill 50 percent of the pipe diameter, which scours the sidewalls in order to transport the solids.

Composter as Filter

In one respect, a composter is simply a large filter that filters solids out of the used water stream and provides an environment for them to be biologically decomposed through composting. It is likely that one day, flush toilets will be commonly used with composters. The remaining solids-free effluent can then be transported and treated at a lower cost.

Using water defeats the water conservation and dry-end-product benefits of composters, but improves the acceptance of the technology. So, until buildings are designed for energy- and waterless transport of excreta, micro-flush toilets will be an intermediate step on the way to a sewerless society.

Mechanical-Trap Toilets

Here is a sampling of systems:

• SeaLand 1- to 3-Pint (.47-1.4 Liter) Flush Toilets

Using an average of one pint per flush (less with urine only, more with solids), these toilets provide all of the advantages of a flush toilet while minimizing water usage, making them popular for use with composters.

Effective rim-wash action keeps the toilet bowl clean. Pressing a foot pedal simultaneously opens a rinsing water valve and a ball valve in the trap, which allows the bowl contents to pass into the pipe below. Releasing the pedal closes the ball-valve and shuts off the rim wash. Pulling up on the foot pedal fills the bowl with water to a level selected by the user. A small amount of water is maintained in the bowl by a self-cleaning ball-valve trap with a Teflon seal. These typically are connected to a pressurized water supply and deliver about one gallon per minute at 50 psig (344.7 kPa) pressure, so you don't want to open the valve for more than a few seconds, to minimize the water flow. If you don't have pressurized water, it can be connected to a gravity-fed water supply from a roof tank which is sufficient to flush a one-pint toilet. In the winter, it can be flushed by pouring water into the bowl.

The SeaLand micro-flush toilet

The two most popular low-flush toilets in North America are the SeaLand Traveller 510 and 910 ceramic toilets. The 510 has a regular-sized toilet seat, and an elongated bowl. The low-profile 511 and 911 models are available for platform-mounted applications; a model for handicapped-accessibility requirements is also available. SeaLand toilets are supplied complete except for a four-bolt floor flange (or a two-bolt flange with two extra holes drilled out). Prices range from $200 to $300.

• Vera Waterless Toilet

Not totally waterless—it uses about a tablespoon of water—the Waterless Toilet by Vera Miljø of Norway is notable because it is a dry toilet stool with a clever plastic rotating eight-inch bowl in the trap, secured with stainless steel hardware. A small soapy-water reservoir within the back of the commode is connected to a manually operated pump which rinses the toilet and the mechanical "bowl" seal with a few ounces of soapy water for cleanliness. The rotating bowl opens only when the user pushes a lever. A spritz of soapy water primes the walls of the toilet to minimize skid marks. Excrement falls into the open soapy bowl (the inside of which is colored black to minimize staining). After use, the user pushes a button that rotates the bowl, which drops the ETPA into the composter. Now the white bottom of the bowl faces up and the toilet opening is sealed. About $700 (F.O.B. Norway)

The Vera Waterless Toilet Stool

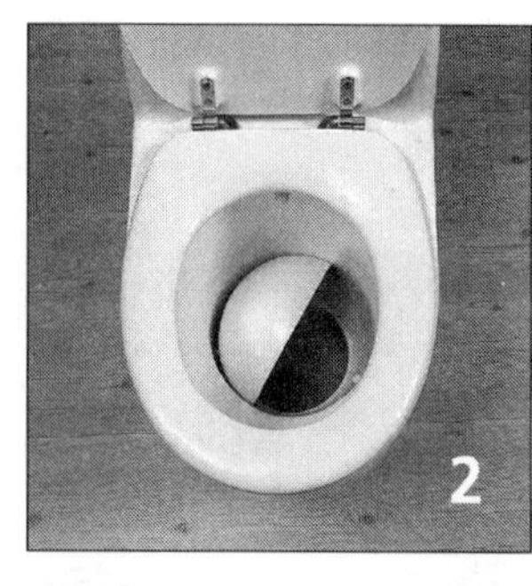

The Vera Waterless Toilet Stool uses a teaspoon of water and this innovative rotating cup system.

• Nepon Pearl Electric Foam Flush Toilet

Another micro-flush toilet used with composters is the Nepon Pearl Foam Flush toilet from Japan. It averages less than one cup of water per flush, and uses foam to carry away waste. In this system, a tiny amount of water mixes with liquid soap or detergent,* held in a removable bottle, with a metering needle that emits one drop of detergent into the mixing chamber every 40 seconds. The water-detergent solution in the tank of the toilet produces a foam created by an adjustable aquarium-style diaphragm air pump and air stone that bubbles air through the mixture and creates a continuous amount of foam that coats and fills the bowl, preventing feces from sticking to the bowl. A unique plastic flap system in the trap holds up the foam, but not ETPA, and prevents odors from backing out into the room. This is best suited for straight-drop toilets, or at least no more than 20 degrees from vertical from the toilet to the composter. An interesting benefit is the addition of the detergent, which seems to precondition the composting solids, making it easier for the microbes to digest it.

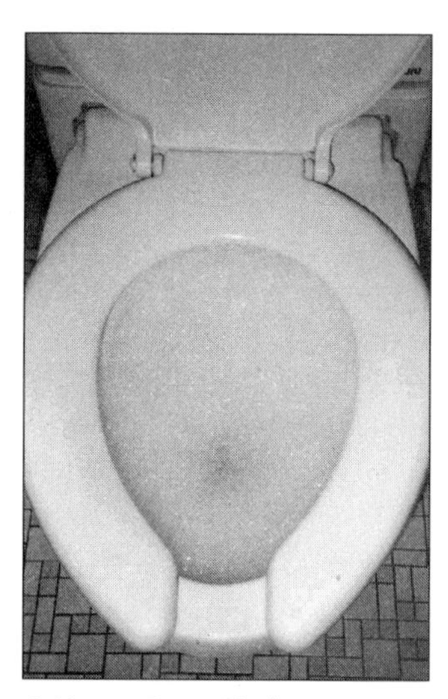

A Nepon Foam-Flush

That small amount of biodegradable soap or detergent does not harm the microbes. Those who have lived with them either love them or hate them. Constant attention is required to provide the right amount of liquid detergent and the right amount of foam flow adjustment. Too little and it dries up; too much causes foam to spill out onto the floor. (And kids love to spend the afternoon flushing the toilet to watch it fill up! One user, who has had one in his home since 1980, reports that his boys delighted in writing their initials in the foam while urinating.) This toilet is very expensive and requires electricity (motor uses 100 VAC at 50 cycles Japanese standard but it works okay on 115 VAC 60 cycles North American standard, grid power). Others have found that parts are difficult to obtain in North America. Price range: $500 to $2,400.

** A 50 percent solution of Liquid All laundry detergent and water works best (a tip from AlasCan's Clint Elston).*

ETPA = Excrement, Toilet Paper and Additive

Vacuum Toilet Systems

Vacuum toilets are very interesting. These are often found in ships, airplanes and trains because of the need to reduce the volume and weight of carried water and wastewater. Space is a costly commodity on ships, planes and trains. Airlines in particular try to minimize the weight of their planes' loads in order to maximize fuel economy.

In a vacuum toilet system, a vacuum pump pulls the material through a small-diameter pipe when a check valve at the toilet is opened. Atmospheric pressure pushes the material through the line to satisfy the vacuum that is created at or on top of the composter. (It works just like the boxes in drive-through banks.) The advantage of the vacuum is that it can literally move solids to an upper floor—in fact, if you have the proper vacuum system, you could even put a composter in your attic or on your roof, and have toilets located anywhere. That's because they do not rely on gravity, which relieves the installation constraints of conventional gravity plumbing systems. For retrofit installations, the 1.5- inch (38 mm) diameter pipe can even be placed to go up and over floor joists and support girders!

There are few composters that are specifically designed to accommodate vacuum technology, because the vacuum chamber has to be airtight. There are in-line vacuum pumps that have a negative pressure on one side and a positive pressure on the other; these literally suck the toilet contents to the vacuum pump and discharge it on the other side. They can be used with any composter. Presently, all of the vacuum systems require electricity to create the vacuum.

Evac of Sweden and SeaLand of Ohio offer vacuum systems for larger boats. Both companies offer stylish color-coordinated porcelain bowls. Evac offers Teflon-lined stainless steel, if you need a vandal-resistant unit for public applications.

SeaLand uses the same toilet as its one-pint-flush system, but it is connected to a vacuum transport system, so you can move waste horizontally. Prices start at around $1,200 for a single-toilet vacuum system.

How a VacuFlush System Works

A vacuum is maintained in the tank and piping system at all times. A small amount of water is in the bowl before use.

2. When the toilet is flushed, a valve in the trap of the toilet opens and the positive atmospheric pressure pushes the contents through the pipes, trying to satisfy the vacuum in the tank. The bowl is instantly (in one or two seconds!) cleared and the waste is moved through the pipes, at seven feet (two meters) per second ,to the vacuum tank, the vacuum pump, and finally to the composter or holding tank (not pictured). The rapid change in pressure in the vacuum tank causes the integral pressure switch to activate the vacuum pump.

3. After the flush lever is released, the vacuum pump continues to run until the vacuum level is recharged in the system. On the SeaLand, recharging the vacuum takes about one minute. Evac systems continuously maintain vacuum in the lines, so there is no waiting.

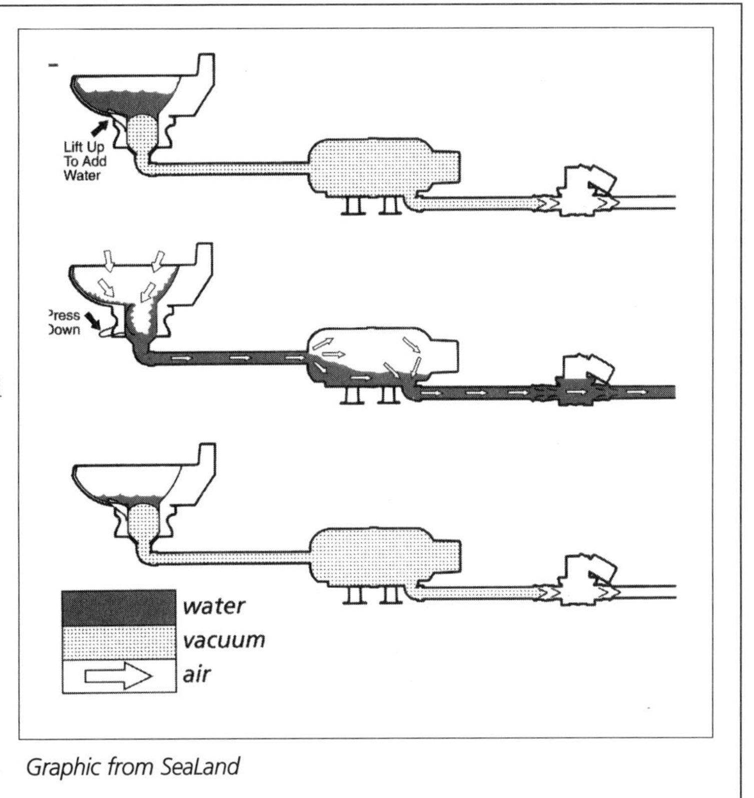

Graphic from SeaLand

• Evac Vacuum Toilet Systems

Evac is the world's leader in vacuum toilet technology. Its toilets can be found on U.S. Navy and Coast Guard ships, Boeing 747 airplanes, Amtrak trains, commercial marinas, hotels, and industrial and municipal sites worldwide. Evac allows the use of toilets with as low as one pint of water per flush, and the flexibility to move wastewater through flexible 1.5-inch (38 mm) pipes. The flush water volume can be reduced by up to 99 percent. We particularly like the new Evac Compact which is available in both stainless steel and porcelain, and is used on trains and planes. Evac allows connecting many toilets to one central composting system. The smallest system can serve eight to nine toilets, and was designed for yachts. It costs $9,000 for the vacuum system alone! Add to that the price of the toilets, and you can see that this system is mainly for public facilities or apartment buildings. The author specified an Evac system for a popular heavily trafficked furniture store in Massachusetts that had extreme constraints on hydraulic capacity of the wastewater system and extreme distances of more than 500 feet from toilet to septic tank. The Evac works flawlessly. (But these toilets can really suck! Don't let that scarf hang too low!)

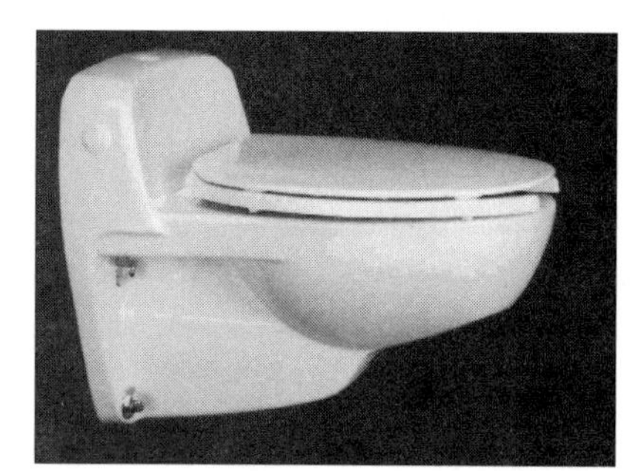

An Evac toilet stool.

There's even the do-it-yourself option. Jeremy Criss, who developed the Bio-Recycler composter, built his own vacuum toilet system, using a compressor and a bowl-cleaning hose spritzer.

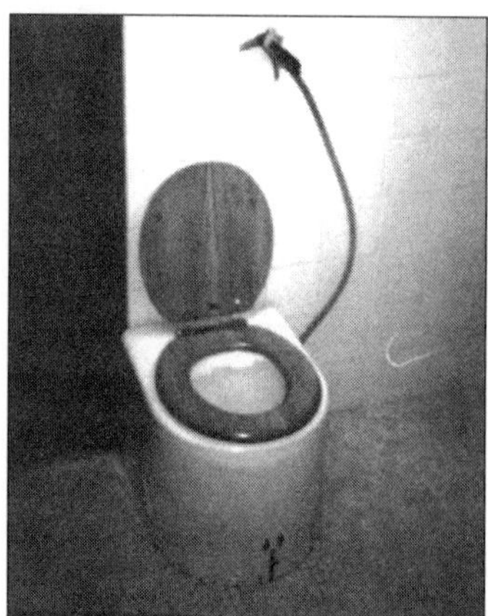

A vacuum toilet made by Jeremy Criss as part of his BioRecycler (Photo: Jeremy Criss)

Dry Toilets

A dry toilet stool is usually a smooth-finished fiberglass or ABS plastic commode with a fitted drop chute or a chair or cabinet built over a toilet chute. Ideally, it is anatomically correct, so the feces do not create too many "skid marks" (streaks on the bowl).

Manufactured Dry Toilet Stools

Nearly every composting toilet manufacturer offers dry toilet stools. BioLet offers a dry toilet with a slide-away plate to hide the composter contents from view; it closes when the toilet is not in use—more manufacturers will likely follow suit.

A black or stainless steel bowl or chute is a good idea, as, yes, there can be skid marks with a dry toilet. The most common question about dry toilet stools is whether one can see the contents of the composter: If the chute is long enough and dark enough (don't install lights directly over the toilet), then you may not. Prices range from $150 to $350.

Above: Five manufactured dry toilet stools by (clockwise from top left) BioSun, Phoenix, Sun-Mar, Clivus Multrum and Vera Miljø

Owner-Built Toilet Stools

The beauty of composting toilets is that they need no water, so you don't need any plumbing. That means that you can easily build your own toilet stool. And forget images of an indoor privy. Site-built toilet commodes allow you to be creative—who says your toilet has to look like the porcelain thrones we grew up with? Considering that this may be the most used piece of furniture in your house, why not design it well?

Building your own can be an opportunity to create a work of art, an attractive piece of furniture. Many composting toilet owners construct their own attractive toilet stools of wood, tile and other materials. Early Clivus Multrums came with kits for site-built wooden box commodes (some owners joke that they belong to the "Birch John Society"). Some, such as the ones designed by the ReSource Institute for use in developing countries, are artfully tiled, sometimes as a continuation of the floor.

Many are made by molding cement around 8- to 12-inch (200-305 mm) sewer pipes or inverted plastic five-gallon buckets with the bottom cut out. Others commonly seen in parks are simply flared stainless steel stools.

Design Considerations

Size of Opening to Composter

When designing your own, consider your anxiety about small children falling in. In 25 years, the authors have never heard such a report, but, for peace of mind, you might consider using an 8-inch opening rather than a 12- or 14-inch opening. Yes, you may use the bowl brush more often, but you'll feel better. Smaller diameter openings also minimize odors in the room.

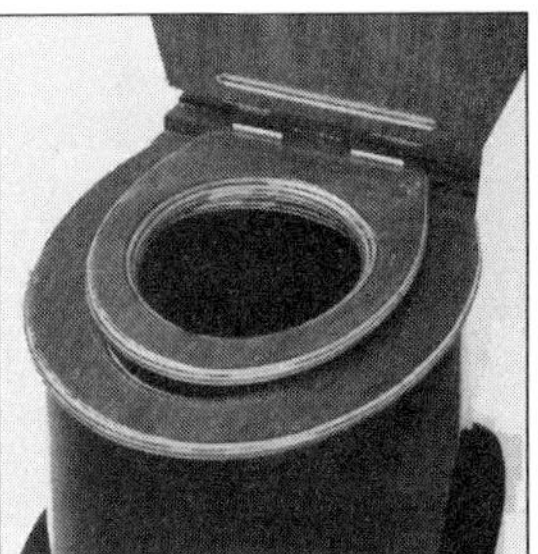

A children's toilet seat hinges right over the full-size seat.

Odors and Insects

If you don't have a power-vented exhaust system, odors will back up into the room under certain weather conditions, so design the seat to make an airtight fit with the commode. A sign alerting users to close the seat and lid after use will help. Some gasket the seat and lid with compressible weather-stripping tape. This also helps keep out flies and other troublesome insects.

Height from Floor to Seat

Keep in mind that the standard height of a conventional toilet is about 15 inches (381 mm) from floor to seat, 18 inches for the physically challenged and elderly.

Toilet Bowls

If the toilet is dry, the bowl should not have much horizontal surface area (where excrement can collect)

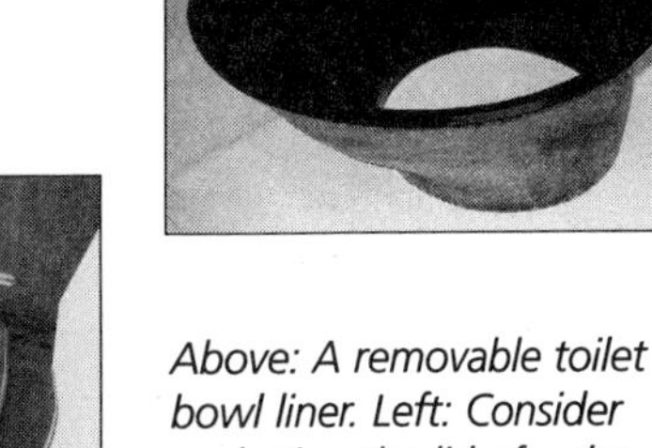

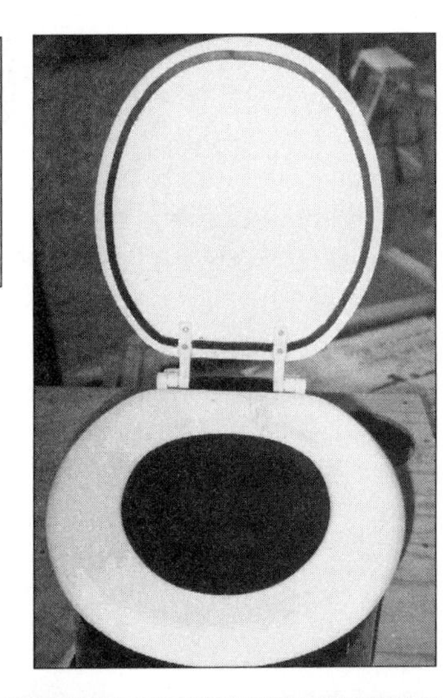

Above: A removable toilet bowl liner. Left: Consider gasketing the lid of a dry toilet stool. Below right: An owner-built toilet stool in California. Below left: A toilet stool constructed with reused tile by the ReSource Institute in Mexico.

Top: A diagram of a site-built toilet stool made of concrete with a wooden lid (Illustration: Center for Ecological Pollution Prevention) Above: A cement-formed toilet in Micronesia and in Mexico. Ω(Photos: David Del Porto)

and it should be as slippery as possible.

EcoTech offers a black fiberglass bowl liner that fits into an eight-inch pipe (about $49), making an ideal bowl for a do-it-yourself toilet stool.

Other Dry or Almost Dry Toilets

• Squat Plates

We also know that the custom in Asia and elsewhere is to use floor-level squat plates. Squat plates are made of typical stain-resistant materials such as vitreous china, plastic, fiberglass, concrete, epoxy-coated wood or ceramic tiles. Many of these are water-flush. Squat plates are sometimes recommended because some say that squatting is the ideal position for elimination. However, North American conventions will probably have you designing the higher commode, as our knees are not used to bending enough to make squatting a comfortable position, nor are many of our clothing styles amenable to this position. As always, take into account who will be using your system.

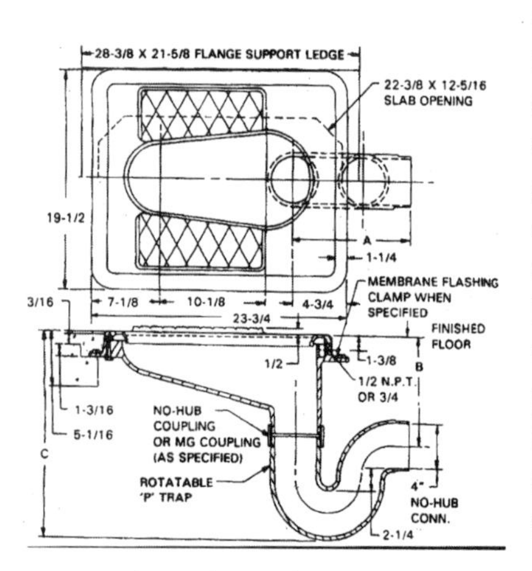

A squat-plate-style pour-flush toilet from Zurn.

Squat plates are available from many plumbing supply companies based in Asia, such as Toto.

• Pour-Flush or Water-Seal Toilets

Consider also flushing your composter's toilet with water from your washbasin. You can design a shallow-trap pour-flush toilet. Often referred to as "water-seal toilets," these pour-flush toilets have been used successfully worldwide, but as of this writing, we have not seen them integrated with composting processes. But why not? As long as the compost is unsaturated, and you are willing to put up with added leachate, a small amount of water poured from the height of two feet should move a lot of ETPA from a shallow-trap toilet that has a 2- to 4-inch (50.8-101.6 mm) pipe.

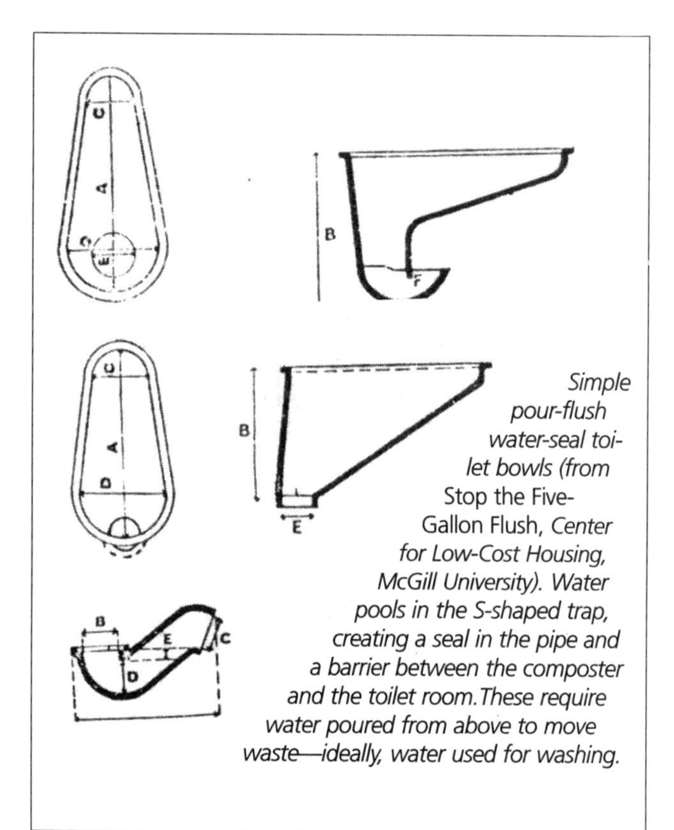

Simple pour-flush water-seal toilet bowls (from Stop the Five-Gallon Flush, *Center for Low-Cost Housing, McGill University). Water pools in the S-shaped trap, creating a seal in the pipe and a barrier between the composter and the toilet room. These require water poured from above to move waste—ideally, water used for washing.*

ETPA = *Excrement, Toilet Paper and Additive*

Urine-Diversion/Separation Toilets and Urinals

Diverting urine from the composter means less leachate to manage and less care. It also provides the opportunity to use the urine as fertilizer. The solids will compost just fine without the urine—in fact, sometimes better and with less likelihood of odors. However, the question is, what do you do with the urine. See "Leachate, the Liquid in Your Composting Toilet System" and "Urine: Managing It Separately" in Chapter 4 for an explanation of how this affects composting.

Urine-separating toilets are perhaps better referred to as urine-diverting toilets, because the urine is diverted before it is combined with the ETPA, not separated from it.

In a urine-diverting toilet, urine is collected in a built-in urinal in the front of the toilet from seated users. It flows via a pipe to a separate container or is combined with graywater and used for direct irrigation or drains to a septic system leachfield. Solids either drop directly to the composter or they are flushed with the standard 1.6 gallons (6 liters) of water. Urine is not flushed (although an occasional rinse may be needed).

Urine should be diluted with water at a rate of at least eight parts water to one part urine (as you may know, undiluted and unoxidized urine can kill plants), and may be used as fertilizer, or it can be taken to a treatment facility. Removing the extra water minimizes the need for energy to evaporate it, making the composting system more efficient. Separating the urine from the fecal matter improves the composting process by diverting the extra water, nitrogen and sodium from the composting process. Urine is also rich in table salt (sodium chloride), and that can cause problems for the beneficial bacteria in a composter.

University studies in Sweden reveal the fertilizer value of human urine is equal to that of chemical fertilizer. It is important to remember, however, that urine can contain significant pathogens (liver flukes, schistomyosis, etc.), but in healthy populations it is usually sterile. Urine should be treated with the same respect as graywater.

Denitrifying at the Toilet

Urine diversion is now being proposed for communities with conventional water and sewer systems to keep the troublesome nitrogen out of receiving waters, where it causes unwanted plant growth and contaminates drinking water supplies. When urine is diverted at the source, the wastewater treatment plants do not require costly denitrifying additions to the facility. Nitrogen from urine currently accounts for 80 to 90 percent of the nitrogen in sewage, so urine diversion offers significant advantages for communities upgrading their facilities to meet modern low-nitrogen requirements. Some states have established new lower levels for the maximum allowable nitrogen in treated effluent in recognition of the trouble it causes.

Left: A urine-diverting toilet by Vera Miljø of Norway.
Below: A diagram by another urine-diverting toilet manufacturer, Ekologen of Sweden, shows how urine can bypass a simple composter, to be stored for use in a tank.

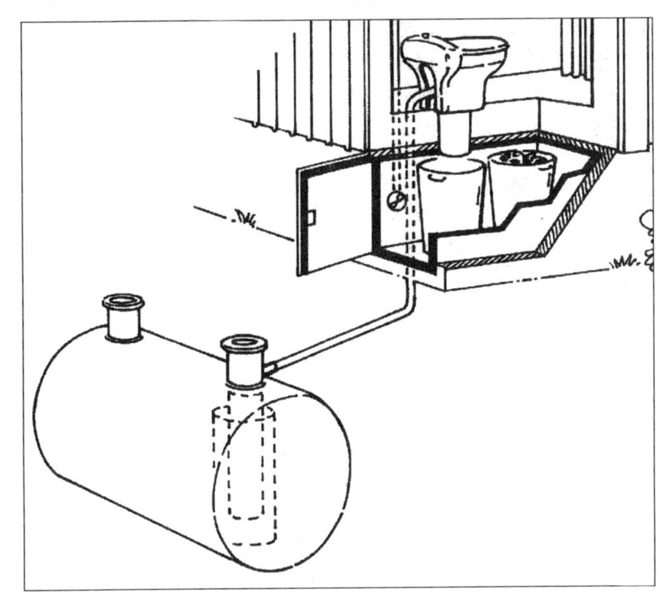

From a composter management point of view, urine diversion makes good sense if you have a place for the urine to go. Otherwise, it is just another task in the maintenance of your system. And as we have reported, if you cannot use it for fertilizer for plants, the nutrient value of urine is usually lost. Experience with urine diversion in Mexico reveals that if there is not a well-managed urine management plan, including a person responsible for urine collection, stabilization, storage, transport, etc., the urine is usually drained to a rock-filled soak pit, where it

ETPA = Excrement, Toilet Paper and Additive

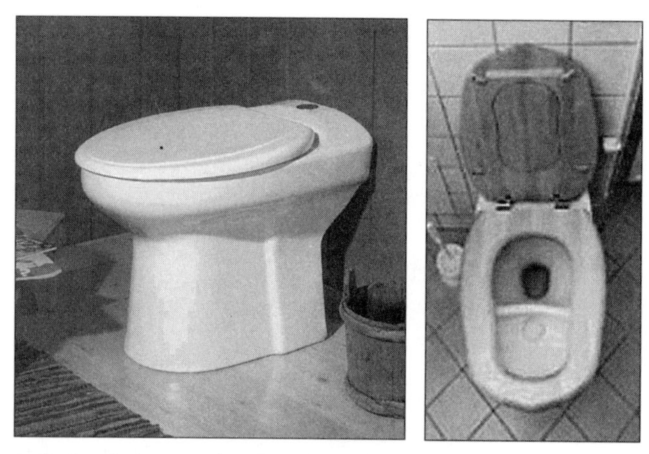

Left: An Ekologen urine-diverting toilet available from Vera Miljø. Right: A Dubbletten urine-diverting toilet in Ekoporten, an ecologically retrofitted apartment building in Sweden.

Urine-diverting toilets made by workshops organized by Cesar Añorve and TESEC in Mexico. This shows the toilet openings that are used interchangeably as the bins of the two-bin system underneath fill up.

can potentially cause problems.

It is best to drain urine directly to shallow, subsurface aerobic planting beds and trenches so that users and others cannot be in direct contact with this liquid. One approach is to combine separated urine with filtered graywater.

In central Mexico, dry urine-diverting toilet stools are made by workshops organized by César Añorvé and TESEC. These are simple ceramic and fiberglass urine-diverting dry toilet stools available in many colors. (See "Composting Toilet Systems in Development Programs," Chapter 8.)

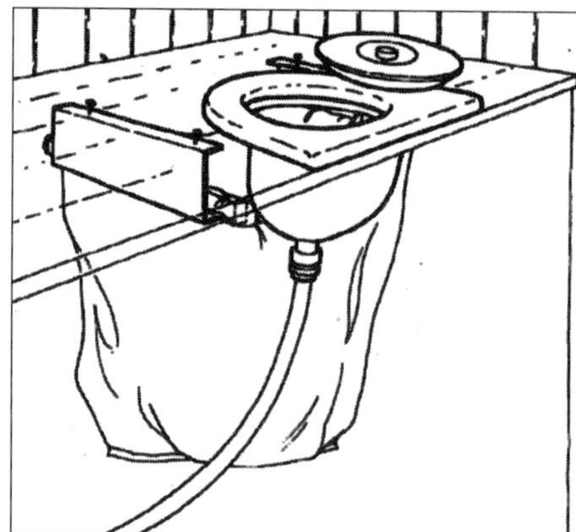

The Septum Torrdass 403 is a urine-diverting toilet bowl. You provide the compost container and urine containment or management system.

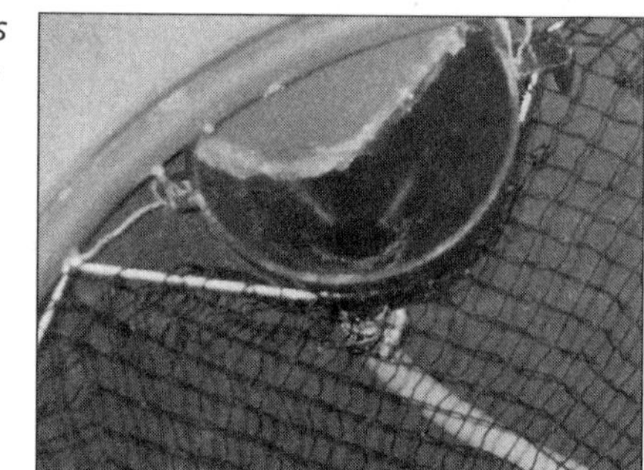

A urine-diverter in this barrel system is made of a funnel trimmed to fit against the wall of the barrel and wired and caulked in place. A garden hose is hose clamped onto it, and drains to a urine composter.

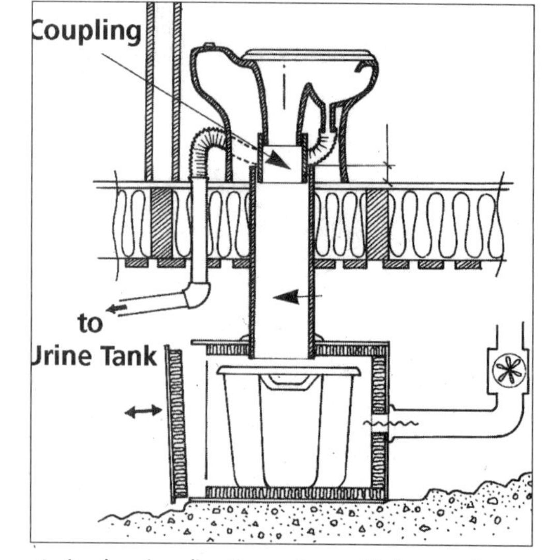

A simple urine-diverting system with feces-only composting. (Graphic: Ekologen)

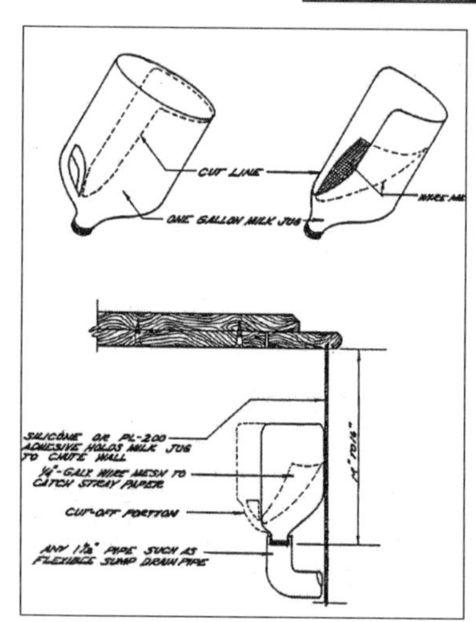

Yes, you can make your own simple urine-diverting toilet or waterless urinal. This one uses a plastic jug. Other inexpensive tools include liquid and motor fuel funnels. (Illustration: National Center for Appropriate Techology)

Toilet Water Separation

The only commercially available toilet water and urine separation system we know of is the Aquatron system, manufactured in Sweden and available with that company's several types of composters.

Aquatron's Separator is a passive solids and liquids separator, requiring no additional energy. When flush water, excrement and toilet paper are flushed down the drain, it enters the top of the polyethylene module, which is constructed to initiate a whirlpool effect in the upper container. There, by centrifugal force, it separates the liquid, which moves to the outer wall and ultimately out of the system. That allows the solids to drop into the composter beneath it. This device can be adapted to any solids composting system, or for that matter, any solids containment and removal system. The liquid, which includes suspended solids from fecal matter, as well as urine and flush water, passes through an ultraviolet light box, which irradiates the liquid, alegedly killing bacteria and viruses. In Sweden, this liquid is legally treated as graywater. The Aquatron is often used with an Ifö Cera adjustable six-liter-flush toilet. It can also be used with a urine-diverting toilets, as it is in Sweden's Ekoporten, a model apartment house retrofitted with ecological systems.

Left: The Ifö Cera adjustable six-liter-flush toilet can be used with the Aquatron toilet-water separation system (below).

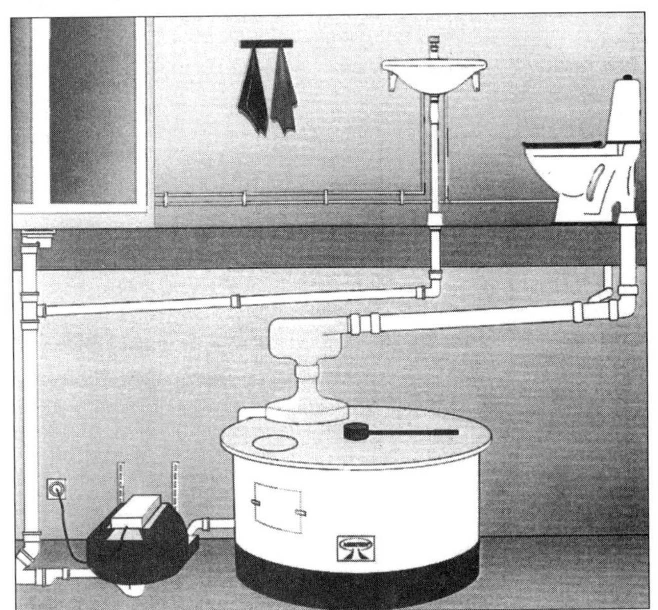

Waterless Urinals

You can also install separate waterless urinals for the males. These can be purchased or made. Just remember to rinse them out with freshwater after use to keep the odors under control. The Choi building in Vancouver simply keeps pitchers full of water next to its urinals.

One novel proprietary urinal, The Waterless Urinal, employs a small amount of oil to provide a trap seal, so that odors do not result.

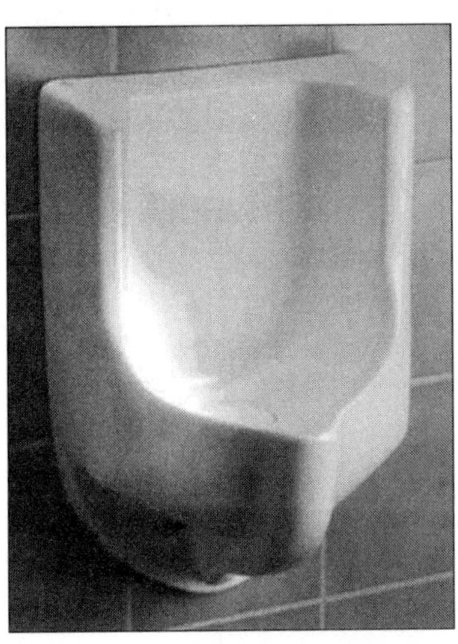

Heres how it works: Urine, which is 96 percent water, flows down the surface of the urinal, draining into a special trap where it emerges through and beneath a floating layer of sealant liquid (biodegradable oil) that is lighter than urine. This oil allows urine to sink through its layer and down the drain, sealing the urine from the atmosphere and eliminating the odors associated with urine decomposition. Both the special trap and liquid must be replaced periodically.

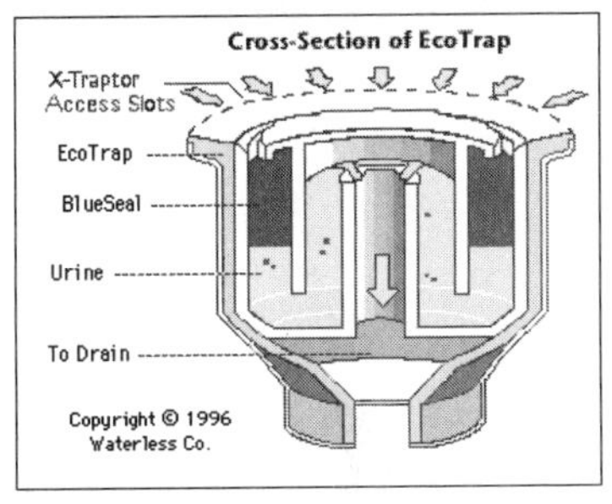

The manufacturer claims the "system requires only 250 ml (7.5 ounces) of sealant liquid per year to operate. On average, a single installation of the trap cartridge and one pouch of sealant liquid will provide sanitary and functional operation for approximately 7,000 uses when maintained properly."

The MisterMiser Urinal is a trim, fold-down low-water (10 ounces!) urinal that fits right between the wall studs,allowing easy installation.

Toilet Sources

Dry Toilet Stools and Risers

Nearly all composting toilet manufacturers sell dry toilet stools made of fiberglass or polyethylene—you'll find their addresses earlier in this chapter.

A source for toilet risers:
Romtec
18240 North Bank Road
Roseburg, OR 97470
www.romtec.com
503/496-3541

Adult/Child Toilet Seats

About $25 from
Bathroom Machineries
495 Main Street
Murphys, CA 95247
www.deabath.com
209/728-2021

Micro-Flush Toilets

Available from most organizations that sell composting toilets, as well as:
SeaLand
P.O. Box 38, Fourth Street
Big Prairie, OH 44611
800-321-9886

Vera Waterless Toilets

In Europe:
Vera Miljø A/S
Postboks 2036
N-3239
Sandefjord, Norway
www.vera.no

In the Americas:
Ecos, Inc.
50 Beharrell Street
Concord, MA 01742
978-369-3951
The price in the U.S. is about $800 to $950

Urine-Diverting Toilets

Ekologen
Box 11162,
S-10061 Stockholm, Sweden
46 8 641 42 50

Vera Porcelain Separation Toilet, about $439
Vera Miljø A/S
Postboks 2036
3202 Sandefjord, Norway

In the U.S., Ekologen urine-diverting toilets and TESEC fiberglass dry urine-diverting toilet stools are available from:
Ecos, Inc.
50 Beharrell Street
Concord, MA 01742
978-369-3951
Ecos also sells

AB Dubbletten
Bromma, Sweden
46-08-87-71-00

Urine-Separator System

Aquatron International AB
Björnnasvagen 21
S-113 47 Stockholm, Sweden
+08-790- 98 95

Squat Plates and Pour-Flush Toilet Components

Zurn Industries, Inc.
1801 Pittsburgh Ave.
P.O. Box 13801
Erie, PA 16514
814/455-0921

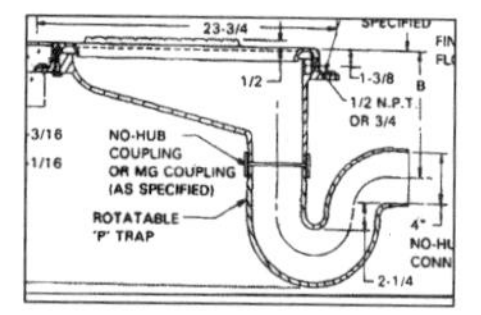

Low-water and Waterless Urinals

Waterless Co.
1223 Camino Del Mar
Del Mar, CA 92014 USA
888-663-5874, 619-793-5393
info@waterless.com

Clivus Multrum, Ifö Sanitar, Toto, Romtec and several other companies also offer waterless urinals.

MisterMiser Urinal
4901 N. 12th Street
Quincy, IL 62301
888--228-6900

Foam-Flush Toilets

Nepon, Inc.
1-4-2 Shibuya-Ku, Shibuya
Tokyo, Japan
In the U.S., try Clivus Multrum, Ecos and AlasCan

Vacuum-Flush Toilet Systems

Evac
1260 Turret Drive
Rockford, IL 61111 USA
815-654-8300, 800-435-6951

SeaLand
P.O. Box 38, Fourth Street
Big Prairie, OH 44611
800-321-9886

CHAPTER EIGHT

Composting Toilets at Work

Learning about composting toilet systems from their manufacturers and designers is one thing. Talking to those who actually live with them and maintain them can be much more enlightening. Who else can tell you about the idiosyncracies of a particular system, potential problems and homegrown solutions?

What follows are profiles of people who own or maintain various composting toilet systems. We've tried to include a representative sampling of systems and their applications. Some of these folks were referred to us by manufacturers.

To identify owners or system operators you can contact for information, call the manufacturers for references, talk to your state's office for on-site systems permitting, or search the World Wide Web for composting toilet information forums. Try to talk to composting toilet owners who are using their systems in the same way you will. The usage and maintenance considerations for public facilities, cottages and year-round residences are quite different.

There are many variables that affect operation. Remember that some owners have not been using their systems in the optimally functioning way or may not understand the science of the process, so be sure to get information from both the source and the users. For larger composters, talk with owners/operators who have had their composters for five years or more or have gone through a few service cycles or about four years.

You will find information about nearly all of the systems featured here in Chapter 7.

Celebrity composting toilet owners: People who have composting toilet systems in their homes often draw curious visitors who want to know more about life with a composting toilet. Here, Dr. Andy shows reporters from the British Broadcasting Corporation some compost from his composting toilet system. The interview was included in a BBC 4 radio show segment called "The Influence of Effluent."

Dr. Andy and Karen: Composting Toilet System Saves $16,000 (*and* Repels Deer)

Andy and Karen's* home in northwestern Connecticut serves as a model of composting toilet success in a conventional, somewhat upscale home to two adults and three children.

The home features two SeaLand toilets connected to a Carousel composting toilet system located in the basement.

In 1993, when Andy, a medical doctor, and Karen, a jewelry designer, looked into expanding their small lakefront ranch-style house, they were told that their septic system had to be upgraded significantly. In fact, about 72 truckloads of sand (900 cubic yards) would have had to be trucked in to expand the leachfield. "It was going to change the topography of the land, and really tear things up," says Karen. It was also going to cost an estimated $16,000. The couple looked at alternatives, and asked their health agent about composting toilet systems. Won't work, they were told. They decided to go directly to the state level. A permitting official told them to give it a try. "The state liked the fact that it was a contained system," says Andy. With the composting system, they were allowed to use their existing septic tank system for gray water and leachate only. Because they agreed not to put blackwater (toilet water) into the septic tank, Karen had to use a diaper service.

"If this were in any way unhealthy in any way, as a physician, I wouldn't install it."

Andy realized after he had installed the system that the basement was too cold for composting. He found that the toilet water from the two SeaLands cooled the composting mass, and the system was not completely composting, resulting in odors. Andy constructed a simple heater: a simple metal-lined box that holds a 150-watt lightbulb. He placed it above the central air intake. After those first hairy months, the system has been operating well. It's been smooth sailing since. "The key was adding some heat," Andy explains. "The composting material generates some of its own heat, but the toilet water is 50 or 60 degrees, which chills the bacteria and makes everything slow. The heater helps evaporate the water and provides a better environment for the bacteria. The lightbulb heater was the key. This works great now."

It was important to Andy and Karen to have attractive bathrooms with toilets with traps. Sealand micro-flush toilets are in both of their bathrooms.

Having flush toilets was crucial for them. "We just didn't want that contact with the biomass," Andy says. Their first micro-flush toilet, a Nepon Pearl, was too much maintenance, they say, as it required frequent adjusting. After installing the second toilet,

*Many people call Andy and Karen to learn about their system. Karen says the average call is about an hour long. For that reason, they asked that we leave out all identifying information, and they would prefer all inquiries to be referred to the manufacturer.

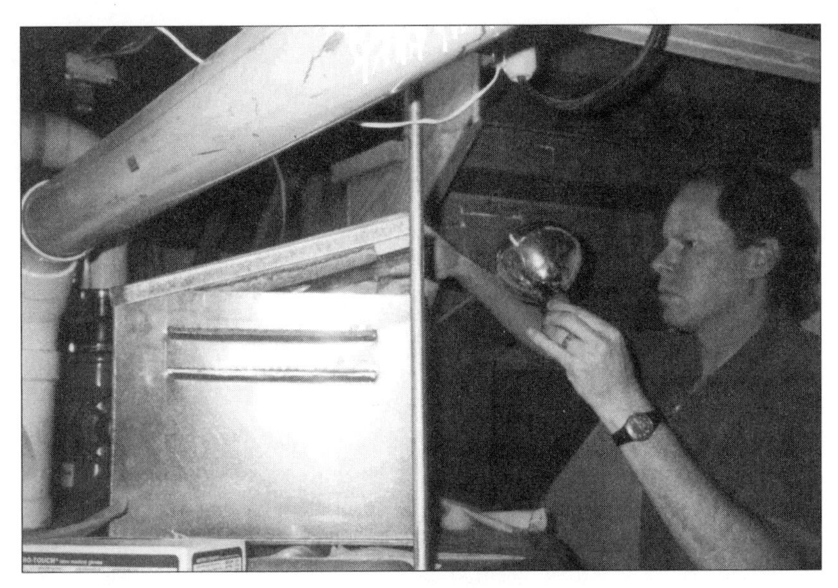

Left: Located in their basement, Andy and Karen's Carousel Composting Toilet System (they lightheartedly call it "the Beast") is loosely wrapped in insulation. Above: Andy changes the lightbulb in his homemade breadbox-style heater, which provides supplementary heat to promote composting.

they also widened the leachate drain opening and added a pump, so the leachate was better drained to the ground-level leachfield.

The SeaLand toilets work well, they say. The only drawbacks they offer are more frequent "skid marks" and an occasional need to replace gaskets around the flush valves. Also, because the second toilet was installed after the initial planning, it is two floors up and about 20 feet away from the composter, occasionally necessitating an extra flush or two.

Maintenance entails adding pine bark chunks to the active compartment every week (they call it "feeding the Beast"). They add vegetable scraps by simply flushing them down the toilet. Every six months, they rotate the inner Carousel to the next compartment, and empty the one with the oldest material. "The material comes out dry, although mostly hard and chunky because it is formed with water." And no bad odors: "It's a little earthy, a little sweet." The couple sprinkle the compost around their yard, where it appears to repel deer. "I think it smells just a little too human for their tastes," Karen muses.

Andy, a medical doctor, notes, "If this was in any way unhealthy, as a physician, I wouldn't install it." He notes that they put the compost only on inedible crops. "But if I had to consider the [potential pathogen] problem at large, there are other, larger health threats."

They say they consider the system fairly easy to maintain and live with. "None of the obstacles were insurmountable at all. They were all solved in a day," says Andy. "It's sort of like taking care of a pet."

The total cost of their system was $5,000 to $6,000, including toilets and installation costs.

Initially, the couple did have some misgivings about the resale value of their home. However, a realtor assured them that the system would be appealing to some people, and an assessor said that it would be an asset from a cost-savings vantage, actually increasing the value of the house.

Andy and Karen also saw benefits beyond money saved in water bills and loads of sand. "It was a little bit of a risk, but it appealed to us in terms of doing our bit with ecology and not being greedy with the Earth's resources," Andy says. "There comes a time when you need to take a stand to make change happen. I was happy to be the first around here to be that person."

Karen adds: "We thought we could be educators in that way and show that this is something that can be lived with. I like that it's ecological and gives me something that keeps deer away and is not so toxic that I have to keep my kids away from it." ✲

An AlasCan in Minnesota

In 1994, Barbara and Earl Bosshardt designed and moved into the new caretakers' house at the 1,332-acre Chesterwoods County Park near Rochester, Minnesota. Because the house is located in an environmentally sensitive rocky area with limited drainage, the Olmstead County environmental department decided to pilot the AlasCan composting toilet and graywater system, with two types of flush toilets. The installed cost was about $20,000.

Barbara was hesitant at first. "I had my reservations," she says. "But now I think it's just great. There's no odor, and it's so clean—I don't have to spend time cleaning." The system includes a Nepon foam-flush toilet directly above the composter and a SeaLand micro-flush toilet two floors up.

To maintain the composter, every two months, Earl opens the upper door and inspects the contents, sometimes giving it a turn with a hoe, and always adding some wood chips. Occasionally he removes about three gallons of compost, which he spreads on the couple's vegetable garden.* "It looks like soil, and there's no odor, not even when you open the top," Barbara reports. The kitchen garbage grinder also drains to the composter.

Besides aerobic organisms, earthworms are at work inside the compost reactor.

Early on, the graywater system has run into some minor problems, she says, due to filters clogging with fibers, however, it now operates well. It drains to a small leachfield.

The entire unit is large—about the size of two refrigerators lying on their sides.

Barbara uses biodegradable cleaners from Amway: LOC for cleaning surfaces and SA8 for laundry detergent.

If any part of the processor overfills, an alarm is signaled at a service provider in Rochester, and a service person notifies the couple. (This has not happened yet.)

She particularly likes the foam-flush toilet. "It saves water and uses just half a cup of water. My grandchildren come and flush it a lot because they like to see it fill up with foam. And I don't have to clean it."

Barbara Bosshardt stands next to the AlasCan composting system located in her basement. Below: She shows the inner workings of the tank of her Nepon Pearl Foam-Flush toilet from Japan.

Although AlasCan recommended that she staggers her washing machine usage, never doing more than a load or two a day, she admits that she has washed up to 10 loads in a day without a problem.

But Barbara considers the system low maintenance, and appreciates the support program she gets from AlasCan. "I think it's great. I like the fact that you get soil fertilizer out the of the system, and we're not polluting the environment." ✲

(Photos by the Bosshardts)

*It's best to use humus on nonedible crops only.

Urine Diversion and Site-Built Composting Toilets in Sweden

On a lake shore outside of Stockholm, is Gebers, a former hospital that was renovated into a 32-unit condominium building using many ecological construction materials. In nearly every unit—which range from studios to three-bedroom units—an Ekologen urine-diverting toilet drains to a composter container located in the cellar. These containers are made of rollaway garbage bins; each is attached to a vent chimney with a fan, and is enclosed in an insulated steel box. (The steel containers were required by fire safety inspector, although fire is not a particular danger with this system.)

When a composter bin fills, it is disconnected, and moved to a finishing area in the corner of the cellar.

Urine drains to two large polyethylene tanks, which are pumped out periodically by a local farmer, who uses the urine to fertilize his grain crops.

Resident Erik Wåhlin says the system was well-received by the residents. The only concern reported by some residents was the appearance of occasional small flies around the toilet opening. These appear to be due to "skid marks" in the narrow toilet chutes. A solution is sprinkling diatomaceous earth or ash in the opening or using some sink or used shower water to rinse the residues.

Nicholas Hort is the Gebers resident who researched and advocated this system. Hort also hopes to ultimately install systems that use graywater for irrigation, and envisions using water from the nearby lake for purposes that do not require potable-water cleanliness, such washing clothes and cars, and watering plants.

A similar system is in use in "Ekoporten," another Swedish multi-unit building.

The system had been in operation for less than a year when this book went to press. It's possible that the composter design will prove too passive for the cool cellar. An easy fix would be suspending a hanging net in each rollaway bin and placing the filled bins in an enclosed ventilated, heated area, perhaps using waste heat from water heaters or dryers.

Nonetheless, this system of urine diversion (and utilization) and feces-only composting appears to be a good use of a new technology and an old approach, and it is worth watching its progress. ☼

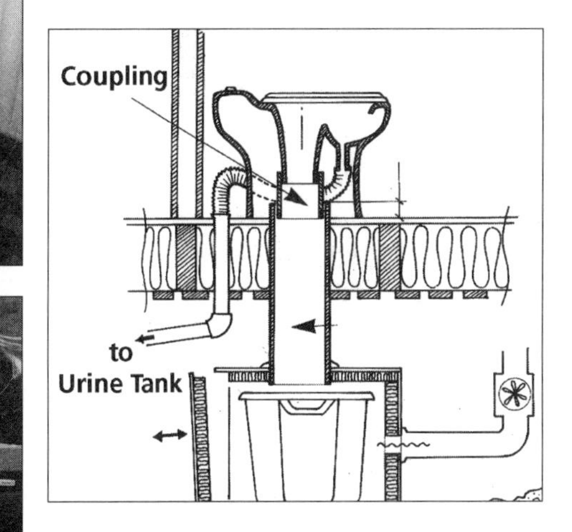

Clockwise from top: Gebers resident Erik Wåhlin next to the urine tanks; a diagram from Ekologen shows how the toilet connects to the composter; the composters are encased in insulated boxes; composter containers not yet in use; an Ekologen urine-diversion toilet.

A Graywater System in the Bedroom

Lush, leafy tropical plants fill an eight-foot-long planter in Heather and Peter Kendall's bedroom, giving the room the feel of a tropical rainforest. But the planter bed offers more than ambiance: It manages the graywater from the home's sinks and single bathroom.

The Kendalls' home also features a Clivus Multrum composting toilet with a fan powered by a photovoltaic panel, as well as a solar hot water heater. For Heather, a decorator specializing in environmentally friendly designs, and Peter, director of an environmental foundation, installing lower-impact technologies was primarily an experiment. "We wanted to have fun with the house," Peter explains.

The composting toilet and graywater system reduce the effluent to the holding tank, cutting their septage pumping costs.

It's also a cost-cutting measure. The house, an older cottage located about an hour outside of Toronto in an environmentally sensitive river area, had an existing holding tank when they purchased it in the mid-1990s. The couple's integrated composting toilet and graywater system reduces the effluent to the tank, cutting their septage pump-out costs.

The graywater from all sinks and their one bathroom is filtered, then drained into the eight-by-three-foot planting bed (formerly a cattle trough) about three feet high. It is planted with large-leafed thirsty tropical plants. The filter—a nylon knee-high stocking—must be changed once a month, as food wastes and fibers clog it. If there's a possibility of an overflow, such as when someone takes a bath, a turn of a valve diverts graywater to the holding tank. Although he is pleased with this system, Peter admits that he would think twice in the future about an indoor graywater system. The drawback, says Peter (who reworked the Clivus design a bit), is that the room can take on an inoffensive damp smell occasionally—"It's like the smell of compost, but not quite," he says.

The couple chose a Clivus Multrum because, they say, it was the only system they knew of that didn't require a heater. They say they are content with the system. Once a month, Peter opens the front hatch and rakes the material forward with a three-pronged claw. He adds a bowl of wood chips through the toilet chute. After two years of use, they had yet to remove more than a few shovelfuls of compost. He pours the leachate into their flower garden. (Note that this is illegal in many places. –Ed.).

In the beginning, they had problems with flies, but that cleared up when they stopped putting food wastes in the composter. "We couldn't master the balance," says Heather. "And besides, I prefer to compost kitchen scraps outside so I can use the compost right away."

For a toilet stool, they chose a plastic one offered by Clivus Multrum. "I actually prefer it," says Heather. "The toilet stool is comfortable, and I like the look—it has European lines." The chute is wide enough that there are few "skid marks," and any waste on the chute dries and flakes off, Peter assures. To clean it, they use a sponge on a stick.

"I don't like using regular toilets now," Peter adds. "With this one, there is no splashing, and I've gotten used to the draft! It's cleaner from a smell viewpoint because of the fan," which pulls odors out of the bathroom.

Because of the existing holding tank, and because the graywater system is indoors (the plumbing code only applies to outdoor systems), they did not have any permitting issues. However, Peter says, "It's a small lot, so if we ever expanded the house or had to have a septic tank, we would have to buy more land." ✲

A graywater bed like the Kendalls'. This one is in Massachusetts. (Photo: Gerry Ives, Architect)

A Clivus Multrum in a Bed-and-Breakfast

A bed-and-breakfast operator* in Boston installed a Clivus Multrum for use in one of his six guest rooms in 1973. "A friend of mine who was working with the company told me about it, and it just seemed like the right thing to do," he says. "It saves water, and it's good for the environment." The bathroom it serves features a wooden box-style dry toilet stool. It is used nearly every day by one to four guests, with an occasional pause of a few days during three winter months. Although most guests are intrigued by the system, every now and then, a guest will refuse the room. "I don't try to talk them out of it. But 99.9 percent of guests' reactions are very favorable."

He is surprised that there are only a few shovelfuls of composted matter to remove every year.

This Clivus, one of the larger models, is perhaps six feet high. Maintenance, he says, entails raking the composting mass every few weeks at top and bottom, and adding a few handfuls of wood chips once a month. He removes material every spring, and says he is surprised that there are only a few shovelfuls of composted matter to remove. This is likely due to the low volume of wood chips he adds. Once a year, he also pumps out some leachate ("fecal treacle," as he calls it) and pours it down a (sewered) drain or sprinkles it on his flower beds after a rainy day. "There's very little care—next to none," he says. "And there are no odors. It works." His system features a fan. The furnace in the cellar keeps the unit warm in the winter.

He says his only troubles were early on when fruit flies appeared. He introduced a parasitic wasp and sprinkled diatomaceous earth in the Clivus. "That was 20 years ago, and they've never come back." ✲

*He chooses to remain anonymous because "I don't need the publicity—I already have enough!"

A CTS Plays Hard to Get

For 16 years, Kate Drum, her husband and two sons lived with a well-performing CTS composting toilet system and dry toilet stool in her Newport, Washington home. But in 1998, she capped off the chute and installed a flush toilet and septic tank system. Her reason is a classic one: The compost reactor was hard to get to, with the removal hatch accessed in a crawl space under the house. "I just asked myself, do I want to be doing this as an old lady? No," says Kate Drum.

Every four to five years, she removed material. "I didn't look forward to it," she says. "I miss it, though. I liked the idea of it, and now I'm just aware of how much water we're wasting. In fact, our flush toilet smells worse than the composting toilet did. The fan worked pretty well."

Maintaining the system entailed raking the top of the mass every one or two months. Because she waited awhile to empty the composter, the material was often quite packed at the removal hatch. Ultimately, she cut a hole in the lower baffle through which she poked a stick to break up the mass. Leachate drained to a soak pit. Drum placed the composted matter in a compost pile, and waited a year before she spread it on her fields.

Although she stopped adding kitchen scraps, every spring, a few flies would appear. She hung a no-pest strip from the inside top of the composter. "That worked real well." She kept a bucket of wood shavings or composted bark next to the toilet, and suggested users "flush" with a cup of it.

One bane of the system for her was the chute, on which excrement would leave skid marks, and salt from urine would build up. "The salt wouldn't come off. That was a drag."

When she capped it off, three people offered to buy the composter. Unfortunately, the house is built around it.

Drum has since seen better-sited composting toilets, and wishes she had the same setup. "I was just thinking about the future," she says. "I wish we had thought about the access issue when we installed it." ✲

A Homegrown Passive Solar Toilet in Colorado

The Sunrise Farm in Loveland, Colorado, is a community-supported agriculture farm as well as an education center. In 1991, the farm built a Sunny John system designed by neighbor John Cruickshank to take the blackwater load off the farmhouse's overtaxed septic system. Another was to be built in 1998 to handle the increasing numbers of visitors and interns.

The toilet system is a passive solar composting/moldering toilet design that employs interchangeable perforated 55-gallon drums in a passive solar structure, and uses no electricity. It is located in its own building with sinks and showers that drain to a settling area, then flow to a reed and rush bed. During gray and cold weather, composting slows to psychrophilic levels, hence they call it a "moldering" toilet.

The building is constructed of clay on timber-frame with bamboo and willow reinforcement. The straw-clay roof (comprised of a rubber liner, a gravel bed, a filter and clay earth) is planted with wildflowers.

When serving 12 people, the barrels are changed every six months, according to David Lynch, the farm's director. When a barrel is full, it is taken to a storage site where grass and leaf clippings, leaf mold and some worms are added, then it is capped. Periodically, they add some water. After a year, the barrel is taken to the nursery, where it is used as fertilizer for nonedible plants. "We could use it earlier, but I like to let it compost as long as possible to be sure," Lynch says. "The staff is trained to handle this."

Flies and odors have never been a problem, he says. "If the wind turbine is working, it never smells," he says, adding that the one time in eight years it produced odors was when a bearing needed replacing. Screening on the toilet's air intake and a rubber gasket around the lid help keep out flies.

Early in the system's use, very little leachate was evident. After the number of users doubled, some sawdust had to be sprinkled on the floor of the composting area to absorb leachate.

He estimates that the cost of the materials to build a Sunny John is $600 to $800. Construction costs for the system and superstructure could run $2,000 to $3,000, if the work was hired out. Sunrise Farm conducts demonstration workshops for building Sunny Johns.

"You have to be willing to handle your own crap. Some people have an aversion to that. But once you do, you have an appreciation for utilizing your waste without flushing. It's rewarding work." ✲

Above: The Sunny John at Sunrise Farm is used by several staff members and visitors daily. Its walls are made of straw and clay, and its roof is planted with wildflowers.

Left: This photo of the interior reveals the straw that makes up the walls. Holes are drilled in the toilet seat lid to improve air flow.

(Photos by Judy Morris)

A Phoenix Back at the Ranch

Mike Evans and his wife, Joyce, have used a Phoenix composting toilet system at their home, Indian Rocks Ranch, in Saratoga, Wyoming since 1991.

Before that, they had a Clivus Multrum for 13 years, initially installing it because water was in short supply, and they didn't want to use a flush toilet.

Mike Evans says excess leachate would build up in the Clivus, requiring periodic draining, which he says was not easy in the older models. When their Clivus dealer told them about his new composting toilet system, the Phoenix, they decided to try it.

The Evanses have been content with the unit, which serves one dry toilet stool. They don't add kitchen wastes because, he says, "it's not recommended. They don't break down very well." Evans adds a handful of wood chips each time it's used. He used to add grass clippings, but found they matted down too much, and didn't decompose. Thanks to the fan, there are no odors, he says.

The Evanses put the composted material on the garden—when they have composted material, that is. Mike reports that he has removed only a few buckets of compost now and then in the past seven years. (Phoenix recommends removing material every two years to avoid potential compaction.)

Evans says he suspects that composting toilet systems like his might not be for everyone. "You have to be okay with the idea of dealing with human waste, and some people can't accept that. Even though you don't have to be directly involved with it, with a composting toilet, it's not a matter of 'flush it and it goes away.' And you have to do a few things other people [with conventional systems] don't do. But it's never been a problem for us and our family." ☼

Above: Mike Evans removes compost from his Phoenix composter.
Left: The family's dry toilet stool, with a pail of sawdust next to it.

(Photos by the Evanses)

Sustainable Demonstration House Saves Big on Water

The Alberta Sustainable House is the home of architects Jorg and Helen Ostrowski. Open to the public periodically, it features solar heating aspects, lower-impact construction materials, a rainwater collection system, a graywater irrigation system, a new Phoenix composting toilet, a worm composter, yard waste composters, solar ovens, and other eco-tech devices. Jorg says the home saves thousands of gallons of water annually over a conventional home. ☼

The graywater system, and, right, the Phoenix composting toilet. (Photos: Jorg Ostrowski)

A Homemade Self-Contained Composting Toilet in Nebraska

For their Nebraska home, Sharon Eby-Martin and her husband Gerry constructed for about $150 a small plywood box version of an Envirolet-style self-contained composting toilet. It features a metal grate that is moved back and forth to sift finished compost into a removal drawer.

Sharon reports that it served them well until their family grew from two to five. "This system works quite well when it comes to low odor and convenience," she says. "However, the toilet cannot keep up with that many people. It fills with urine and the vent holes clog."

The system is basically a 2'x2'x2' box (open at the bottom) set on a 2'x2'x1' high box (open at the top) with a piece of sheet metal drilled with many large holes and laid flat between them. Under the sheet metal are three PVC plastic aeration pipes drilled with holes, which face down. On top of the sheet metal is an old grate or wire oven rack which can be slid back and forth (with handles outside the unit) to help mix the material and sift the drier, more processed material down into the removal drawer. It features a 4-inch PVC exhaust pipe and is heated with a 100-watt bulb surrounded by a half-cut-off coffee can (as a splash shield) at the bottom rear of the box. The interior was coated with epoxy.

Jeremy Eby-Martin demonstrates the family's site-built composting toilet. *(Photo: Sharon Eby-Martin)*

They add a cupful of sawdust (free from the local lumberyard) after every use. They plan to construct a 55-gallon drum batch system for their new strawbale home. ✲

Life with Composting Toilet: The Feel-Good Factor

Colleen and Glenn Caffery and their two daughters had lived with a Phoenix composting toilet system for only six months in their new western Massachusetts home when we spoke with them. They could have installed a conventional septic system for their house. "But, from the beginning, we knew we wanted to do something like this," Glenn Caffery says. "We wanted to be true to our principles."

Their system features one toilet, a SeaLand micro-flush. They note a little more "streaking" with the SeaLand than with a standard toilet. To prevent this, they sometimes place a few pieces of toilet paper on the inside bowl. "But it's a great toilet. We're happy with the aesthetics of the system. We really wanted a flush toilet so it would be mother-in-law friendly! But the biggest kicker for us was flies," he says, noting that the flush toilet prevents any flies that might be in the system from entering the house. So far, he has noticed only a few tiny gnats outside the composter, and has taped some fine screen onto the air intake.

Glenn adds planer shavings from his workshop, and rakes the top of the pile once a week. He plans to remove the end-product in a year, and bury it per Massachusetts law, perhaps near tree roots. "I treat this seriously. We're not going to put this on our tomato plants," he says. "Why take a chance?"

"It is something you have to be involved with, and maybe that wouldn't be acceptable to some people," Caffery says. "It actually feels nice to have this in the house. There's a connectedness to nature's cycle that we really like. And we're not producing anything that has to be pumped out and taken away to be treated." (Note: A professional maintainance service would cost about $30 per visit. –Ed.) ✲

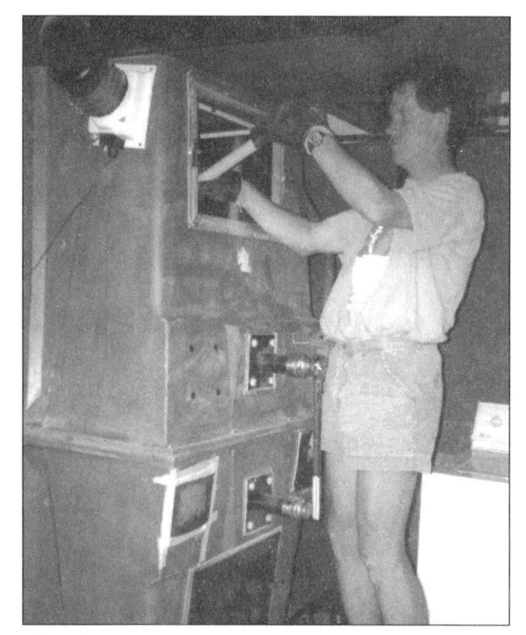

Glenn Caffery rakes the contents of his Phoenix system.

A Carousel Composter in the Kitchen Pantry

In 1981, when Patti Nesbitt was designing her passive solar Virginia house, there was no question that it would have a composting toilet. "The bottom line for me was not to poop in my drinking water," says Nesbitt, a former consultant to the EPA on small flows and alternative wastewater systems and a contributor to the 1977 book *Goodbye to the Flush Toilet* (Rodale Press, out of print).

In her home, a Carousel composting toilet system serves one dry toilet on the second floor. Installing the system allowed her to have a reduced-size septic system, which manages the rest of the home's wastewater, including a standard low-flush toilet on the first floor.

What's unusual about Patti's system is that the composting reactor is in her kitchen pantry. That location allows it to be directly underneath the upstairs dry toilet stool *and* get the benefit of the ambient temperature in the house, improving the processing rate. A leachate line drains to her septic tank. Nesbitt, her husband and two children use the toilet, which is on the second floor near the bedrooms.

Having a composting toilet reactor in the pantry (she stores kitchen cleaners on top of it) has not been a detriment, she says. "There's so much retention and no odor. I always thought I'd put a door on the pantry at a later date, but it's been 16 years that we've lived without one."

Patti empties her Carousel every two years, usually emptying two or three chambers at a time. "It's dry compost, and sometimes I have to chop it up and pull it out in chunks," she says. "It's no dirtier than working in the garden." She usually removes six to seven five-gallon pails full of finished compost per chamber, and mulches her flower beds with it.

Patti is comfortable with the safety of what she takes out of the composter. "The retention time in the Carousel is so long, it shouldn't be a problem. Beyond 12 months, it's not an issue. Parasites, bacteria and viruses get broken down. The ovum cannot survive outside the body for a long time. You don't get that in a [continuous composter]. It assures a safety factor."

When she removes material, she also pours some soapy warm water into the outer case to flush out any salts that may have built up from urine. For additive, she keeps a container full of untreated wood mulch (purchased from the garden supply store) next to the toilet stool. Her formula: "a scoop per poop." Some years, a few gnats appear. She treats the problem by

Top: Patti Nesbitt and her Carousel in her kitchen pantry. Above: Removing compost from a compartment to give the authors a sample. (Usually she would have a bucket and spade on hand.)

adding more wood mulch and Precor, a juvenile growth-inhibiting hormone.

Patti notes a little-considered advantage of the dry toilet stool: No sound. "You can get up to use it in the middle of the night, and there's no flushing sound. With a water toilet, everyone can hear you!"

Allowing her children to use the toilet was never a problem. "I was always worried that my babies would fall down it. They never did." When her children were young, she threw diaper waste into the toilet, although the diaper liners do not compost. "I'd pull them out when I was emptying it," she says.

"I really like it. But it's not for everybody, because you have to be willing to manage it, and some people just won't," she says. "It's a wonderful toilet, a great system. It makes sense engineering-wise, biology-wise and environmentally." ✲

A Clivus Multrum in a Cohousing Community

Lee Ketelsen of Acton, Massachusetts, keeps her Clivus Multrum (model M-3, installed 1996) in a room in the basement with her solar hot water heater tank to help keep both systems warm.

Ketelsen, a professional water conservation advocate, says the composting toilet system ultimately saved her money over the cost of a conventional septic tank and absorbtion field system, as it allowed her to install a reduced-size leach field and save on water costs.

Ketelsen's unit is serviced regularly under a Clivus service contract. Every few months, a service person removes material and troubleshoots the system. Under Massachusetts law, leachate must be removed by a septage hauling company (or treated in some other approved way). If she had to do it all over again, Ketelsen says she would locate the unit closer to a window or door, so the septage hauling company would have easier access to the leachate container. Leigh and her husband rake the composting mass, leveling the pile at the top under the toilet and raking it forward at the removal hatch, to improve the throughput.

According to Clivus New England, the unit was a little undersized for her family of four (the company has since introduced a slightly larger model, the M-10).

On the day of our visit, some gnats buzzed around a light (but not in the bathrooms). She planned to sprinkle some pyrethrums in the composter or spray the bugs with soapy water. She adds wood chips periodically as additive. ✲

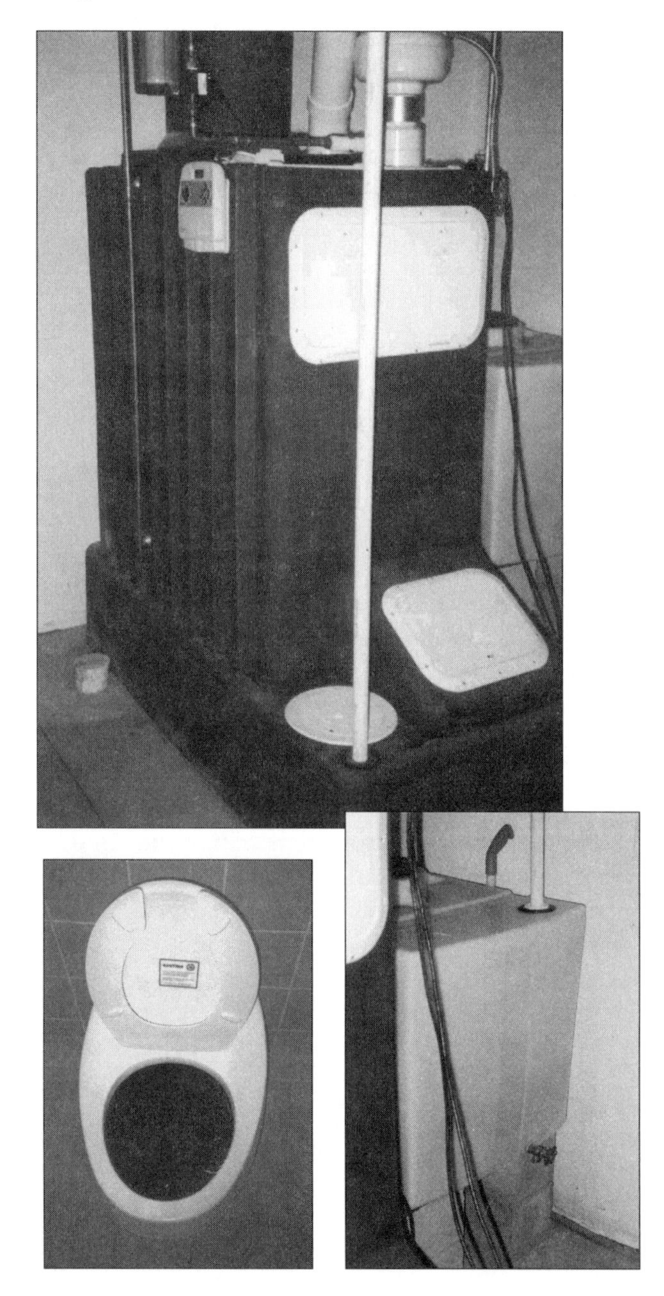

Clockwise from top: The Clivus Multrum M-3. The leachate collection tank. A dry toilet in the first floor bathroom.

A Clivus Minimus in Australia

Leigh Davison, a water and waste studies lecturer at Southern Cross University in New South Wales, Australia, built his Clivus Minimus in 1982 for the house he shares with his wife. He suspects that his Minimus, a concrete-block sloped-bottom floor composter, may be the first in Australia. He and his wife live in an ecologically oriented community on 260 acres in Lismore.

There are a total of five Minimuses, three Farralones and a Wheelibatch in his community. A larger Minimus serves the community house. Davison reports: "It gets a lot of use, including a 10-liter bucket of food scraps every day. No real problems. There was a colony of cockroaches in there, but that has been controlled using sprinkles of borax or sprays of pyrethrum (a botanical insecticide that breaks down within one day). The pyrethrum did not seem to affect the vitality of the heap. It makes me wonder about the suggestion that people taking antibiotics should not use compost loos. Certainly this does not seem to affect the larger loos such as Minimus and Farallones. I would imagine that with a large heap, any antibiotic damage can be rectified by recolonization from an untouched zone. [See the Appendix for more on the effects—or lack of effects—antibiotics on composting systems. –Ed.]

"The Minimus is a wonderful totally low-tech no-moving-parts device that works fine in our subtropical climate. However, I have had good reports of them working in cool temperate climates in Victoria and Tasmania. As with any alternative technology, I believe that you start with an idea or a prototype and then depend on human ingenuity to adapt it to local conditions."

Davison usually periodically adds straw, although he and his wife have been experimenting with "adding a twisted-up page of the *Sydney Morning Herald* and dropping it down after each visit. This seems to work well as we have been doing this for a few months and no nasty smells as yet. Only other maintenance is removing material a couple of times per year and knocking the top off the mountain with a plank kept just outside the door." ✲

Above: Leigh Davison removes compost from his Clivus Minimus, conveniently located under and outside of his home. Left: The removal hatch. Above, far left: Davison's bathroom is a large, sunny room, an enjoyable setting for passing a few moments of the day... (Photos: Leigh Davison)

Compost Happens in Olympia

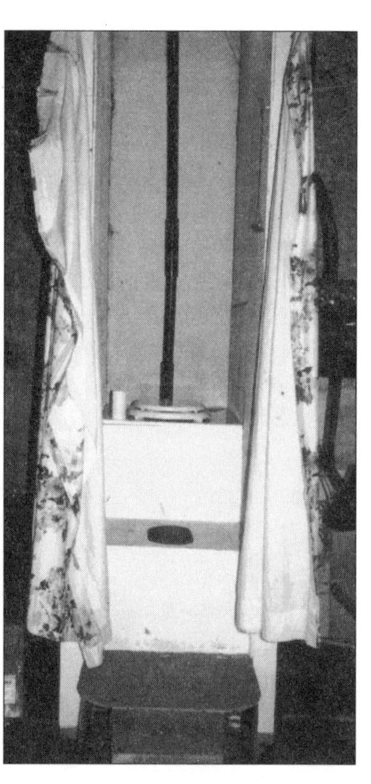

Steve Rentmeester of Olympia, Washington uses a batch composting system consisting of a set of interchangeable 55-gallon (208-liter) drums (one metal, two plastic) with urine diversion. Seven people use the system, which supplements a failing septic system.

A lower seat funnels urine to a five-gallon jug. "The funnel is attached to a hose, which has a J-shaped bend in it to create a water (or urine) seal," he explains. "This almost completely removes the smell associated with urine collection." Rentmeester mixes the urine with equal parts water, and uses it to fertilize trees.

Above left: Steve Rentmeester empties a composting drum after five months of processing. Above right: A homemade urine-separating toilet. Left: Rentmeester periodically checks the temperature of his composting drums. (Photos: Abby Thurston)

There is a vent behind the upper seat and a roof vent in the top of the garage were the toilet is located. When a drum fills up, he covers it with a plywood cap, which he finds breathes better than metal. "My ideal lid would be similar to a chinaman's cap, like what you would use to cover a vent extending from your roof with a screen to keep out the bugs," he says. Unlike many drum batch systems, his drums contain no aeration devices.

Rentmeester has measured temperatures as high as 160°F at the top center of the composting mass after two weeks. The temperatures were far lower in the center and bottom of the composter. (Ambient temperature averages 61.5°F/16.4°C) Both active and full composters are kept in his garage.

He says several people in Olympia have similar toilet systems. Unlike their systems, which produce compost after one to one and a half years, Rentmeester says his only takes four to five months. He suggests two reasons: "Red worms. I always add a couple of shovels full of composted horse manure, which is crawling with red worms [and also inoculates the compost with microbes], to the bottom and top of every drum. And I use a very coarse sawdust, planer shavings from the local hardware store."

Rentmeester's system has not been approved by a local health agent; however, he knows of owners of similar composters in the area who do have permits. ✲

Composting Cooperatively

The Olympia-based Green Frog Co-op began to build and test 55-gallon drum batch composters. The group hopes to sell the composters and offer a service plan. Green Frog hopes that the composters will be used to supplement the overtaxed septic systems of homes around Puget Sound, as well as help ease the load on Olympia's central sewage plant, which is suffering growing pains.

One of Green Frog's experimental composting reactors. (Photo: Philip Vandemann)

Solviva's Composting Toilets

Anna Edey is widely known for her bio-intensive solar greenhouse at her home on Martha's Vineyard, an island off the Massachusetts coast. Heated in part by keeping chickens and rabbits, the greenhouse features raised beds, hanging planter tubes, water walls and a variety of other innovations. Edey's innovations in lower-impact living are detailed in her colorful and informative book, *Solviva: How to Grow $500,000 on One Acre & Peace on Earth* (Trailblazer Press, 1998).

Edey uses two types of composting toilet systems in her home: a small alternating-container system and a two-chamber system with a flush toilet.

The 20-Gallon Alternating-Container Composting Toilet

In the first system, a 20-gallon pail is used as a collection container in an attractive polished mahogany cabinet toilet stool. Adjacent to the toilet seat is a hinged compartment containing what she calls "bio-carbon," a mixture of sawdust or leaf mold and some compost that is added after each deposit. A leachate drain line is attached to each bucket, and flows to an outdoor yard waste composter. Edey usually collects her own urine in four-gallon jugs via a homemade urinal. She dilutes it with 10 parts water and adds it to her compost pile and plants. Although this part of the toilet system is unvented, Edey says there are no odors. When the pail is three-quarters full, which usually takes about three weeks for one person and occasional guests, Edey takes it outside to a sunny area. There, she adds some finished compost to inoculate it ("like adding yeast to bread," she says), and caps it. The material inside appears very composted after about three weeks in the summer and five to seven weeks in the winter. To more fully compost it, she empties the composting material into an outdoor classic three-bin composter, and adds more wood shavings and garden clippings.

Edey says those who use this design might consider building a "solar shed," a small shed with a south-facing large-windowed wall, in which to place the buckets. Also, if a bathroom is against an outside wall, a removal hatch/door can be added, so that the collection container could be removed from outside and taken directly to the solar shed. (One could also build a solar shed against the wall of the bathroom.)

The Double-Chamber "Compostfilter" with Flush Toilet

Edey knows that many Americans will only accept a flush toilet system, so, in 1995, she installed a composting system with a standard 1.6-gallon (6-liter) flush toilet. The toilet flushes to a double-chamber composting box located outside on a sunny side of the house. It is three feet tall by four feet wide, insulated with two-inch foam board and made watertight with a plastic lining. Inside, it is divided into two compartments to which the flush pipe can be directed. It contains several drainage holes that drain to Edey's "Greenfilter" system. To start up the composter, she filled the containers halfway with wood shavings, compost and about two pounds of earthworms. Edey says the temperature has never dropped below 55°F/12.8°C, even on the coldest winter days, and, most surprising, the first chamber never seems to fill up. When she finally decided to switch to the second chamber, the material in the first appeared to completely compost in one week. She calls this reactor a "compostfilter," as it "filters" the solids from the toilet and composts them. However, in three years, she has removed only two five-gallon (19-liter) buckets of compost—"and that only because I needed the compost for the garden," she says.

Greenfilter

The leachate from the composter drains through a sloped pipe to a series of "Greenfilters" designed by Edey. These are shallow growing beds lined with plastic membrane and filled with wood shavings, compost, sandy topsoil and healthy plants. From there, any remaining liquid drains to a second bed, and then to a 20-gallon (76-liter) pump chamber equipped with a float switch-controlled sump pump that periodically pumps 15 gallons (57 liters) of effluent into a perforated pipe in a third Greenfilter that includes six inches of wood chips. After that, it perks through the subsoil in a landscaped area. Tests of the effluent show a 90 percent reduction in BOD, Edey says.

Compost Pasteurization

To use composted human excrement on food crops, especially greens and root crops that touch the soil, Edey recommends pasteurizing it to be sure that any pathogens are killed. (Ultimately, though, she recommends using only composted animal manures on edible crops.) Edey built an experimental solar pasteurizer, a simple wooden box, about three feet long by two feet wide and six inches deep. The bottom is lined with one inch of rigid foam insulation. She fills it five inches deep with composted human waste, and inserts two min./max. thermometers. She covers it with a double-glazed lid, and tilts it to face the afternoon sun. On sunny days, it can reached 150°F/65.5°C inside within three hours. At night, it cools off. Over a series of sunny days, she finds the pasteurizer can reach temperatures over 170°F/76.7°C throughout the compost, "more than enough to kill the toughest pathogens," she says. "With concentrating collectors, far higher temperatures could be attained."

Edey's graywater system is a long pipe that runs first to a shallow area filled with wood chips and a shallow leach field planted with grasses, wildflowers, dogwood trees, honeysuckle bushes and even some loosestrife. To prove that this is a feasible system for mainstream Americans, Edey uses standard soaps and cleaners, even though these are not best suited for use by plants. "I've seen no bad effects. The plants are thriving," she says.

Her ecological wastewater systems are usually well received by guests, who are inspired by the idea that their "wastes" will ultimately feed plants—and that the system does not smell. "When people see this system, they get so excited," she says. "They ask a lot of questions."

For her, developing these systems was a matter of trying out a concept. "I get ideas, I try them, and many of them work out better than my highest expectations."

Her book, Solviva, *is available for $35 + $3 shipping from Solviva Company, RFD 1, Box 582, Vineyard Haven, MA 02568. Solviva@vineyard.net* ☼

Putting the Sun to Work in New Mexico

John Cobb, M.D., M.P.H., a professor emeritus at Humboldt State College in California, built his "Sol-Latrine" in 1994 at his home in sunny Corrales, New Mexico. Housed in a passive solar building, it is a batch system with interchangeable 55-gallon (208-liter) barrels with an evaporation chamber. The whole structure cost about $1,000.

In his system, a tank of solar-heated water is used to wash hands, then used to flush the toilet. A warm water shower is also possible. Wash water flows to a solar pasteurizer designed by Cobb and is then released to the vegetable garden.

Cobb says the system works well, with no odors. And thanks to the dry New Mexican climate, usually all leachate evaporates away, except when too much flushing water accumulates. When that happens, a simple bilge pump is used to pump the leachate into another compartment to evaporate. The system uses no electricity. ☼

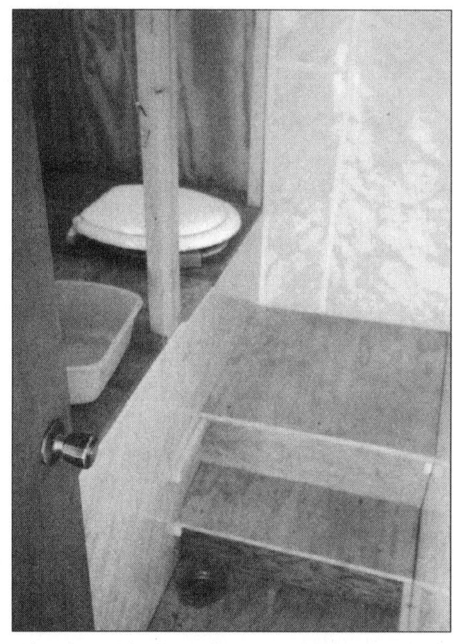

Above: Inside the Sol-Latrine. Right: Outside.

A Humanure Toilet System in Oregon

Ole Ersson and his family have a standard flush toilet in their downstairs bathroom, typically used by guests in their Portland, Oregon home. But upstairs is their "Humanure-style" five-gallon bucket toilet, a system Ersson calls his "$1 toilet" because nearly all of the materials were scrounged for no cost.

Ersson, who is a medical doctor, was inspired to install the system after reading Joe Jenkins's *The Humanure Handbook* (Jenkins Publishing, 1994), which details Jenkins's simple sawdust toilet system. Ersson purchased an oak toilet seat at a garage sale, and screwed it onto the top of a five-gallon bucket. He then cut off the bottom three-quarters of the bucket. This becomes the fixed toilet stool. The active collection pail, another five-gallon pail, is placed underneath it, so that the edges are inside the active pail. The family adds a cupful of additive after each use. Every two days, Ersson takes the pail out to an active three-bin composter, made of shipping pallets. He empties it into an active bin and covers it with kitchen scraps and three to four inches of shredded wood chips and brush. Just like a garden composter, it is turned periodically. Ersson has found that the piles reach temperatures high enough to kill pathogens in the warmer months. He usually lets the material actively compost for four to six months before using it.

Maitri Ersson adds shredded plant material to the sawdust toilet.

Ole Ersson empties the contents of a sawdust toilet system collection bucket in his three-bin composter made of pallets. (Photos: Ole Ersson)

For additive, the Erssons use chopped-up leaves and brush, which is brought to his house free of charge by a tree pruning company. "We can get all we want delivered to the house," says Ersson. In fact, this mix often contains evergreens, which provides "a wonderful frangrance," Ersson says.

"There's no offensive odor. We've never had a complaint from the neighbors," he says. Ersson says, however, that he may purchase an in-house composting system, so that he can get municipal approval for his excrement-composting efforts.

At one time, the Erssons even used their compost to help heat their water in a solar straw-bale greenhouse, running water tubes through 30 cubic yards (23 m³) of compost. They found they could heat water to as high as 130°F/54°C. (The family has subsequently dismantled the greenhouse to reclaim the yard for gardening and recreation.)

Ersson has a Web site featuring many of his experiments in lower-impact living:

http://www.rdrop.com/users/krishna

Two Sun-Mars in a Canadian Cottage

Richard Kraushaar has had a Sun-Mar Excel NE (non-electric) at his lakefront vacation cabin in Canada's northern Georgian Bay area since 1988 and a Centrex NE with SeaLand micro-flush toilet since 1997.

Both are located in a small building next to the cabin. "We use the Excel until it's relatively full, then let it sit a week. Then it's odor-free and easy to empty," he says. "Sometimes we go six weeks before we empty it, sometimes every two weeks if we've had a lot of people over. It's certainly not obnoxious at all to empty. There's no smell." It takes him about 40 minutes to remove material. He says that anywhere from one to seven people use the cabin from May to October.

What he takes out of the composters is usually not fully composted, so he lets it continue processing in one of two outdoor composting piles. One is for the current year; the other is the previous year's compost, which he uses on his flower plants. "We've got some amazing flowers up there," he says. "There are more blooms and they're healthier and much stronger than what grows at home in Ohio."

Above: Richard Kraushaar tends his flower beds, where he trenches in the humus from his composting toilet system.

A Sun-Mar Centrex outside of a cottage

The cabin's graywater drains to a leachfield.

For additive, he recently switched from peat moss to pine wood shavings from the pet store, because the peat was sifting out of the rotating drum. He also adds stale bread to help absorb liquid and encourage composting. He stopped adding kitchen scraps to the composter when doing this drew fruit flies. Every other day, he rinses any skid marks off the inside of the bowl of the Excel. Leachate and flushing water from the micro-flush toilet evaporate. He installed solar-powered fans on both units, and noticed an improvement in the thoroughness of the composting. Kraushaar says he finds that rotating the "bio-drum" every two days offers the best compost results and discourages flies.

He is pleased with the systems, and has had no complaints from visitors. "People are astounded when they go into the building and there's no odor," he reports. ☼

A BioLet in a Texas Cabin

George Kronke installed a BioLet XL in his Greenville, Texas cabin in 1996. Located "in the middle of the woods" on 70 acres, the cabin has electricity but no water service. Water must be hauled in, and water drained from the sinks is brought to town.

He and his wife and occasional guests use their BioLet nearly every weekend. For gatherings of up to 12 people, Kronke says he provides a "porta-potty" to supplement it.

For additive, Kronke uses a mix of vermiculite and Canadian peat moss, and is playing with the ratio: "We're just trying to figure out how to get the starter right so the compost mix is best." He empties the toilet every two and a half to three months, taking the tray outside, and spreading the material around trees. He says the material looks composted and has no odor. "I never mind doing it," he assures.

Kronke has encountered flies only once—"just a a couple"—and odors only when mud daubers built a nest in his exhaust pipe. Placing a screen over the pipe, however, blocked air flow, so he monitors the pipe.

"It works really well. It's a fabulous little machine for what it does," Kronke says. "It almost looks like a regular toilet. A lot of people thought we were nuts when they heard about us intalling one. But it's a great system for places without water." ☼

An Office Building Goes Sewer-Free in Vancouver

Built in 1996, the C.K. Choi Building in Vancouver is one of a few office buildings that employ composting toilets and waterless urinals. Built to serve up to 300 people, the three-story building is home to the University of British Columbia's Institute of Asian Research.

The 30,000-square-foot building uses just 500 liters of water per day. Comparably sized conventionally designed buildings consume an average of 7,000 liters daily (650 for irrigation, 500 for handwashing and 6,500 for toilet flushing), according to University reports on the project.

Three dry toilets on each of three floors connect via stainless steel chutes to five Clivus Multrum composting toilet systems in the building's basement. Rainwater from the roof is collected and held in a 7,000-gallon tank below a staircase. It is used to irrigate the site's landscape, which is bordered by thirsty ginkgo trees. Leachate and graywater from washbasins are filtered and pumped to a 300-foot-long outdoor contained planter bed planted with lillies. The final discharge is used to irrigate plants. A test by the city of Vancouver of the fecal coliform counts of this end-product showed that it contained less than 10 CFU per 100 mililiters. (to compare: swimming is permitted in waters with up to 200 CFU per 100 mililiters.).

(Photo by Mike Sherman, Matsuzaki Wright Architects, Inc.)

Designed by Matsuzaki Wright Architects of Vancouver, the building also includes reused materials, photovoltaics, and natural lighting and ventilation features.

Architect Eva Matsuzaki says the decision to use composting toilets resulted from a meeting of planners, engineers, and university staff members to plan the design of the building. "Out of that came the decision to use half the electricity and half the water [of a conventionally designed building]," Matsusaki says. "And one of the decisions was to have no sewer connection." She anticipated some regulatory hurdles from the city of Vancouver, but the system was approved with relative ease.

She estimates that the building's composting toilet system cost about $35,000. Matsuzaki says the toilets have been well received after an initial period of skepticism. The only problems were a brief period of odors when a fan stopped working and again when the level of the composting mass dropped below the air-intake vent. Also, a pitcher of water is placed next to the waterless urinals so users can give them an occasional rinse when urine pools.

For the first few years, the systems are to be raked, emptied and checked by Clivus Multrum through a service contract. In the future, the maintenance staff will take over. Every day, the university maintenance staff wipes down the toilets and adds a can of wood chips or bark mulch to each toilet.

"Vancouver's sewer is straining its capacity, and there's controversy about how well things are treated before they are discharged into the sea," says Matsuzaki. "We wanted to show that this could be done in an urban environment." ✲

Composting Toilets in Communities

There are few communities and developments with composting toilets in every home. Most are in eco-villages—planned communities built along ecological design principles. (Exceptions include a few communities in Asia and Latin America described in the following section, "Composting Toilets in Development Programs.") Occasionally, they are considered for more conventional developments, but usually the classic issues of acceptance, maintenance and regulations factor them out. That's too bad, because besides the cost-saving potential, this application offers the best way to assure that users are supported and the systems are maintained in a planned and accountable way.

Sweden

Not surprisingly, Sweden, the origin of so many composters, is home to some community composting toilet initiatives. In northern Sweden, the municipality of Tanum mandated that all homes must switch to dry toilets by the year 2000. A coastal area too rocky for effective septic systems, Tanum is home primarily to fishing industry and vacation cottages.

In Norrköping, the Ekoporten (Eco-House) is a three-story building with 18 apartments. In 1996, it was outfitted with water- and energy-saving features, including urine-diverting flush toilets. Urine and toilet water drain to a tank, where it sits for six months, then is removed for use by a farmer. Feces are separated from toilet water in a separator and composted with kitchen garbage. The compost is used on the site's gardens. A UV water purifier cleans the toilet-flush water, which is then combined with graywater, filtered, and drained to a nearby constructed wetland system. (Drangert, 1997)

Several reports on the performance and acceptance of the systems in the Ekoporten were published in 1999.

Other community projects underway in Sweden that involve urine separation and either on-site or central composting include Hammarby Sjöstad, Bergjön and Hamburgsund.

An eco-village in Toarp was the subject of a critical study that followed this village's experiment with composting toilet systems. Three types of composting systems were installed in 37 houses. Three years later, when homeowners were given the option of installing flush toilets, all but four of them chose to switch. (The Vera Carousel had the highest satisfaction rate.) The authors reported that the homeowners did not always understand how to care for the systems, and did not add sufficient additive. The composters were poorly ventilated, hard to access and drew flies. The toilet systems were installed as a public initiative, and not chosen by the home occupants. This is nearly always a major cause of lack of acceptance.

Another eco-village, Solbyn ("Sun Village"), undertook much the same composting toilet installation plan, with much the same results. A coordinator of this initiative says that poor design, installation and explanation of the operation ultimately led to many of the systems being replaced with conventional systems (again, the Carousel rated high).

Sources:

Jan-Olof Drangert, "Perceptions, Urine Blindness and Urban Agriculture." Ecological Alternatives in Sanitation, Proceedings from the Sida Sanitation Workshop, 1997.

I. Fittschen and J. Niemczynowicz, "Experiences with Dry Sanitation and Grey Water treatment in the Ecovillage Toarp, Sweden." *Water Science and Technology*, 1996.

Nils Nyberg, in correspondence, 1998.

An Aquatron separator and composter of the sort used in the Ekoporten in Sweden.

Australia

Crystal Waters, an eco-village in northwestern Australia, uses various innovative composting toilets and graywater planter systems, according to spokesperson Max Lindegger. Dowmus Resource Recovery tests its composting toilet systems there.

Lindegger says that the community has dealt with and overcome problems of insects, moisture control and fan breakdowns. "Possibly the biggest problem is user acceptance of the dry composting system. People don't seem to enjoy looking down a dark hole."

Lindegger's home features a Dowmus system into which *all* of the home's blackwater and graywater flows—including shower water, washing machine and dishwashing water produced by six to eight people. They also add organic wastes such as vegetable scraps, dead animals, bones and paper into the system. "Worms (red wrigglers and tiger worms) and microorganisms process the material at the rate of about 1 kilogram per kilogram of worms," he says. "This no doubt is related to climate, and I believe that the ideal temperature is above 19°C.... The outlet from the 'wet' composting tank constantly discharges into a storage tank which is below ground. The liquid is nutrient rich, but I am told it is pathogen free.... I use the water to irrigate fruit trees, and the results to date have been excellent. As I use fine sprinklers, I have some blockage problems, and plan to install a sand filter before the storage tank to filter out organic matter." (Using a composting system as a biological solids filter in this way makes a lot of sense, however, the powerful flow rate of so much graywater might be too strong for sufficient detention periods, resulting in inadequate microbial processing. We hope to hear more about this. -Eds.)

Max Lindegger
Global Eco-Village Network (GEN) Oceania
59 Crystal Waters, MS 16, Maleny Qld 4552, Australia
lindegger@gen-oceania.org
http://www.gaia.org/thegen/genoceania/index.html

In Lismore, in the northeast corner of New South Wales, several homeowners have installed composting toilets in the past 20 years. The subtropical climate is ideal for effective composting.

To regulate these systems, Lismore's city council issued guidelines for composting toilets in 1997, and approved 89 of these composting systems on a trial basis.

The experiences of 29 of the owners of these composting toilet systems were compared in a report compiled by Rachel Pollard and Leigh Davison, a student and a professor respectively at Southern Cross University, with an environmental health officer. Of these, 24 had constructed their own systems, most using the Clivus Minimus (inclined vault) design; the rest were Wheelibatch (interchangeable roll-away garbage bins), Pickle Barrel Batch (interchangeable 55-gallon drums) and Farallones Batch (double-vault). The manufactured systems were Dowmus (single round chamber), Rota-Loo (rotating carousel style) and Natureloo (alternating round chambers). All of the owners rated their systems "satisfactory" or better.

Problems they listed included fan breakdowns (sometimes due to using fine sawdust) and occasional odors during peak usage. Due to a high number of variables—from number of users to design aspects—the authors noted that they could not make firm conclusions about the general effectiveness of each kind of toilet. They report that each design appeared to have its strengths and weaknesses: "The smaller batch systems require less space but require more maintenance [monitoring and changing the containers]. The larger systems were more forgiving to peak usage."

Design and maintenance adaptations included changing the types of additive used, increasing the incline of the Minimus to better drain leachate, and changing the door design of the Farallones system, so that material did not tumble out when it was being inspected.

They concluded, "Owners showed a high degree of satisfaction with the technology. Any management problems that had arisen were overcome relatively easily by minor structural or behavioural modification. Owners also showed a generally high degree of interest in the technology and a willingness to experiment in order to optimise system performance."

Source:

"Effectiveness and User Acceptance of Composting Toilet Technology in Lismore, NSW," Rachel Pollard, Applied Science student and Leigh Davison, Lecturer, Southern Cross University, Lismore, NSW; Tony Kohlenberg, Senior Environmental Health Officer, Lismore City Council, NSW.

United States

In the United States, some eco-villages and cohousing developments report plans to install composters in all homes. Stay tuned. Waterfront areas are also looking to the compost solution. In upstate New York, a plan to install composting toilet systems in 100 lakefront cottages began in 1998 by piloting three systems. The cottages are currently served with a bucket collection system by the Syracuse Water Department. In some seacoast areas of Maine, property owners who replace their flush toilets with composting toilet systems can receive rebates from the state, which hopes to prevent further pollution of the clam flats. ✲

A Northern California Town's Management Plan

In a small Marin County town in California, a group of homeowners found that being accountable for successfully maintaining their composting toilet systems was the only way they could avoid costs of more than $30,000 to install improved septic and mound systems.

Located on the rural outskirts of the town, these seven homeowners had built composting toilet privies using the Farallones design in the 1970s.

"This was back when Jerry Brown was governor," explains Pat, one of the homeowners. "There was a state directive about how they should be built, and those systems built using the Farallones design had tacit approval." Over the years, however, composting systems fell out of favor with some regulators.

Their homes were built on communally owned agricultural land, across which flows a creek. The homeowners installed composting toilets because they did not want to use farmland for leachfields. When one of them built on to her house in 1993, she upgraded her system—and triggered a county review of the systems in the area.

Marin County officials told them that their composting toilet days were over: They had to install advanced septic systems, specifically, mound systems, to prevent pollution of the creek. The estimated cost of this system would range from $30,000 to $50,000 per home. Pat checked with other county environmental health departments, and found that they did not regulate composting toilets.

Pat then joined forces with the homeowners and the community public utility board to figure out a way to prove that the composting toilets were being maintained properly and not polluting the ground or water. With help from utility board member Eugenia McNaughton, an EPA environmental scientist, they developed a program, which included holding meetings with composting toilet-owning homeowners, as well as administering for homeowners to demonstrate their understanding of the operation and maintenance requirements of their systems.

Pat says county officials seem to be appeased by the program, but she is uncertain about the future. "The county seems to have turned its back on the issue," she says. "It's been a long struggle here. We're living on the rural fringes of the richest county in America. People almost lost their homes over this issue." For that reason, she asks that the name of her town not be used until the matter is firmly resolved.

Their first step in putting the plan into action was to test all of the systems' end -product for fecal coliform. "We found some privies had coliform, but four or five of the houses were coming out with a very clean product." Pat and Eugenia wrote a simple 12-page manual with simple standards for design and instructions for maintenance. They conducted workshops with homeowners and other interested locals, demonstrating proper maintenance and handling of the humus. In a follow-up test, the results were better, she says. One of the systems is pumped by a septage hauler.

Pat says that in the systems with higher fecal coliform rates, owners were not stirring the material, not adding enough dry material, and urinating in the systems, resulting in too much accumulated leachate that led to anaerobic conditions. With the Farallones design, excrement in the active chamber is shoveled into a "curing" chamber to continue processing. The shoveling helps aerate it, she says. Pat and some of the other owners even allow a third phase, shoveling it out of the second chamber and into an outdoor composting bin. Then, per law, it must be buried six inches underground, although Pat feels it is best spread thin around the root zones of trees. The homeowners have small septic systems to manage their graywater.

Pat herself adds a scoop of sawdust to hers and stirs it once a week. The back of the system is vented; the idea is that air comes through through the back vents, over the pile and up and out through a vent pipe.

Pat also helped train utility company personnel, and

The removal doors to Pat's Farralones-style composting toilet system. (Photos by Pat)

attends monitoring visits with the utility department staff person. One or two times a year, Pat and the utility staff person visit each system. The monitoring visit is also to ensure that the same homeowner is maintaining the system and operating it according to the guidelines.

"Our goal was to make operating these kosher and to really teach people how to make these work," she says.

Pat knows several people who successfully use electric composting toilets to supplement their septic systems. "I hear they work great," she reports. "A lot of people around here have them, because all around are failing septic systems."

Pat thinks composting toilets can be used more widely. "Our culture is afraid of poop. We overprocess it and kill it. In a healthy home, a lot of this is overkill. We have dogs pooping all over the place, and no one seems very worried about that."

Eugenia notes that the increasingly high cost of wastewater treatment plants makes alternative systems worth considering. In Los Osos, California, the estimated cost of a treatment plant is $71 million. Serving 10,000 residents, it will cost $7,100 per person—not including costs for operation, maintenance and hook-ups. "Composting toilets deserve more consideration than they are getting," she says. "Here, the monitoring plan resulted in a group that gets together and works out their problems." ✲

Pat's site-built toilet stool, tiled in terra cotta, includes a built-in compartment for additive.

A Composting Toilet at the Office

A secondhand BioLet XL self-contained composting toilet serves the office of the Center for Ecological Pollution Prevention, where this book was written.

It is used by two to four people daily. Because it is used only during the day, it mostly manages urine. We routinely add plant trimmings, and every two or three weeks, a cupful of additive (soil-less growing media) and a tablespoon of composting microbes. The removal tray is emptied every six weeks, and occasionally some of the contents from the top are vacuumed out with a wet-dry vacuum because the material dries into pieces too large to pass through the metal grate to the removal drawer. This over-drying effect is due to the powerful fan—adding a fan-speed controller would allow adjusting it to exahust odors while only minimally drying the contents. Toilet paper, too, tends to dry out before it can decompose, and wraps around the turning arm (which ultimately jammed the system and broke a shear pin). To remove it, we spray it with water and pick at it with a knife to loosen it. A better, preventive solution is to surround the turning arm with a small section of two-inch plastic pipe cut in half lengthwise. We rarely turn on the system's heater, as that would dry the contents further. Occasionally, we pour a basin full of handwashing water into it. Turning off the power periodically allows the material to stay moist, but sometimes results in some odor coming into the room.

Open windows and the plentiful moist organic activity at the office—thanks to a multitude of plants, a fountain and a worm composter—draw occasional flies during the summer, and they sometimes make their way to the composting toilet.

Diatomaceous earth, fly paper, pyrethrins spray and soap spray all work, but a few days of keeping them out of the toilet works best. We made a fly screen insert for the toilet opening by cutting fly screen in the shape of the opening, and tucking it in. A wire handle is attached to it, so it can be removed when the toilet is to be used.

Carol Steinfeld takes the removal tray outside.

(The CEPP office was moving as this book went to press. The new location will feature a Net-Batch 55-Gallon Drum system.) ✲

Composting Toilet Systems in Parks and Recreation Areas

Composting toilet systems should be a natural for parks and nature conservation areas—after all, these areas are usually far from sewer systems, hard to access with pumping trucks, environmentally sensitive and conservation oriented, and they have maintenance staffs. Alas, issues of capacity, maintenance, access and vandalism have proven sticking points to more widespread use of composters in parks.

Sometimes it's a matter of lack of knowledge about what maintenance is required or an unwillingness to perform the maintenance. Sometimes the construction considerations seem too onerous. At times, composting toilets simply get an undeserved bad rap because of a few mismanaged systems (the authors have seen some malodorous holding tank systems labeled "composting toilets" despite the absence of ventilation systems). And some poorly designed or poorly chosen composting toilet systems in parks indeed deserve their bad rap. Many parks simply use latrines or vaults that must be pumped out.

In parks that do have composting toilet systems, it is key to make the operation and maintenance procedures consistent and well documented, so that knowledge of the systems doesn't leave as staff members move on to other jobs, a common occurrence).

"It comes down to dedication," says Lisa Walker. "I am sold on composting toilet systems, but you've got to maintain them on a weekly basis and follow the manufacturers' instructions." For 10 years, Walker, who is Operations and Maintenance Foreman for Mt. St. Helens National Park, has overseen the operation of 16 Clivus Multrum and 10 Phoenix composters located throughout the park. The park also uses holding-tank vault systems in its highest-usage areas.

Top: A single composter with two toilet rooms on the Lakes Trail in the Gifford-Pinchot National Forest, Vancouver, Wash. All materials were taken in by boat. Access is on the other side. With its walk-in basement, "this is the design and setup we really like," says Jeff Hull, maintenance coordinator. (Photo: Jeff Hull)
Left: A toilet facility at the Wolf Education and Research Center in Winchester, Idaho, features a Phoenix composter. (Photo: Phoenix)

Every week, Walker's crew adds a two-gallon bucket of wood chips and rakes the material in each of the composters. Getting to the composters, however, is a challenge, as most of them are located as much as 10 to 15 feet (3-4.6 m) below ground in basements. Staff members must be lowered in by rope. These confined spaces also affect air quality. The maintenance crew tests the air and pumps in fresh air periodically while someone is working with the composter.

When we spoke with her in 1998, Walker had yet to remove any end-product from the composters, some of which had been in operation for 10 years! (Do not try this at home, folks! Manufacturers usually recommend removing material at least every two years or so to avoid compaction—otherwise you may need a longhandled chisel to remove bio-concrete.) Removing material will require hoisting garbage pails into the basement, filling them, and pulling them out, she says.

In 1997, Walker was invited to Yosemite National Park to conduct a seminar for staff members on composting toilet success factors after the park had some composting toilet systems fail. Walker suspects the problem was mistaken capacity calculations. "When the manufacturer says a system can handle 25,000 uses, it probably means over a year, not in three months!" she says.

Yosemite has also developed a secondary composting process, whereby material removed from eight composters is taken to a black box where it processes for four to five months at temperatures of 100°F/37.8C and higher.

At Mt. St. Helens, Walker says that park visitors appear to respect the composters more than the vaults. "The vaults stink, so visitors fill them up with trash," she says. "With the composters, people ask questions and even want to see the basements."

She occasionally experiments by adding semi-composted horse manure (which likely inoculates the mass with local microorganisms –Ed.) and different kinds of carbonaceous materials, although she hasn't done it consistently enough to reach firm conclusions about success. Counters on the doors give her a rough idea of how many people use each unit. Fans are powered by propane generators or small photovoltaic panels. Leachate is pumped to a separate leachfield. Occasional flies ("mostly in the basement, not the composter," she notes) are managed with vapona-impregnated pest strips.

Walker thinks more parks should use composting toilets. For one thing, they can save money in some situations. Estimating that the large composters cost about $20,000 each, and six 1,000-gallon vaults cost $4,000 a year to pump out, the composters should pay for themselves after just a few years.

"I think a lot of people are afraid of the maintenance aspects," she says. "It takes some dedication. You have got to want to make them work. If you don't, they'll just be giant stinking vaults like those in parks everywhere."

However, she adds, "There needs to be more research of these systems. There's more involved than the manufacturers can present."

Help from San Dimas

Issues of maintenance, capacity, design and staff commitment are also emphasized by Brenda Land, sanitary engineer for the USDA San Dimas Technology and Development Center, which investigates systems and solutions for the maintenance of federal lands and forests.

"When they are sized correctly and maintained properly, these can be good systems," says Land. Maintenance, she says, can flag when staff people find that the systems involve more work than they thought or they are not enthusiastic about the systems in the first place. For that reason, she recommends placing composters where no other system would be appropriate, such as where a pumper truck cannot access the site. "When you have to remove material yourself instead of using a pumper truck, you have an incentive to make composting work," she notes. She also recommends getting staff members involved in the decision process when choosing and designing a toilet facility. "The staff has to buy into it," she says.

Her agency is still looking into composting in cold-weather locations, and is assisting with a study of the SIRDO double-chamber sloped-bottom composting system from Mexico.

Yosemite National Park

Yosemite National Park has about 10 composting toilet facilities featuring both manufactured composting toilet systems as well as a site-built design of its own. Some of these require a day's hike to access them and "mule power" to haul out the end product.

Ed Walls, chief of maintenance at Yosemite National Parks, oversees the maintenance of these systems. His department hosted a conference in 1995 on the viability of composting toilet systems for parks. It was attended by people worldwide.

Yosemite developed its own system, which is a sloped-bottom composter with a false-bottom grate. The park has also developed a secondary composting process, whereby material removed from eight composters is taken to a black box where it processes for four to five months at temperatures of 100°F and higher. "We feel we're getting an inert pathogen-free compost," Walls says. Closing the organic loop is important to Walls, and he usually turns over the compost to another agency, which uses it as a soil amendment. "If you close the loop, you've achieved a good product and a good facility." ✲

Resources

Brenda Land has authored an informative 39-page report on issues of composting toilet systems in parks. Based on field visits to 30 systems, the report includes information on five manufactured composting and evaporation systems.

Composting Toilet Systems, Planning, Design & Maintenance, July 1995, #9523 1803-SDTDC, San Dimas Technology & Development Center, San Dimas, CA 91773

Yosemite National Park, with the Department of Interior, produced a report that compiled as much scientific information as they could find on composting human waste. Unfortunately, it contains little anecdotal information about Yosemite's own adventures in composting. However, it can serve as a basic primer for public facilities.

Backcountry Human Waste Management: A Guidance Manual for Public Composting Toilet Facilities, 1996. Department of the Interior, National Park Service, Yosemite National Park.

$330,000 Composting Toilet Facilities?

Park composting toilet systems received a publicity black eye in 1997, when it was reported that a composting toilet facility in the Delaware Gap National Recreation Area cost taxpayers $330,000. Yet, the facility's great expense had little to do with the composters. Most of the money was spent on the handsome superstructure, with its tile roof and specially chosen paint and moldings, as well as the involvement of 12 Park Service designers, architects and engineers.

Comparing Composting Systems in Tasmanian Parks

Australia-based Leonie Crennan, Ph.D, has completed a number of government-sponsored studies on the effectiveness of composting toilet systems in Tasmanian parks. She produced a series of reports that detail problems of incomplete processing, underestimated capacity and inadequate management of leachate, mainly in sloped-bottom composting toilet systems.

Her examinations and conclusions about the systems lead her to variations on both fixed and movable batch systems.

Based on her experiences, Crennan says she prefers batch composting systems. "They are easier to maintain and more likely than the continuous systems to give a pathogen-free compost. All my work these days is with the batch system."

For both parks and development initiatives, Crennan has organized installations of batch-style composters: a double-bin with wire mesh internal walls which she calls the "cage batch composter." She reports that the cage composter in the World Heritage Area of Tasmania has been trialed since 1992 in one location and since 1994 in another location, and has performed well. "The National Parks Service plans to replace all their [single-chamber sloped-bottom composters] with this unit as funds become available. There will be a few design adjustments based on the trial, mainly to the drainage system, to accommodate the wet, cold climate," she says.

The double-vault systems have performed well. "For six years, there has been no necessity to remove compost from the toilets despite usage of about 6,000 people per year at each site," Crennan reports. "When the second batch is ready to be changed, the first batch has decomposed enough that the first bin can be reused. This pleases staff members, as the [sloped-bottom composters] require the removal of uncomposted material by helicopter every year—very expensive and a health risk." ✲

Source:

Composting Toilet Project, Reports Nos. 1-7, 1992-1993. Centre for Environmental Studies, University of Tasmania, Hobart, Australia.

Nature's Way: Bio-Sun at Mianus River Gorge

The Mianus River Gorge is a 710-acre preserve located one hour outside of New York City and featuring some of the last old-growth forest in the Northeast. It is open to the public eight months a year, and gets 12,000 to 15,000 visitors.

Until 1992, restroom facilities consisted of chemical toilets that filled up every few days and had to be emptied. Installing a septic system was going to be disruptive of the area, so it was decided to install a Bio-Sun tank composting toilet system with a solar-powered fan. "Ultimately, flush toilets and a septic system were going to be more expensive to maintain," explains executive director Anne French. "And I liked the idea of using the same processes that occur on the forest floor, the microbes, the energy from the sun... It's a great ecological education tool, too. It's made people more aware of the whole issue of human waste and what happens to it." The system uses four dry toilets and a four-by-four-foot plastic tank.

Warden John Shull maintains the system, occasionally raking the material, and annually turning and removing it in the spring right before the season opens. "You literally go down into the tank, open it and scoop it out. It's pretty easy," he says. He places the compost around trees and plants. During one rainy year, he found that leachate was building up, so he called in a septage pumper to drain it.

"It's amazing how little odor there is," Shull says. "People comment that it's better than a conventional system." ✲

Walden Pond: If Thoreau Could See It Now...

Walden Pond in Concord, Massachusetts is known to people worldwide thanks to the book *Walden*, in which Henry David Thoreau chronicled his two years of living in a tiny two-room cabin there in the 19th century. Today, the large pond and the surrounding 411 acres are part of a state reservation with 6,000 visitors a year.

In 1995, two Clivus M-35 composters serving four dry toilets were installed in a public bathroom in the parking lot. (A seasonal bathhouse next to the pond includes several flush toilets and a septic system.) Through a maintenance contract, a Clivus staff person visits once a year to remove material and check the system. Graywater from four sinks is filtered, dosed and sent to a planted area next to the building.

The reservation's assistant supervisor reports that "skid marks" can appear on the toilet chutes occasionally, but the bathrooms are cleaned every hour, so these do not build up. One cannot see the contents of the tank through the toilet chute. "It doesn't smell. It's been hassle-free," she says. "That's important." ✲

Above: Leachate tanks and graywater filter. Left: Two composters serve four dry toilets.

Ecotourism Helps Keep the Waters Clean

Increasingly popular, "ecotourism" projects and "eco-resorts" are a newer approach in the travel industry, whereby efforts are made to reduce the environmental impact of tourism by simplifying facilities, using more sustainable technologies, and emphasizing local architecture, materials, food and culture.

A few eco-resorts use composting toilet systems, and their numbers are increasing as awareness of pollution issues grows and aesthetically acceptable systems become available. One of the better-known is Concordia Eco-Resorts, five tent cabins at St. John, Virgin Islands. Affiliated with the renowned Maho Bay resort, Concordia's five tent-cabins rent for $100 a day in season and offer "luxurious camping," says general manager Michael Caudill. They use solar power, solar-heated hot water, water pumped by guests, and Sun-Mar composting toilets. Caudill maintains the toilet systems—two Centrexes with SeaLand flush toilets and three Excels (self-contained toilets). Installed in 1995, the systems use no power—the climate is warm and small turbines at the top of the exhaust stacks catch the nearly nonstop trade winds, aiding venting. Leachate drains directly into the ground. Extra leachate can be a problem in the Centrexes due to the flush water, he says.

During high season, the Sun-Mars need to be emptied before the contents have finished composting, so Caudill is considering creating a secondary composting area. Another 120 tent-cabins are planned for the area, and Caudill plans to install composting toilets, perhaps another kind, in these, too. "We've had nearly every problem, from mongooses crawling into them to guests falling off the high Excels," he says. For a lot of people, using a composting toilet is "a stretch," he says; however, he has never received a complaint about them. "We work these really hard, and they work well. I plan to get one myself."

Above: In the Virgin Islands, Concordia Eco-Resort's five tent-cabins feature composting toilets. (Photo: Harmony)

Left: A bathroom building featuring a ReSource Institute for Low-Entropy Systems (RILES) composting toilet at an eco-resort in Mexico. (Photo: RILES)

Engineering firm Sustainable Strategies designed systems for the Republic of Palau in the western Pacific. Several composters were constructed on eight of its Rock Islands to replace pit latrines, which were producing measurable water pollution.

Very likely, more composting toilets will appear in tourist facilities and conservation areas worldwide as planners understand their value and learn how to construct and maintain inexpensive systems. ✲

Far left: Eight composting toilet facilities were built on Palau's Rock Islands, popular stopping spots for divers. They replaced polluting pit latrines. Here, a college group poses with the system they are helping to construct. (Photo: Carol Steinfeld)
Left: A twin-bin system in Mexico. (Photo: Tad Montgomery)

Students learn to build three types of composting toilet systems at EcoSan Camp in Baja California Sur, Mexico. Clockwise from left: Students visit a CEPP twin-bin with net system built in 1999 by The Global Classroom and CEPP. The system is used by kayakers, tourists and fishermen. Built large, the system only needs to be serviced every three years. Its two toilet stools (top left) are made of concrete. Top center: Two workshop participants make an alternating aerated barrel system. Right: Carol Steinfeld and a student show the side of a movable Baja Box Batch system with a Mexican urine-diverting toilet. This system was made to serve guest bungalows. Urine mixes with shower water to irrigate planted beds. (Photos: Carol Steinfeld and David Park)

Composting Toilets and Graywater Gardens at a Fijian Eco-Resort

Richard and Linda Kwasny operate Lalati, an upscale eco-resort on the Fijian island, Beqa, a spot popular with divers.

The pair contracted engineering firm Sustainable Strategies to design and engineer site-built composting toilets (shipping manufactured systems from North America to Fiji was too expensive) and graywater systems, which were installed in 1999.

Each Fijian-style thatch bungalow, called a "bure," features a SeaLand micro-flush toilet that flushes to a composter underneath, made of a rollaway trash container fitted with a hanging net (to assist aeration), a vent/exhaust chimney, a viewing porthole, an air-intake, and a leachate outflow line. Leachate from the composter and graywater from a sink and shower flow to a Washwater Garden located on the side of the bure and covered with a transparent Lexan roof overhang to keep out rain. These gardens are planted with broad-leafed thirsty plants.

Left: A typical guest house at Lalati, an eco-resort in Fiji. Each guest house features a rollaway composter, a micro-flush toilet and a Washwater Garden for graywater.

Top left: A rollaway composter underneath a guest house serves a microflush toilet. Leachate drains to a Washwater Garden (right), as does water from a sink and a shower. Left: Microflush toilets are used with this system. (Photos: Linda Kwasny)

Guests are supplied with locally made natural coconut soap and a biodegradable shampoo, although Linda Kwasny continues to use Dial antibacterial soap, and is monitoring the effects of that on her Washwater Garden. For laundry, they use a citrus oil-based detergent.

She reports, "In the beginning, we had a problem eliminating the odor in the composter. We definitely had to play around with the composting additive, and now think we've got it right. Our additive currently is a daily dose of about two cups of woodchips, popcorn, sugar, and a good deal of peat moss. It took about two months for the composting to really start working well. [It may have occurred faster if composting microbes were added. -Ed.] We added worms to all of the composters and that has probably helped." They first changed a composter after six months in use by a house inhabitated by male staff members. "There is no smell, so we must be doing something right," Linda says. When a composter is full, it is disconnected from the toilet, exhaust and leachate pipes, and taken to a sunny spot covered with transparent Lexan (to keep out rain). The openings are securely covered with two thicknesses of insect screening. After a minimum of six months of processing, the composters can be emptied.

Larger Washwater Gardens use up graywater from two clothes washing machines and the kitchen sinks. "We planted the Washwater Gardens with different varieties of gingers and lots of canna lilies, which really love it there," says Linda. "The gardens have flooded on occasion with heavy rains; we alleviated that by building rock berms to divert the water. In time, they will be just integral part of the surrounding gardens as the gardens spread. The plants all seem to be thriving well in the gardens."

Lalati has won an award for this system from the World Health Organization (WHO) for best ecotourism practices. Inspired by the success of Lalati's composting toilet and graywater system, WHO plans to introduce a composting toilet installation plan to Fiji as a method of reducing fecal coliform levels in the water, which have been traced to septic systems.

Ultimately, that is perhaps the best that ecotourism facilities can do: not only lower their impact and highlight local aspects, but model lower-impact technologies appropriate for both developed and developing countries. ✲

Composting Toilets in Development Programs

In developing countries, the need for low-cost, low-water, nonpolluting and sanitary excreta disposal makes composting toilet systems a strong candidate for use.

According to the World Health Organization, nearly one-half of the world's population does not have access to sanitation—essentially clean water and ways to deal with human waste without polluting or spreading disease. Flush toilets are often not an option: At least 25 percent of the world does not have sufficient water for water-carriage plumbing.

Perhaps humankind's oldest method of dealing with excreta—simply walking out into the forest, off a path, to the edge of a field or into the water—becomes a problem when the population density of an area increases. Too much waste in contact with too many people results in diseases and contamination of water and soils.

About 95 percent of the developing world discharges its wastewater to a river or sea. (Winblad, WHO). Unfortunately, that same water is often a source of food, and water for drinking and washing. One of the most common on-site solutions is the pit latrine. This method often simply entails digging a pit, using it until it fills up, then covering it with soil and digging another one. However, usually only slow anaerobic decomposition takes place in the pits, because they get little air (even with a vent pipe) or heat. This attracts "vectors"—organisms such as flies that carry diseases from unprocessed excrement to human food or surfaces from which diseases can be transmitted. Also, bacteria, viruses and nutrients can migrate from pit latrines—some testing has shown migration of nitrates, phosphorus and pathogens as far as 50 feet away from a pit latrine—contaminating drinking water and soils. (USEPA)

Peace Corps volunteer Steven Herbert explains the operation of a composting toilet system to Chief Marabout Cheikh Ndiaye in Senegal.

Some areas, often those with Asian influences, employ the "benjo," simply a toilet enclosure over a stream. In other areas, sewage is collected much as it was in pre-20th-century European and North American cities: It is poured into street gutters. In some countries, both animal and human wastes are dried and burned as cookstove or heating fuel. Using methane digesters to produce fuel from combined animal and human waste has been successful in other places. And, often, raw human excreta is collected and placed directly on the fields.

There are some serious problems with these methods, however:

- There is too much direct handling of the excrement.
- Pathogens are not killed.
- The nutrient value of the excrement is not utilized.

The Revered Flush Toilet

In much of the world, flush toilets are the idealized technology for wastewater carriage. However, when governments and development organizations introduce flush toilets, septic tanks and wastewater treatment plants, they bring with them those systems' attendant technical and management problems. Importing the developed world's ideal has already proven an unsustainable method. Besides the significant water required for flushing toilets, most developing countries simply do not have the money, training or human resources to maintain wastewater treatment plants—which often provide poor treatment anyway. They may not have the water—especially the huge amount of water needed to move this waste along. And the world just can't afford it.

The good news is that several government- and non-government-organization-sponsored programs are promoting and installing dry composting toilets all over the world, from Latin America to India.

Composting and dry toilets are especially effective in warmer climates, as decomposition occurs faster at higher temperatures. Many programs employ the Vietnamese Double-Vault Privy design, a batch system whereby one vault is used at a time. About 15 years ago in Tanzania, a vent pipe was added to this design, and it was thereafter referred to as the Ventilated Improved Privy, or VIP Toilet.

Efforts to introduce composting or dry toilets often run into two problems: lack of funding and low acceptance. Ultimately, as in much of the world, people want that shining example of upscale bathrooms and conspicuous consumption, the flush toilet.

Needed: Better Sanitation And Water Supplies

Much of the world still needs a safe, affordable and sustainable method of managing human exreta. Consider these figures from the World Health Organization (WHO) and the United States Environmental Protection Agency:

- Nearly a quarter of humanity still remains today without proper access to both water and sanitation. About 1.2 billion people in the world have no access to safe drinking water, while 2.9 billion people do not have access to proper sanitation systems.
- Compared to 95 percent of the First World nations, only 40 percent of Third World nations have adequate water supply and only 25 percent have water for sanitation needs.
- 80 percent of illnesses and 33 percent of deaths in developing countries are due to polluted water and lack of hygiene.
- Every eight seconds, a child dies of a water-related disease. Every year, more than 5 million human beings die from illnesses linked to unsafe drinking water, unclean domestic environments and improper excreta disposal. At any given time perhaps one-half of all peoples in the developing world are suffering from one or more of the six main diseases associated with water supply and sanitation (diarrhea, ascaris, dracunculiasis, hookworm, schistosomiasis and trachoma).
- According to WHO, providing safe drinking water and proper disposal of human excreta would do more to improve the public health of humanity than any other intervention. The public health value of improving water and sanitation services is most clearly seen in water-related diseases, which arise from the ingestion of pathogens in contaminated water or food and from insects or other vectors associated with water. Improved water and sanitation can reduce morbidity and mortality rates of some of the most serious of these diseases by 20 percent to 80 percent.

(Who Facts, 1996)

Make the Tech Transferable

In addition to the usual technical and cost issues, consider local social, political, cultural, religious, gender and privacy issues when designing or recommending a sanitation method or practice. Often the enthusiastic promoters of these technologies have dealt only with public policy officials during the planning of systems, without looking into local customs and traditions regarding a very private activity.

For example, on the island nation of Kosrae, four years after a composting toilet was installed there as a Greenpeace demonstration model, its owner related to the authors that local custom dictates that it is considered disrespectful to handle one's father's excrement, making the maintenance of the system difficult for him. On the island of Yap, we learned that brothers and sisters cannot use the same toilet. In many places in the world, the quiet private moments in a remote field, jungle or warm lagoon comprise the only time a woman has to herself, and putting a composting privy in the middle of a village where there is no privacy may mean the toilet will never be used.

A benjo on Yap island deposits excrement right into a stream.

When choosing a design, also consider local climate,

A Twin-Bin Net Composting Toilet System in Fiji, built as part of a demonstration workshop for public health officers. (Photo: David Del Porto)

"It's like remote rural water treatment plants. They don't work. They require biological and chemical expertise beyond that of the local schoolteacher. Once one breaks down, it's expected that whoever was responsible for the initial financing will fix it."

—*Paolo Lugari, in* Gaviotas: A Village to Reinvent the World (Chelsea Green Publishing, 1998)

building materials, skills and other conditions, and adapt the design accordingly.

It is also critical to remember that the most successful programs are those backed up by technical and public education support programs.

In developing countries, as in the rest of the world, composting toilets represent a different way of doing things, and their use must be integrated into individuals', families' and communities' ways of life. Ideally, a full cycle of use and maintenance is demonstrated to a community, with compost removed and properly managed—either buried or used on planting beds perhaps specifically designated for that purpose. Users and operators should fully understand how and why the systems work. When possible, systems should be demonstrated in the homes of officials and other high-standing members of the community, as well as in schools and churches and other public sites that have enthusiastic people accountable for maintenance.

Revamping the Exported Development Model

Major progress in acceptance of these technologies may only occur when industrialized nations clean up their own acts and show that ecological sanitation systems and practices are anything but backward. Maybe one day, the World Bank, the World Health Organization and other foreign aid programs will install composting toilet systems in their office buildings. In fact, all wise-water-use advocates should remember that the charge "Think globally, act locally" starts in our own homes.

What follows are some examples of ecological sanitation initiatives in developing regions of the world. Some of the systems used in arid Latin American countries are not technically *composting* toilets but *drying* toilets, because the excrement does not get moist nor aerated enough to actively compost. In these systems, the excrement is simply dried, as urine is diverted and water is not available to keep it moist. Presumably, some composting occurs. Urine is diverted in these systems to help prevent odors, and occasionally to use as fertilizer, although this use is still uncommon. (See "Composting versus Drying" in Chapter 3.) Some of the organizations promoting dry toilets add caustic lime and/or wood ash to the drying excrement to discourage flies, odors and pathogens. However, this also inhibits biological processes, and can make the material too alkaline to use as fertilizer. Adding a vent pipe—more air—to these systems could eliminate the lime (which is an added cost and can hurt handlers' skin) and allow the material to actively compost and be transformed into an innocuous, usable product.

Note that Espacio de Salud and Grupo de Tecnologia Alternativa have been involved in thousands of ecological toilet installations in Mexico. Perhaps Mexico will ultimately serve as a model of a better way for the rest of the world.

Waste Reuse: Setting Standards Too High?

Diseases often have several means of transmission besides that of human excrement reuse. Also, there are many ways to reduce potential disease transmission from handling processed human waste. Until recently, however, sanitation officials assumed that the only safe precaution was to require treatment of the waste to remove *all* of the pathogens, thereby eliminating the potential disease transmission from the reuse of excrement. However, according to World Bank sanitation specialist Richard Feachem, this led to setting treatment standards that were "practically impossible to achieve." Worse, when these unrealistic standards were observed, the fertilizer value was considered uneconomical and therefore communities would resort to using the excrement with no treatment at all. Subsequently, it was realized that a "potential risk" might not necessarily become an actual risk, and that a particular disease may have other pathways of transmission within a community other than using excrement. That's especially true of ascaris, trichuris and hookworm, all intestinal nematodes which in warm climates may be more important than bacteria and virus issues. So, many measures should be used to protect the health of farm workers and others, such as wearing protective clothing, immunizing and deworming, in addition to choosing the correct method of applying excrement-derived fertilizer (EDF) to crops.

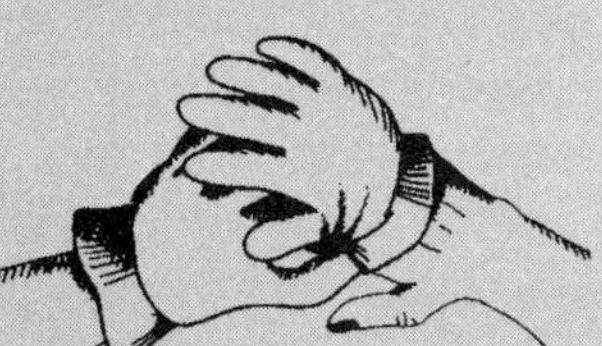

(Illustration: The Yap Non-Polluting Bathroom, Center for Clean Development

Composting Toilets and Washwater Gardens in the Western Pacific Islands

Top: Sustainable Strategies Twin-Bin Net system with a integrated planted leachate system. (Adapted from an illustration by Center for Clean Development) Above: A twin-bin system on the island of Yap. Right: Making a concrete toilet stool in Fiji. (Photos: Center for Ecological Pollution Prevention)

Since we know it well, we'll start with the work of one of the authors (Del Porto) with composting toilet demonstration projects in the Micronesian islands and Fiji.

Pollution from human waste in these islands has been identified by the South Pacific Regional Environmental Programme (SPREP) as "perhaps the foremost regional environmental problem of the decade." Unmanaged waste results in the dying off of coral reefs and eutrophication of waterways due to nutrient pollution, as well as contaminated drinking wells and the spread of diseases such as cholera and gastrointestinal diseases.

In the past few decades, the United States Environmental Protection Agency (USEPA) had constructed central sewer plants on some of the islands. On some islands, the systems failed. On others, they are overtaxed. These facilities require continuing resources, training and parts that must be imported at great expense. The USEPA had also installed septic tanks (although at least one island had no septage pumping truck), pit latrines and central sewage facilities. The septic tanks were often located in environmentally sensitive areas, such as just yards from the seashore or next to mangrove swamps. Often, there was not enough capacity in the central systems for septage. Also, some islanders not accustomed to on-demand water would leave water running and remove faucet aerators and showerheads, overloading the systems, which caused break-out failure.

Since 1992, three types of composting toilet systems have been introduced through demonstration projects sponsored by Greenpeace, Center for Clean Development, the South Pacific Commission, Koror State, the Republic of Palau and the Federated States of Micronesia:

- Carousel and Rota-Loo rotating four-chamber batch composters;
- twin-bin composters with net aerators (called Soltran Non-Polluting Toilets); and
- twin-bin composters with Wastewater Gardens, which are covered planted evapotranspiration beds to manage leachate and graywater, while keeping out the tropical rain.

These systems were designed by Sustainable Strategies, Del Porto's design and engineering firm. Although all of the systems appear to function well, the system that has been replicated by islanders is the double-vault net system, which can be made with local materials. This variation on the double-vault system features a super-strong fishing net hung inside, which suspends the excreta as it composts and serves as a large surface-to-volume aerator. No mixing or turning of the pile is required.

In Palau, eight Carousel composting toilets with covered Wastewater Gardens for leachate were installed on the Rock Islands, popular stops for tourists, as part of an ecotourism initiative.

In Fiji, the firm conducted a hands-on composting

toilet design and construction workshop at the Fiji School of Medicine for future South Pacific Island health officers, who took this knowledge back to their island homes.

For more information:
Sustainable Strategies
P.O. Box 1313
Concord, MA 01742-01313 USA
978/369-9440
www.ecological-engineering.com
sustainable@aics.net

Double-Bin Composting Toilets with Urine/Wastewater Gardens in India

When Paul Calvert arrived to teach boat-building in the fishing villages of Kerala, in southern India, he discovered a very apparent lack of sanitation. Calvert reports, "Open defecation was the norm, the habitation crowded, the water table high and use of open wells for all water needs widespread. Periodic flooding in the seasonal rains only made matters worse. Conventional pit latrines, VIPs (ventilated improved privies) and septic tanks—although few had tried building them—were not the answer. They only added to the pollution of the wells and they overflowed in the rainy season. Open defecation, it seemed, was here to stay."

In response, he researched systems and developed a composting toilet that separates feces, urine and anal washwater (the mostly Muslim population of Kerala uses water for anal cleansing, not paper).

In its simplest form, he says, "the system is a raised slab over two vaults. The vaults are built above ground to keep the contents out of the high water table and periodic flooding. Over each vault there is a hole in the slab for feces and a funnel to receive the urine. In the center of the slab between the two vaults is a trough over which anal cleansing is performed. The anal cleansing water trough and urine funnel are interconnected and flow to an aerobic planted bed outside and adjacent to the composter base.

"Each vault is primed with a bed of straw prior to each cycle of use. After each use, the user sprinkles some cooking ashes down the feces hole then closes it with the cover provided.

"The first vault of the compost toilet is opened and emptied when the second is full, typically after one year or more of operation. The end-product looks and smells like good garden soil. A toilet used by a family of six has a cycle of at least one year to one and a half years before removal of the first end-product. There is no fly nuisance or any odor problem, and the toilets remain clean and pleasant to use. If there is no interest in productive use of the urine, the plant bed needs almost no maintenance; all that is required is to cut back excessive growth. This can be chopped up and added to the compost pit. However the plant bed can be planted with vegetables or flowers for sale or family use."

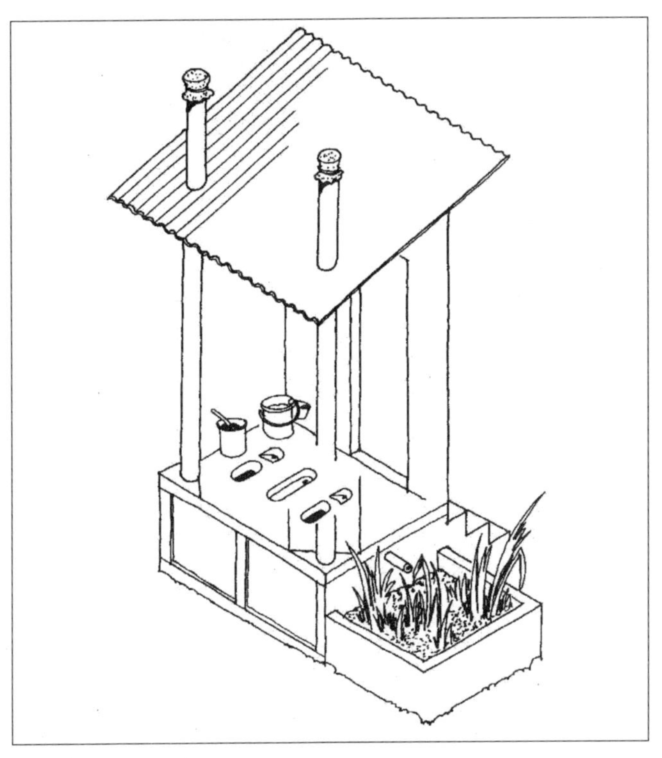

(Drawing: Ecological Sanitation*)*

Nearly 150 of these toilets are in operation in southern Kerala. Each costs about US$100. Some of them, Calvert says, "are built as part of the house and are accessed from within the house—a major step in confidence in traditional communities. However it should not be thought that this is a technology only for the rural poor. It is the urban middle class and elite that are polluting at a far greater pace than the villagers. The design is now developed for use in urban homes and apartments and can be built in simple or luxury form to suit all tastes and pockets."

For more information:
Paul Calvert, designer/builder
Sustainable Technologies in the Community
Pulari, TC42/937(11), Asan Nagar, Vallakadavu
Trivandrum 695008, Kerala, India
paulc@md2.vsnl.net.in

Drying Toilets and Urine-Diverting Toilet Stools in Central Mexico

In Mexico, half of the population has no sewage service, and more than 30 percent has no indoor plumbing. Gastrointestinal infections are the second leading cause of infant mortality, which many ascribe to a lack of sanitation provision. On a national scale, only 13 percent of wastewater is treated, and only 2.6 percent is processed in treatment plants that function adequately. At the same time, conventional treatment is causing pollution (sound familiar?), and water is scarce and expensive.

In the Cuernavaca region in central Mexico, dry sanitation is critically needed. The water source for the city is an aquifer contaminated by latrines and inadequate septic systems located over it. Sewer systems are inadequate as well, with pipes often disintegrating and mixing water and wastewater, which has led to a cholera epidemic.

In response, architect and entrepreneur César Añorvé developed and began promoting double-vault dry toilets with urine diversion (see diagram) in the early 1980s. Añorvé also designed an attractive urine-diverting toilet stool made of cement and sand, and available in many colors. These are sold for 130 pesos (about $18 U.S.). His workshop has produced more than 6,000 of these.

Like most double-vault systems, one side is used at a time. With this one, the dried excrement is removed, usually after a year or two, and occasionally used on farm fields. The diverted urine is usually drained into the earth next to the toilet, although ESAC is now working on some schemes for utilizing it.

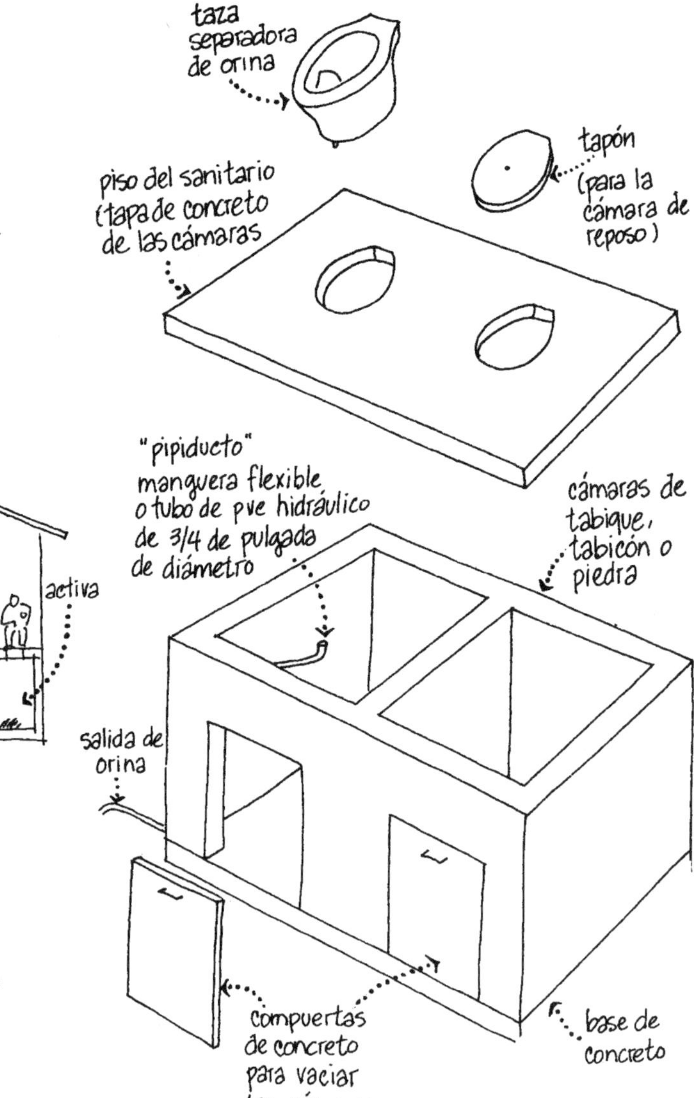

This diagram of ESAC's double-vault dry toilet with urine diverter shows how the vaults are used interchangeably. (César Añorvé and ESAC)

When one side fills up, it is "capped," and the urine-diverting toilet is placed onto the other chamber. (ESAC, drawing from Ecological Sanitation. *Sida, 1997)*

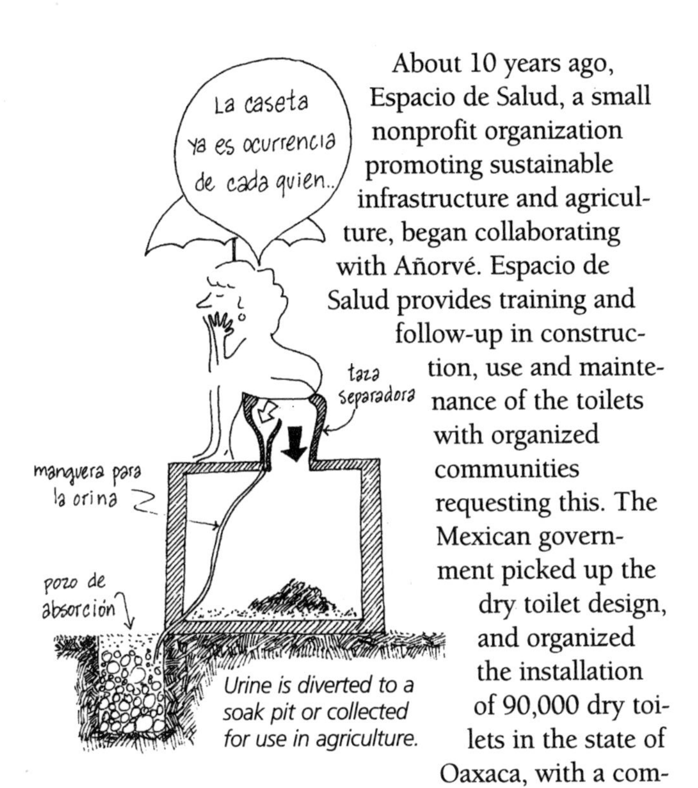

Urine is diverted to a soak pit or collected for use in agriculture.

About 10 years ago, Espacio de Salud, a small nonprofit organization promoting sustainable infrastructure and agriculture, began collaborating with Añorvé. Espacio de Salud provides training and follow-up in construction, use and maintenance of the toilets with organized communities requesting this. The Mexican government picked up the dry toilet design, and organized the installation of 90,000 dry toilets in the state of Oaxaca, with a commitment to build 30,000 more annually. A plan to install 1 million dry toilets throughout Mexico is also in the works.

But ESAC's coordinator of environmental initiatives, George Anna Clark, reports that the government-sponsored dry toilet installations are not always well received as they are "usually constructed without the homeowners' request, and with inadequate, incorrect or complete absence of instructions regarding their proper use and maintenance." She says that the toilets are best accepted, used and maintained when they are voluntarily adopted by homeowners who fully understand the need for the systems and receive maintenance support from a local organization.

Clark says she hopes central Mexico's growing numbers of decentralized neighborhood composting centers can take on the role of servicing the dry toilets, and perhaps collect the urine for farming.

ESAC has produced some colorful posters and excellent graphic materials depicting the systems and explaining their operation.

For more information:
George Anna Clark
Espacio de Salud, A.C.
A.P. 1-1576, Cuernavaca, Morelos
62001 Mexico
esac@laneta.apc.org

GTASC: Promoting Several Ecological Systems in Mexico

Grupo de Technologia Alternativa S.C. (GTASC), a for-profit cooperative, calls its wastewater systems "SIRDOs," an acronym which translates to Integrated System to Recycle Organic Waste. SIRDOs are basically three types of composting/dry toilets and one saturated (wet) system to service graywater and flush toilets.

GTASC has installed 1,000 SIRDOs in 16 states of Mexico. GTASC's composting toilet designs include a polyethylene inclined vault toilet (Clivus style) with a window in the rear to collect passive solar heat, and air- intake holes. The lowest point is layered with sand, ashes and lime to de-acidify the leachate that pools there. A two-chamber version of this is also available (think double mini-Clivuses with solar windows). It includes a toilet riser with a chute flap inside that is directed to one side or the other as the vaults fill up. Another model employs two plastic interchangeable containers used with a urine-separating toilet seat, which diverts urine to a rock-filled soak pit. These composting toilets are made of recycled

When one vault fills, a flap in the toilet riser is redirected to the new active vault.

A SIRDO with plastic enclosure in Mexico. (Photo: GTASC)

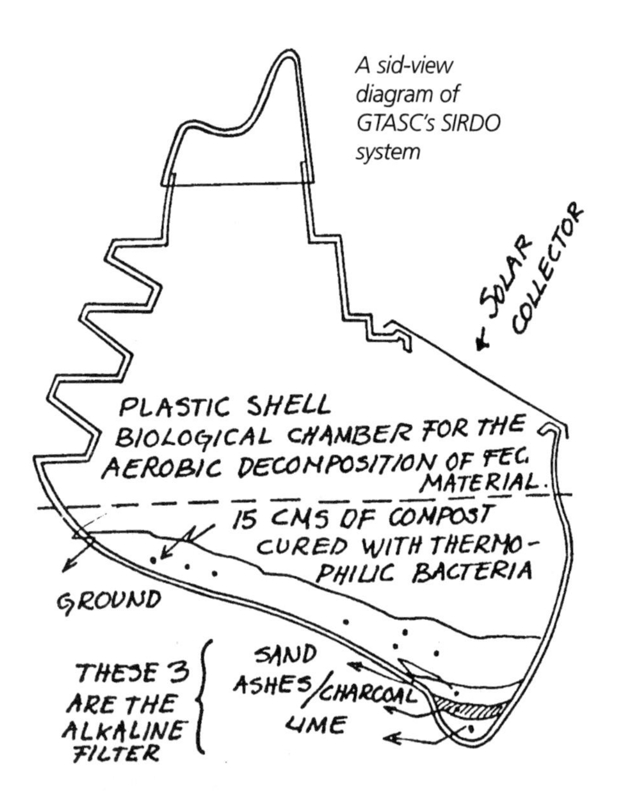

A sid-view diagram of GTASC's SIRDO system

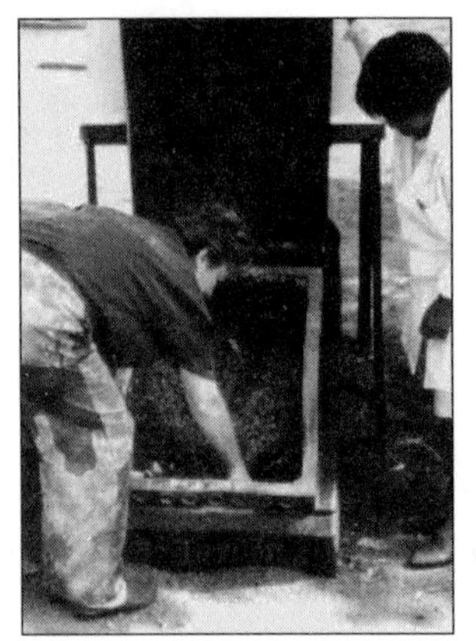
Removing material from a SIRDO (wear gloves!)

polyethylene by women's cooperatives. The third model is the classic concrete two-vault toilet system.

The "Wet SIRDO," a central system which can be used with flush toilets, is essentially a packaged treatment plant. Apparently, this system separates and composts the solids, and allows graywater, urine and leachate to drain through composting waste, and then to a sand and gravel chamber for primary filtration treatment and some evaporation. It is then reclaimed and used on non-food crops. (No test data is provided.)

GTASC maintains that the fertilizer value of the compost ultimately pays for the system. However, all the costs, including purchase and managing of additives, collecting, stabilizing, and transporting of the end-product, are not quantified. The National Wildlife Federation (U.S.) is working with GTASC to test SIRDOs in the United States to better promote their merits to development organizations as a viable sanitation solution.

For more information:
Josefina Mena Abraham
Grupo de Tecnologia Alternativa S.C.
Alamo #8-16, Col Los Alamos
San Mateo Naucalpan, Edo de Mex. 53230

GTASC's two-vault system

Dry Toilets in El Salvador

In El Salvador, double-vault dry toilet systems, called in Spanish "Letrina Abonera Seca Familiar" (LASF), were adopted by El Salvador's health department and UNICEF after witnessing their successful introduction in Guatemala in the late 1970s by Centro Mesoamerciano de Estudios sobre Tecnologia Apropiada (CEMAT).

An example of a typical project is an installation of about 130 in homes in a high-density squatter community in Hermosa Provincia. The dry excrement is used in a nursery garden. With these systems, toilet paper is typically collected and burned per local tradition. Projects organizers recommend stirring the vault's contents with a stick and adding more wood ashes weekly.

Jean Gough of UNICEF reports, "Studies by CEMAT and the federal Ministry of Health indicate that after eight months of storage, there is a rapid decline in ascaris eggs, reaching zero after 10 to 12 months. Harmful bacteria are less hardy than ascaris eggs and are not likely to survive in conditions where ascaris eggs cannot." (Sida conference proceedings, 1997)

UNICEF with Sida and the Ministry of Health experimented with the design, attempting to make it a one-vault system with urine diversion to bring down the cost of the units. Results were mixed. (*Ecological Sanitation*, 1997)

Jean Gough
Apartado Postal 1114
San Salvador, El Salvador
Or contact UNICEF

Sloped-Bottom Vault Composters in Mexico and Central America

"Beauty, function and cost" are the factors that the ReSource Institute for Low-Entropy Systems (RILES) considers when working out a wastewater system design for a client. ReSource organizes the installation of site-built Clivus Multrum-style (sloped-bottom single-vault) toilet systems, usually through nongovernmental organizations. As of this printing, more than 600 have been built in Mexico and Nicaragua for both homes and some eco-tourism accommodations. "We go to a place and work with local folks and train builders through an apprentice program," says Laura Orlando, RILES's director. The organization also trains a local expert in installation, design, public policy and outreach. Usually, they construct an entire bathroom and wastewater system, from composter and graywater irrigation system to the superstructure and the toilet stool.

Depending on what is available, the systems have been made of cinder block, cut stone (in Nicaragua) and ferro-cement. The system can cost the equivalent of $300 to $800 U.S., depending on the extent of the structure. The owner pays for materials and labor.

Building ecologically beneficial bathrooms is not RILES's only goal. "We try to make them beautiful, no matter how rich or poor the family," says Orlando. "That way, it does its own marketing." RILES has constructed mosaic-style toilet stools, using broken tile donated by hotels. RILES's website features several beautifully finished bathrooms and structures.

In Mexico, about 225 ReSource systems have been constructed on the Yucatan peninsula in the state of Quintana Roo, and another 200 or so in other states. Future plans include manufacturing fiberglass toilet stools.

For more information, contact: Laura Orlando
ReSource Institute for Low-Entropy Systems (RILES)
179 Boylston Street, 4th floor
Boston, MA 02130
617/522-7258
resource@riles.org
www.riles.org

Top: A tiled toilet commode with a built-in compartment for additive.
Above: The removal door of a RILES system.

Left: A toilet room structure in Mexico.

Why "Child-Friendly" Is Critical

In terms of preventing disease, the first population to consider is children. Children under 10 years old are the main sufferers of oral-fecal diseases. They are also the main excreters of the pathogens that cause these diseases. That means that the composting toilet should be as child-friendly as possible—accessible, usable, safe—and kept clean. (Feacham and Cairncross, *Environmental Health Engineering in the Tropics.* Wiley, 1993) Systems might include toilet seats for children, and be sited within children's walking distances from their homes.

A children's toilet seat insert

This first half of a two-vault toilet system shows how urine is diverted from a squat plate to a soak pit.

A Peace Corps Project in Senegal

While serving as a Peace Corps volunteer in Senegal, West Africa, from 1992 to 1994, Steven Herbert found that sanitation was a significant problem, as most families simply used the fields or fly-infested pit latrines. He adapted composting toilet designs he found in the Peace Corps library to create a two-vault system with a urine diverter. As in many of the aforementioned designs, one vault is used at a time. Urine drains to a small buried rock- and lime-filled "soak pit." At his own expense, Herbert constructed a prototype for Peace Corps workers' use, then led the construction of several of these systems, mostly for use by schools. "I figured schoolchildren would be more accepting of these than the adults," he says. The vaults are half buried; finished compost is dug out through an upper side panel. In Senegal, anal cleansing is done with water, so the vaults each have a small drain, in case leachate builds up (although that is uncommon in the dry, hot climate of Senegal). Per Senegalese custom, his design featured a squatting plate, not a stool. The Peace Corps provided the concrete and other materials for toilets; villagers provided some labor and constructed privacy fences around them. Herbert recommended villagers put the composted excrement into the composters he designed, along with brush and other compostables. "That way, there was another stage of processing," he says. As he was leaving, material was just beginning to be removed from one of the systems. Reports from Senegal in 1998 were that all the systems were successfully operating.

Steven Herbert
15 Whispering Pines
Colchester, VT 05446
Or contact the Peace Corps

Two Batch-Composting Toilet Designs for Kiribati and Tonga

Several batch-style composting toilet systems were tested and installed by the University of Tasmania's Centre for Environmental Studies and local counterparts for a pilot project for the Pacific island nation Kiribati and Tonga. One of the designs was a "cage batch" toilet originally developed for the Tasmania National Parks. This is essentially a twin-bin system with a metal wire mesh cage aerator on the sides and bottom. Another design uses modified 240-liter wheeled garbage bins as the composting chambers, which are interchangeably used under the toilet chutes. Each bin features a mesh false floor. Air is drawn in through an opening in the base and vents through perforated aeration channels along the interior walls of the bin.

The designer of the systems, public health specialist Dr. Leonie Crennan, reports, "The cage batch has produced an acceptable pathogen-free compost every six months, and the system has proved more durable than the Wheelibatch there. The Wheelibatch works well in Australian domestic locations where people have retrofitted below their bathrooms, and are willing to change the bins and fiddle with reconnections. However, in the Pacific, where maintenance is a dirty word, the fixed-batch system is more acceptable and durable. Also with the fixed-batch system, they can be built with local materials, which reduces long-term dependence on imported goods and aid."

Crennan and her associates conducted a well-thought-out technology transfer campaign, which included showing a video of composting toilets used in upscale public facilities, offering T-shirts emblazoned with a project logo, teaching a song about the project, conducting community focus groups to decide how the systems would best be introduced, and, a few months later, holding a public event around the removal of the finished compost. "That's essential," she says. "Seeing a soil-like material produced as end-product has a very powerful impact. Unfortunately, that kind of time and patience isn't usually written into aid projects." (See her report in WHO's

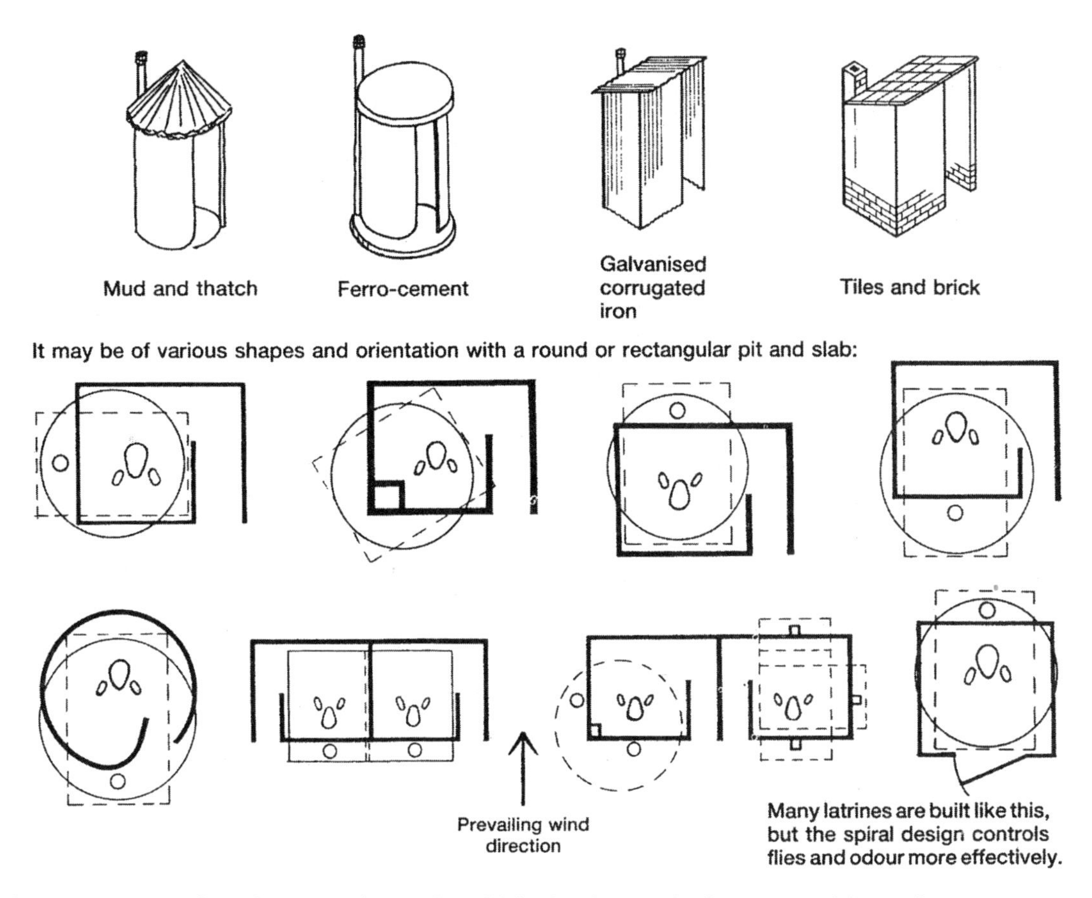

Examples of privacy structures for toilet rooms. (Originally published in The Worth of Water, *IT Publications)*

Sanitation Promotion Kit for more details.)

A three-year trial of cage batch composting toilets in Tonga, examining both acceptance and technology issues, has proven successful. Several have built them on their own, including a school. "They hope to be able to decommission most of the existing septic tank flush toilets. The main incentive has been huge savings on water bills since using the CTs. Some teachers are also concerned about pollution of groundwater by pit latrines and septics," she says.

"It is always a matter of balancing what is culturally acceptable (in terms of taboos, economics, political and traditional vested interests, attitudes to maintenance and hygiene) with what is biologically effective (in the context of climate, geography, hydrogeology, available biomass, resilience of local pathogens)."

For more information, contact:

Dr. Leonie Crennan
85 Dunbar Street
Stockton, 2295, Australia

Other grassroots and government organizations are promoting ecological sanitation in places such as India and Southeast Asia. A reported 5 million Chinese use a "double-urn latrine," a system of earthenware jars in which one either defecates or urinates. The urine is diluted and used in agriculture, and the feces may be composted in a central location before being used on fields.

Surprisingly little information about composting toilet systems for developing countries is available. Most books on appropriate technology and sanitation focus mostly on septic tanks and pit latrines. However, more agencies and nongovernmental organizations are getting into the composting act, and this technology is increasingly a topic at conferences and the subject of agency publications.

Left: A concrete toilet stool in Yap (Photo: David Del Porto). Above: Adding bulking material (Illustration: The Yap Non-Polluting Bathroom, *Center for Clean Development)*

Resources

Ecological Sanitation
Uno Winblad et al., eds.
Stockholm: Sida, 1998. Presents composting and drying toilet options for developing countries, with an emphasis on drying toilets and urine diversion. Contains excellent and informative graphics and overview information about approaches to ecological toilet systems. Available free for the downloading on the World Wide Web at www.gwpforum.org. Click on "publications" then on "Ecological Sanitation." Copies can also be ordered for 100 SKR plus postage on a "print-on-demand" basis from Arkitektkopia, Attn: Lars Lundberg, Box 3436, 103 68, Stockholm, Sweden. Email: LLU@akopia.se

"Ecological Alternatives in Sanitation"
Jan-Olof Drangert, Jennifer Bew, Uno Winblad, eds. (proceedings from Sida Sanitation Workshop, 1997). For more in-depth accounts of the systems described in *Ecological Sanitation,* read these reports presented at a 1997 conference on the topic of dry sanitation and urine diversion for developing countries, organized by Sida, Sweden's international development aid program.

Sanitation Promotion Kit
Mayling Simpson-Hebert
New York: WHO, 1998. A collection of case studies and articles written by practitioners about working with political entities and improving technology acceptance for sanitation systems in general. Available in North America for $41 postpaid from Rural Environmental Health Unit, WHO Publications Center USA, 49 Sheridan Avenue, Albany, NY 12210

Environmental Health Engineering in the Tropics
Sandy Cairncross and Richard Feachem. England: 2nd ed., 1993. Features excellent information about a variety of issues relating to sources of disease in communities, water chemistry, water quality, water supply technology, planning a sanitation program, and various methods of managing excreta and other organic wastes. Only briefly touches on composting toilets. Available from John Wiley & Sons Ltd., Baffins Lane, Chichester, West Sussex PO19 IUD England. (Also www.amazon.com)

"Water for Big Cities: Big Problems, Easy Solutions?"
Peter Rogers, Harvard University: Cambridge, Mass., 1997. This paper by a Harvard environmental engineering professor describes the current and impending water crisis in big cities in the developing world, and proposes better approaches to water management.

The Worth of Water, Technical Briefs on Health, Water and Sanitation. London: IT Publications, 1991. This illustration-packed book doesn't cover composting toilets, however, it is a great model of a how-to book for sanitation and building simple water management systems in villages and communities, with simple introductions and how-to illustrations of rainwater harvesting, household water storage, soap making, siting and choosing toilet systems, and more. Order from: IT Publications, 103-105 Southampton Row, London WC1B 4HH, UK. Email: orders@itpubs.org.uk. In the U.S.: Stylus Publishing, P.O. Box 605, Herndon, VA 20172-0605. 800-232-0223. Email: Styluspub@aol.com

A Guide to the Development of On-Site Sanitation
R. Franceys, J. Pickford, and R. Reed. Geneva: WHO, 1992. Mostly focusing on pit latrines (lined and unlined) and varieties of septic systems, this book does include good background information on water and sanitation issues in developing countries, as well as some diagrams of squat plates and insect control traps for latrines.

"Composting toilets differ from the latrine in that they actively treat the excreta, i.e., kill pathogenic organisms within the unit....

"The batch-type [composting toilets] are more appropriate for use in developing countries because they are simpler to operate than the continuous type. The latter requires careful control of waste decomposition and frequent removal of the composted material in order to keep the process going. In addition, because the length of the composting period determines disease vector die-off, process control for the batch-type composting is less important than for the continuous process composting [toilet] with its shorter residence time."

—Appropriate Technology for Sanitation

The Perils of the Pit Latrine

A commonly recommended system for developing countries, the pit latrine usually requires pumping out vaults (lined) or holes in the ground (unlined; most common). Great care must be taken to avoid groundwater contamination.

Often, pit latrines are open-bottomed, making them cesspits. Increasingly, the travel of viable human pathogens through saturated soils reaching connected ground- and surface waters are sources of diseases.

In fact, a pit latrine can be more polluting than a septic system, as it almost directly deposits fecal coliform and viruses into the soil and groundwater.

CHAPTER NINE

What About Graywater?

For many, a discussion of composting toilet systems must include information about what to do with the rest of the domestic wastewater equation: the water used for washing, or "graywater."

Graywater captures our imaginations because it seems like merely soapy water. Why can't we flush our toilets with it, spray it on our lawns, treat it on-site and recycle it in the house? The answer is, you can! *However,* these applications require a level of pretreatment that many of us aren't prepared to undertake or pay for—and that our regulators are not going to approve without system certification by a registered engineer or a testing laboratory.

But there are ways to use it. This chapter describes what's in graywater, how to pretreat it, and some general descriptions of systems for which you can get permits, with an emphasis on utilization systems.

"Water reuse" is the term government agencies, researchers and engineering firms more commonly use to discuss issues of and system designs for irrigation, groundwater recharge and flushing toilets with "used water." How it is used depends on its characteristics (strength, temperature, etc.) and constituents (what's in it).

Many advocates of using graywater assume that it is benign, because excrement, or blackwater, is the primary contributor of potential pathogens to the wastewater mix. But it is not that simple. Think about our propensity to put all sorts of toxic chemicals down our drains (solvents, paint thinners, drain cleaners, etc.), rinse diapers in the washbasin and drain blood and fluids from raw poultry and meat down the kitchen sink. Graywater is much more than soapy water and must be carefully and responsibly managed!

A mound of marigolds and other decorative plants are part of a graywater system at a park in Toronto. (Photo: Alfred Bernhart)

Know Your Graywater

What Is Graywater?

Graywater is washing water from bathtubs, showers, bathroom washbasins, clothes washing machines and laundry tubs, kitchen sinks and dishwashers. Water from toilets, urinals and bidets is considered "blackwater." Some states' regulations consider diaper washing water and water from kitchen sinks to be blackwater because it may contain excrement, animal products and greases. If this is the case, you may need a separate holding tank or septic system.

Graywater accounts for 50 percent to 82 percent of all combined residential effluent volume (sewage), and could supply most, if not all, of a home's landscape irrigation requirements that are not met by average rainfall in many non-arid communities.

More states are recognizing graywater as a wastewater component that can be managed separately from blackwater. In several states, a septic system used only for graywater can be 40 percent smaller than a standard system.

Graywater: Variations on a Theme

Graywater varies from source to source. Since household washing products, lifestyles and customs differ considerably among household and commercial settings, there is a wide variation in the volumes, constituents and chemical and microbiological characteristics of graywater. For example, in an office setting where only hand washing occurs, one would not expect to find the same graywater characteristics as those from a hotel or residence.

For that reason, when planning a graywater utilization system, the very first step is to identify its constituents and the microbiological, chemical and physical characteristics of the graywater streams. Many studies have identified fecal organisms in the graywater from homes and public rest areas, so it is prudent to conclude that graywater may routinely be contaminated by fecal matter. (Enferadi, et al. 1980, and others) Treat it with the respect that this warrants.

Classifications of Graywater

There are many subcategories within the general graywater classification, such as:

1. Effluent from kitchen or food-preparation sink drains and dishwashers, which contain significant quantities of grease, oil and food particles and pathogens, such as salmonella from animal and poultry blood
2. Discharge water from clothes washing machines in the laundry in which large volumes of lint particles from nonbiodegradable fabrics are present, plus fecal bacteria if diapers are washed
3. Wastewater from slop sink and floor drains in garages and buildings that contain many toxic chemical compounds that have accidentally spilled on the floor and are mopped up and disposed of along with unwanted solids and liquids that fit down the drain
4. All other nontoilet washing and processing effluent, including water used in bathing, which may contain fecal bacteria and pathogens washed from the skin
5. Nearly clean residual water that has little or no constituents or pathogens, such as condensate from dehumidifiers and air conditioners, warming-up water from showers (the water that flows while we are waiting for it to warm up)

What We Put into It

A Short Course on Soaps, Detergents and Bleaches

After water, the major component of graywater is soaps, detergents, bleaches, etc., followed by the oils, dirt and grime that this water was used to remove. Shampoo, laundry and dishwasher powders, floor cleaners, face and body wash gels, and some toothpastes and mouthwashes contain detergents.

Soaps

Soaps are alkali salts of long-chain fatty acids. They emulsify soils, microbes and liquids and detach them from surfaces, and act as a surface action agent (surfactant) to reduce the surface tension of water, making it "wetter." They break the bonds that hold these constituents to skin and fabrics, allowing them to be suspended in a form that can be rinsed away.

Soaps differ significantly from detergents in certain important performance properties, and are making a strong comeback in the cleanser department. Hard soap and hard water can create tenacious soap scum. If you are fortunate enough to have a soft water supply (characterized by reduced calcium and magnesium ions) or use naturally distilled rainwater, then you may prefer soaps to detergents. They are more gentle, and some say they do a better cleaning job in soft water than detergents. The gradual switch to detergent came when we started to use hard water from wells and rivers, then found that hard soap left rings in bathtubs but detergents did not. Most hard soaps (usually bars or flakes) are made by reacting sodium hydroxide, a strong alkali, with clarified animal fat, lanolin, glycerin or vegetable and citrus oils.

Soaps That Are Good for Gardens

Most soaps are made with sodium hydroxide. The use of sodium-based soaps increases the amount of sodium in the graywater, while the hydroxide raises the pH, or alkalinity.

Sodium can, in some plants, inhibit the water and nutrient transport in the plant xylem and phloem cells (it affects the ability of a cell membrane to transport water and nutrients and changes the osmotic balance, which causes trouble). In other words, sodium causes hypertension in plants, just as it can in humans. Potassium-based soaps are better to use with a graywater irrigation system, because potassium is a fertilizer and a beneficial nutrient. Most liquid soaps are made with potassium hydroxide. Potassium hydroxide alone serves as an excellent grease remover; it works by turning grease into soap! Tri-potassium phosphate (TKP) is a powerful cleaner that is used on sewer pipes and automotive engines.

Sodium causes hypertension in plants, just as it can in humans. Potassium-based soaps are better to use with a graywater irrigation system, because potassium is a fertilizer and a beneficial nutrient.

Detergents

Detergents have the same effects as soap. They contain surfactants, as well as a variety of chemicals to provide special functions: (l) a builder, normally an ingredient that will chelate (sequester) or precipitate polyvalent metal ions such as calcium and magnesium ions present in the cleaning solution, which give "hard water" its name; (2) bleaches; (3) corrosion inhibitors; (4) sudsing modifiers; (5) fluorescent whitening agents; (6) enzymes; (7) antideposition agents; and (8) inert solid fillers. (*Van Nostrand's Scientific Encyclopedia,, New York.* They will penetrate and wet soiled surfaces, and displace, solubilize or emulsify various substances, especially oils and greases. They also disperse or suspend substances in liquid to prevent their redeposition on the surfaces or fabrics to be cleaned.

A planted graywater system will be happier if you leave out the toxics, solvents and disinfectants. (Illustration: Center for Clean Development)

Degreasing Cleaners

D-Limonene, a monoterpene, is the active ingredient in relatively new multi-purpose concentrate cleaners, such as Citrasolv™. It is an organic extract derived from orange, grapefruit and lemon oils. It is is 100 percent biodegradable, and emulsifies oil, grease, tar and cosmoline; it is noncorrosive and contains no petroleum solvents. (One of the authors used it full strength to degrease an automobile engine. It worked very well and provided a non-polluting nutritious meal for the microbes in a Washwater Garden. –Ed.)

Disinfectants in Cleaning Products

Disinfectants will kill beneficial organisms as well as pathogens. Consider this carefully when designing a biological graywater system. You may need to compensate for the disinfectant death of the beneficial organisms that are transforming your effluent for safe reuse.

Due to the growing concern over salmonella and *E. coli* bacteria contamination from beef and poultry operations, many cleaning and personal hygiene products now include disinfectants. While this may be a simple scheme to sell products with fear-based marketing strategies (they offer no real advantage over washing hands thoroughly with hot soapy water), it is of concern to designers and operators of biological treatment systems. The overuse of disinfectants can reduce or upset the beneficial microbial populations both on our bodies and in our treatment systems.

This is especially true with dishwashers in commercial kitchens, because the effluent may contain large amounts of chlorine bleach used as a disinfectant. During the energy crisis in the mid-1970s, some state health departments and local engineers recommended substituting chlorine bleach for high-temperature water as a means to kill microorganisms and reduce energy costs. This resulted in the failure of septic tanks, biological packaged treatment plants and soil absorption systems. With little or no biological processes in the treatment device, fats, oils and grease quickly accumulate and clog the system.

Phosphorous

Before phosphate laundry detergents were banned by most states, graywater contributed one-half to two-thirds of the phosphorous in combined wastewater. But those percentages are dropping now. Phosphorous can pollute surface water, but it is good for your garden.

Characteristics of Graywater

How Much Is Produced?

Following are estimates of the relative proportions of toilet or blackwater and wash-water or graywater to total household effluent water. It is very important to understand that the relative percentages given here will shift dramatically when more water-efficient appliances are used. For example, as toilet flush volumes go down, percentages from other sources increase. That means that the strengths of the constituents of graywater will be greater when you use modern water-conserving fixtures and appliances. Less water use also affects the diameters and and lengths of waterpipe drain lines in buildings. (Konen, 1998.)

Graywater volumes are measured by water usage, and vary widely. The present formula for urban/suburban sources used by Massachusetts Department of Environmental Protection to calculate design flow is 33 gallons (125 liters)per person per day (gcd) or 66 gallons (250 liters) per bedroom. More will be generated in hot communities where showering is more frequent and less in rural communities with unpressurized water service.

T. Winneberger discovered that rural Native Americans on an Arizona reservation used only 5 gallons (19 liters) per person per day—the lowest in the United States! The increasing use of high-efficiency appliances and fixtures will reduce graywater volumes in the future.

According to a 1998 study of indoor water consumption by the American Water Works Association, changes to the national water conservation code that was adopted in 1994 have caused a shift in water-use patterns. The chart below shows its data on homes with current best water conservation measures, such as 1.6-gallon (6-liter) toilets, front-loading clothes washers, etc.

Typical Single-Family Home Water Use
With Conservation

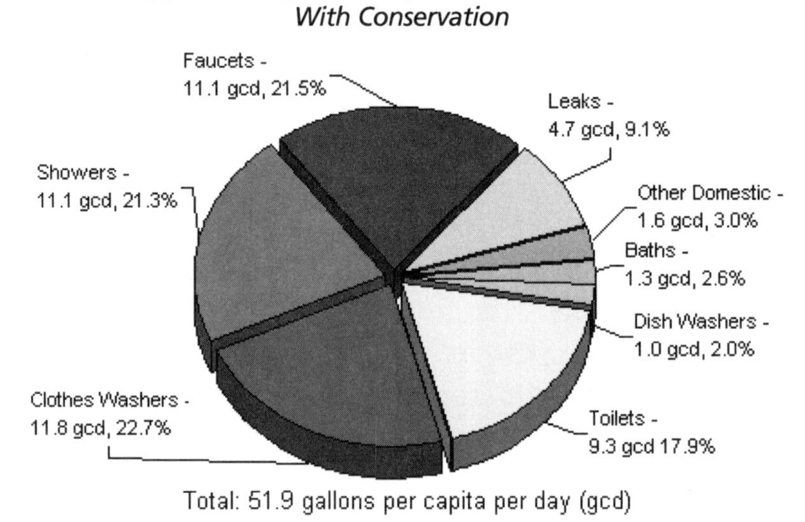

Total: 51.9 gallons per capita per day (gcd)

American Water Works, 1998

Household End Use of Water (gcd*)
Without and With Conservation

	Without Conservation**			With Conservation			Savings		
End Use	Share	gcd*	lcd**	Share	gcd*	lcd	%	gcd	lcd
Toilets	26.1%	19.3	7.3	17.9%	9.3	35	52%	10.0	5
Clothes Washers	22.7%	16.8	64	22.7%	11.8	44.7	30%	5.0	19
Showers	17.8%	13.2	50	21.3%	11.1	42	16%	2.1	8
Faucets	15.4%	11.4	43	21.5%	11.1	42	2%	0.3	1.2
Leaks	12.7%	9.4	36	9.1%	4.7	17.8	50%	4.7	17.8
Other Domestic	2.1%	1.6	6	3.0%	1.6	6	0%	0	
Baths	1.8%	1.3	5	2.6%	1.3	5	0%	0	
Dish Washers	1.4%	1.0	3.8	2.0%	1.0	3.8	0%	0	
Inside Total	100%	74.0	280.0	100%	51.9	196.4	30%	22.1	83.6

* gcd = gallons per capita per average day of the year ** lcd = liters per capita

** Based on the average inside uses measured in 1,188 homes in twelve North American cities including an additional 6% to account for estimated "in place" savings due to existing conservation.

For WaterWiser by John Olaf Nelson Water Resources Management.
Copyright © 1998 American Water Works Association

Graywater's Nitrogen Deficit

The carbon-to-nitrogen ratio of graywater is not ideal for efficient microbial breakdown in planted irrigation systems. The detergents, soaps, soils, fats, oils and grease in graywater add a lot of carbon to it, but it has relatively little nitrogen (most of the nitrogen in combined wastewater is contributed by urine). This deficiency may require adding nitrogen to graywater to balance the large amounts of carbon. As in composting, and from a microbial nutrition perspective, a diet of 25 parts carbon to one part nitrogen (C:N = 25:1) is ideal. All living cells require digestible carbon for growth as an energy source and nitrogen and other nutrients such as phosphorous and potassium for protein synthesis necessary for building cell walls and other structures.

Graywater Is More Digestible!

Many studies indicate that graywater is easier to digest by microbes than is septic tank effluent or sewage. This may be because graywater's five-day biological oxygen demand (BOD5, a measure of its carbon content used by water planners) is 90 percent of the ultimate oxygen demand. This means that graywater has a faster rate of stabilization than blackwater—it breaks down to an oxidized plant-usable form faster. One study determined that the five-day BOD of blackwater only consumed 40 percent of the ultimate oxygen demand, which means blackwater will consume oxygen for a much longer time.
(Carl Lindstrom, "Key Graywater Characteristics," privately published. Cambridge, Mass., May, 1997)

Graywater Characteristics as Reported by Various Studies*

The following is a summary of graywater characteristics from various studies. Due to the diversity of individual households, there is a wide range between many of the characteristics. Also, these data do not report the toxic chemicals, heavy metals and other dubious compounds such as floor washing water, solvents, paint thinner, drain cleaner, disposed contents of chemistry sets, etc.

	Source of Data					
Parameter	Siegrist	Rose	Boyle	Sherman	Karpiscak	EMR
Volume (gcd)**	20-30 (75.7-114 liters)	29.4-59 (111-123 l)	21 (79.5 l)	-	-	45.3*** (171.5 l)
Ratio to total household wastewater	-	65%	-	70%	69%	53-81%

Parameter	Unit	Rose	Enferadi	Brandes	Boyle	Sherman
Turbidity	NTU	20-140	-	-	42-67	-
Phosphate	mg/l	4-35	-	1.4	-	3.4
Sulfate	mg/l	12-40	-	-	0.3-11.9	-
Ammonia	mg/l	0.15-3.2	-	-	0.6-4.5	-
Nitrate Nitrogen	mg/l	0-4.9	-	0.12	0.1-0.6	-
Total Kjeldahl Nitrogen	mg/l	0.6-5.2	2-50	11.3	5.7-18.4	1.9
Chloride	mg/l	3.1-12.0	-	-	-	-
Suspended Solids	mg/l	-	20-1500	162	36-160	-
BOD	mg/l	-	40-620	149	125-291	33
COD	mg/l	-	60-1610	366	242-622	52
Total Dissolved Solids	mg/l	-	420-1700	-	686-925	-
Alkalinity	mg/l	149-198	-	443	382	-
Electric Conductivity	mho/cm	-	443	-	-	-
Total Coliform	MPN/100ml	-	102-109	2.4x106	-	-
Fecal Coliform	MPN/100ml	-	101-106	1.4x106	-	-
pH		5-7	-	6.8	7.1-8.7	-

** Source references are cited at the end of this chapter, by author*
*** gpcd = gallons per capita (person) per day*
**** Reported figure is the average of data from six different sources.*

Source: Graywater Pilot Project-Final Report, City of Los Angeles, California November 1992
City of Los Angeles Office of Water Reclamation, November, 1992

Carbon Clogging

In any biological system, the first carbon-containing molecules to be consumed are the easy-to-metabolize carbohydrates, such as sugars and starches; then the difficult compounds, such as grease and cellulose, follow. Too little nitrogen will limit microbial processing, resulting in a buildup of undigested carbon-containing fats, oils, grease, soaps, detergents, soil, biodegradable fabrics, dust and soot from exhausts. This buildup of organic matter may clog the soil and the distribution and drain pipes.

Where Does the Nitrogen Come From?

The little nitrogen that is in graywater is usually from ammonia and ammonia-containing cleaners; proteins in animal and fish blood and flesh; vegetable proteins; protein-containing shampoos, conditioners and cosmetics, as well as urea- and nitrate-containing houseplant fertilizers.

Where to Find More Nitrogen

An easy way to increase nitrogen is to make more use of ammonia-based cleaning compounds. Also, look for soaps and shampoos that contain the detergent ammonium laureth sulfate or ammonium lauryl sulfate, such as Neutrogena and Pantene Pro V products. Another method is to flush down the sink drain a small amount of acidic fertilizer, such as urea, Miracid (a well-known plant fertilizer) or other nitrogen-rich fertilizers. These mild acid formulations will buffer the high pH from alkaline soaps and detergents. Alternatively, one could attempt to reduce the carbon content, but this is not a practical suggestion on which to base a system design.

Diverting Urine

Or, you can simply add urine. Installing urine-diverting toilets and urinals can add to graywater this normally sterile, nitrogen-rich liquid after graywater filtration. Periodic flushing of any urinal with 3 percent hydrogen peroxide solution (available in drugstores and grocery stores) will control odors. And, while we do not publicly advocate using a bathroom washbasin as a urinal, doing so would put the nitrogen in the right place with no change in the plumbing. (See Chapter 4 for a discussion of urine-diversion options.)

Graywater Sources

As with any reuse plan, you should match the source and characteristics of the graywater with the treatment necessary for the reuse you have in mind. Analyze each graywater source, and recombine it only after pretreatment. Consider whether it contains pathogens (laundry and kitchen) or fats, oils and grease (kitchen) or non-biodegradable lint (laundry). Let's walk through the house to decide how to manage each source for utilization:

The Kitchen

The kitchen is a particularly challenging source because of lipids (fats and oils), food particles and uncooked animal and poultry blood (a potential salmonella carrier) that are washed down the drain. Grease from cooked animal fat and solid (hydrogenated) vegetable oils are often flushed down drains as hot liquids, but they solidify when cooled. In large quantities, grease and solidified oils can clog the soil, preventing it from absorbing water (a common problem in restaurant leachfields. Food particles may attract insects and rodents. Fats and oils are usually "emulsified" (a state in which two or more mutually insoluble liquids are contained in a mixture where one is dispersed as droplets throughout the other).

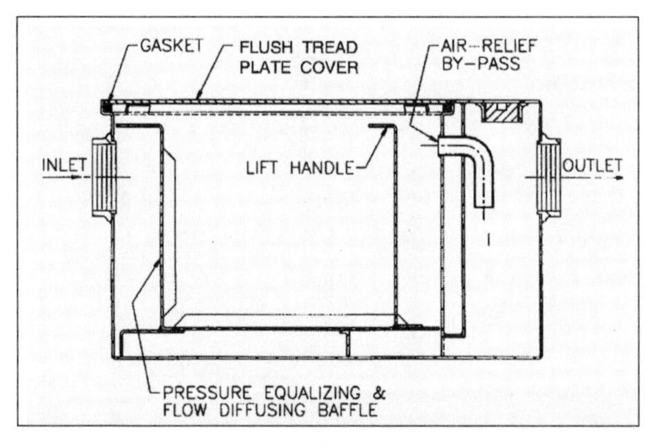

Top: A Zurn Grease Interceptor. Above: A grease-trapping graywater filter from Clivus Multrum.

How to Calculate the Size of a Grease Trap

Many manufactured interceptor/traps give specifications for the flow rate alone. For a two-person home, the smallest commercial grease trap will do. For example, the Zurn Model 100 4-gpm grease trap holds 3 gallons of water and 8 pounds of grease, and will fit under your sink. It costs more than $500. If you want to build one yourself or have special issues, here's what you need to know:

1. Determine the total amount of water used to prepare food and wash your dishes at each meal (G).

Here more then ever, conservation counts. Infrared or self-closing faucets with flow-restricting nozzles will save water and the fuel needed to heat it, too. A new Bosch dishwasher only uses 4.4 gallons per regular wash cycle and 3.6 when used on the "Economy" setting! Given the number of pots, pans and dishes that the dishwasher can hold, it uses less water than if you washed the same load by hand!

Add 1 gallon for processing and cooking.

In this example, total G equals 5 gallons (19 liters)

2. Determine the flow rate of the water as it goes down the drain in gallons per minute (GPM).

Use 6 gallons per minute for the drain. The dishwasher can pump it much faster. This is important if you are cooking many meals in a short period of time, and would overload the trap, allowing unsettled material to flow out. A 4 gpm flow restrictor on the trap may cause your wash-water to drain more slowly.

In this example, GPM equals 4.0 gallons (15 liters)

3. Determine the detention time needed to trap the grease, oils, grit, etc. (DT).

For homemade traps, use 2.0 hours to cool the hot water and separate the solids. Sophisticated (and expensive) manufactured traps can do it much quicker. A heat recovery system using cold water as a heat exchange fluid could significantly reduce the detention time. So, for design purposes you need to know the temperature (T) of the wash water and heat exchange fluid as well.

In this example DT equals 2.0 hours

4. Make sure your interceptor/trap has the capacity (C) to hold all the wash water (G) for the required detention time (DT). Make sure the flow rate (GPM) does not interfere with the detention time.

So multiply 5.0 (G) x 2.0 (DT) equals 10.0 gallons (37.85 liters) or 1.4 cubic feet (C) (or 1m^3).

In this example, C equals 10.0 gallons (37.85 liters).

Maybe a Grease Trap

A "grease trap" or "interceptor" is a tank that retains the hot emulsified kitchen sink effluent for a specific period of time to allow the food particles, fats and oils to separate from the remaining liquid. These particles, fats and oils are periodically removed, and the clarified effluent is drained to be treated and/or reused. Think of it as a sort of septic tank for a kitchen sink. Grease traps or interceptors can be placed under sinks. Larger ones are used with dishwashers, which drain a greater volume of water all at once. Use a flow-control fitting to control the flow rate. Some include a special portal for adding grease-digesting enzymes. Many states require grease traps for commercial or institutional kitchens to protect the waste treatment plant, but homeowners who utilize their graywater will benefit as well.

Obviously, garbage grinders cannot be connected to your graywater system. If you want keep this convenience, have your plumber connect the grinder to flow to your indoor or outdoor composter.

A kitchen waste chute connects to a Bio-Sun composter below. (Photo: Bio-Sun Systems)

The Laundry

In addition to the bleach issues described earlier, the laundry presents its own special challenges for management.

If you are planning to recycle the graywater, do not wash baby diapers in the sink! Choose biodegradable diapers and either send them out with the trash or compost them. Or hire a diaper service or manage this effluent separately. The baby's feces may contain pathogens, requiring that this effluent to be disinfected prior to reuse. Keep in mind that washing machines are responsible for many leachfield failures, due to clogging by the nonbiodegradable fibers from

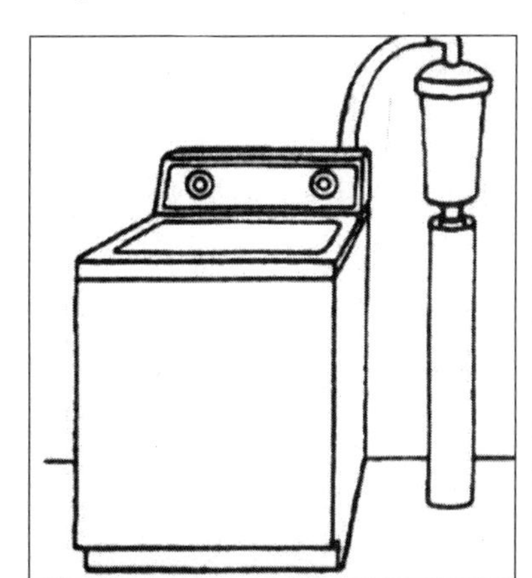

A Septic Protector filters out fibers from clothes washing water—a major cause of clogged leachfields.

clothing—especially polyester, nylon and polyethylene. Other solids that can cause clogging include powdered detergents and soaps and insoluble grit from dirts and construction materials. Even if only small amounts make it through a septic tank or graywater filter, these can eventually block the holes in distribution pipes, drip irrigation emitters and ejectors, and clog the void spaces in filters, soils and sand filters. Filter and use liquid detergents, not powdered versions.

The Bathroom

Issues for Reuse for Flushing Toilets and Urinals

Many people ask about using graywater to flush toilets. It's a reasonable reuse, but certain considerations must be taken into account.

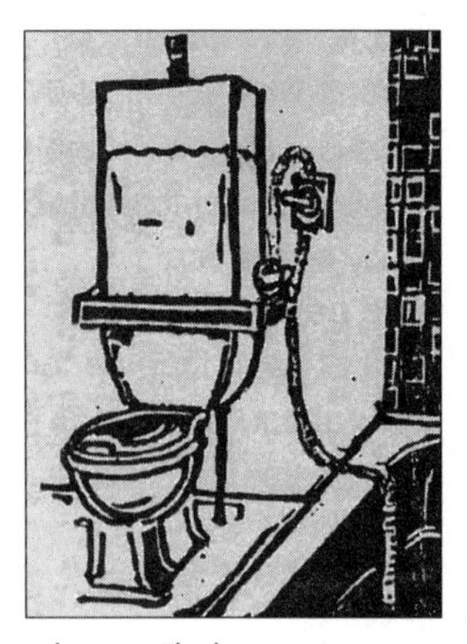

The Eco-Flush, a graywater-flush toilet

The first is that graywater cannot be stored very long without disinfection or (a) the dissolved oxygen in the normally warm effluent will soon be used up, and anaerobic bacteria will rapidly multiply, causing foul odors; and (b) human viruses and bacteria can cause health risks. Second is that untreated graywater contains greases, oils and particles that will collect on the inside of the toilet tank, the distribution holes in the bowl and the drain pipes. This may cause problems and maintenance costs.

Not Quite So "Gray" Water for Special Uses

Some graywater was not used for washing, and poses few or no potential pathogen, nutrient or clogging problems. For example, "warm-up" water from showers can be collected while waiting for the water to warm to acceptable temperatures. Condensate from air conditioners and dehumidifiers can also be reused.

You can use graywater to flush a toilet simply by collecting it in a bucket and pouring it directly into the toilet bowl.

Some products have been designed to use "pre-flush" water. "The Lid" replaces the cover on your toilet tank with a plastic basin with an integrated water spout. When you wash, the water drains directly into the toilet tank and either replaces or supplements the flush water. No treatment occurs with The Lid, but it provides a close-coupled connection for toilet flushing with washwater.

A graywater flush tank was developed by a Wisconsin firm, but never marketed. It required pumping graywater from a tank underneath the bathtub/shower into a large tank above the toilet.

Remember: Graywater is more than soapy water. However, with proper treatment and use, it is a free and valuable resource.

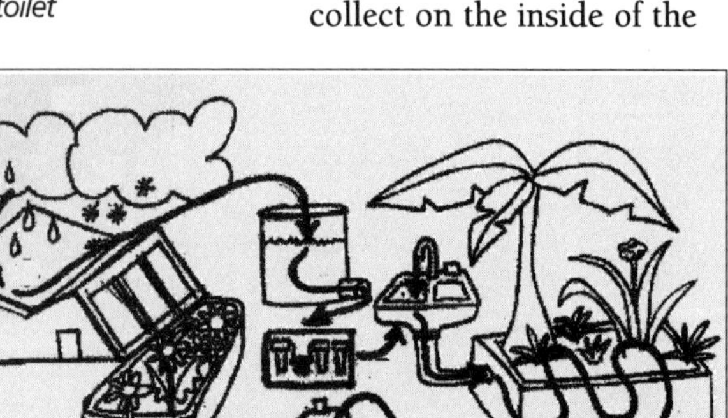

The Eco-Bathroom: This system designed by Earthship collects rain and snow on a roof designed for that purpose, stores it in a cistern, then (via solar and wind power) filters it and pumps it to the household for use. On the other end, household graywater drains to a graywater treatment planter, then is pumped and filtered and used for toilet flushing. (Illustration: Solar Survival Architecture)

Simple Ways to Use Graywater

Here are some simple ways to use graywater, but they are unlikely to be approved by a health agent. Do them responsibly and with careful attention to what is in the graywater you are using:

- ☐ Mini-graywater systems: Insert a basin in your sink when you rinse your vegetables, and pour this relatively uncontaminated water onto your water-loving plants. Hibiscus is an especially thirsty and beautiful plant that will appreciate this effort.
- ☐ Do the same with condensate from your air conditioner or dehumidifier, or use it to pour-flush your toilet.
- ☐ Collect "warm-up" water in a bucket, while waiting for shower/bath water to get hot. Pour-flush your toilet with this.

Filtering Graywater

If you reuse or recycle graywater, you need to remove the particles in it that could clog the pipes in an irrigation system, jam the valves or pumps, and prematurely fill up the void spaces in the soils of your planting beds.

Simple filters at the source can accomplish the job quite well. However, nonbiodegradable and buoyancy-neutral particles, such as the plastic lint from polyester, nylon and polypropylene fabrics, will eventually accumulate and clog up an irrigation bed or leachfield, requiring the replacement of the entire field. The more plastic particles that get through—no matter what size—the sooner they will build up and cause trouble.

Today's microfiber and microfleece fibers, such as those by Polartec, which are made from recycled plastic bottles, require using a very, very fine filter screen.

Washing machines agitate the fabrics and break the plastic fibers into pieces that are often 30 to 40 microns in diameter and length. One micron equals one thousandth (0.001) of a millimeter or 0.00004 inches. The ideal solution is to keep out all nonbiodegradable particles, but that is impractical these days. In the past, when our fabrics were made of only biodegradable fabrics, such as cotton, wool or cellulose-based rayon and acetate, a simple nylon stocking was an inexpensive, adequate filter. If plastic fabrics are never washed in your home, this stocking system will be adequate to use.

Filtration is the process whereby particles of a certain size are separated from particles of smaller size by screens which allow the smaller particles to pass through and the larger to be captured for removal. Filters are graded by "pore size," which tells you the size of particles they let through.

For graywater filtration, 160 microns is minimum and 30 microns is preferred. It is best to use filters in series, so the finer filter does not get clogged every day. If you use a 30-micron filter as your only filter, it will quickly fill up with all particles bigger than 30 microns, and that could mean cleaning it several times a week. Instead, use two filters: first a 160-micron filter, then a 30-micron filter. Empty both only once or twice a month by removing the filter bag or fiber cartridge and brushing out and rinsing off the lint into a container (not back down the drain!).

The Septic Protector

Filters large enough to reduce the frequency of emptying are better. Although there are many brands of filters on the market, available from swimming pool or plumbing suppliers, we like the Septic Protector, because it is designed to be connected to your washing machine, where most of the trouble occurs. The Septic Protector is inexpensive ($149) and easy to install without professional help. It features either a 160- or optional 30-micron removable filter cartridge inside a transparent plastic case with a screw-off lid. According to the manufacturer, the Septic Protector's 30-micron cartridge filter will remove more than 99 percent of solids (lint, hair, pet fur, sand, clay, etc.). It will also reduce oils and soap scum.

Other filter methods:

■ **Insect screens and septic tank effluent filters:** These pore sizes are too large (3/16th inch) for filtering out the nonbiodegradable fibers. (As a final filter, I use a Zabel 1800 septic filter, which catches hair, rings, coins, and so forth, just before the graywater enters my Washwater Garden. –Ed.)

■ **Diatom swimming pool filters** are excellent but too expensive to purchase and maintain for household use. It's like cutting butter with a chainsaw, because they are sized for swimming pools, not washing machines, so they would be too large and expensive to install and typically require professional servicing.

■ **Granular activated carbon (GAC)** used for water filters can be very effective as polishing filters, but are expensive to install and the GAC has to be replaced often.

■ **Sand filters:** Whether they are up-flow, down-flow, recirculating or fluidized bed or the tank types sold in catalogs, these can be effective, depending on sand grain size (smaller is better). However, they need to be back-washed periodically to clean them. Most will not filter out the nonbiodegradable 30-micron particles, because the sand grain size is too large. A series of two Septic Protectors could accomplish much more for much less.

■ **Peat filters:** The recent use of a special grade of peat (yes, it's the same stuff you use in your garden) holds promise for soil and gravel substitutes in leachfields, and may prove adequate for graywater, too. Compacted fine peat provides small pore spaces and has a unique ability to allow contaminants to adhere (adsorb) to the peat fibers.

■ **Utrafiltration and reverse osmosis filtration:** The U.S. Navy is developing a graywater-recycling system using these pressure-driven separation processes by which graywater is moved across a semi-permeable membrane. They can remove bacteria, salts, and more. This is for recycling graywater as higher-quality water such as that for cleaning; for irrigation it is overkill. (A good information source is the Navy Shipboard Environmental Information Clearinghouse, which maintains a web page: http://www.navyseic.com/liquid/html/abstract.htm)

Some Graywater Systems

This section generally describes the graywater systems that you are most likely to get approved by your town or city. These include:

- Septic tank and leachfield
- Pretreatment and subsurface disposal
- Pretreatment and irrigation
- Other natural systems

Treatment, Utilization and Disposal Systems

There are as many variations for the treatment and recycling or disposal of graywater as there are for combined wastewater. After all, it has many of the same characteristics as combined black- and graywater, but in different proportions.

However, by diverting graywater from the solids, nutrients and potential pathogens found in combined wastewaters, it is far less of a treatment challenge and offers more opportunities for use.

Graywater, though, still must be filtered, then treated. Utilization systems are a natural method for this, and managing graywater can range from pouring a bucket of washroom sink water around your shrubs, to draining it to a lined system planted with tropical plants in a greenhouse.

Graywater from the author's kitchen sink drains to a greenhouse, first through two ponds with a fountain and waterfall, then to irrigate plants. (Photo: David Del Porto)

The systems and technology for treating graywater and plumbing it for use or irrigation are not new. These technologies have been used in the water and wastewater industry for decades, and are highly evolved.

In this section, we describe approaches generally. Unlike some books on the topic, which show a multitude of filter-and-irrigation schemes, we cannot provide cookie-cutter solutions. The variables of each site are too great, and we don't want you calling us when your health inspector finds your homemade system and orders you to remove it or revokes your occupancy permit.

The size and type of responsible graywater system you can install depend on:

- **Flow rate:** how much graywater your building produces and how often (volume and rate)
- **Geological conditions:** the characteristics of your site's soil, including percolation rate (how much water your soil will absorb over time), depth of groundwater, and proximity to wells and surface water, to name a few
- **Characteristics and constituents of the graywater:** Does it contain softened water with sodium from an ion-exchange resin water softener? Kitchen water? Diaper washwater?
- **Plumbing specifics of your home**
- **And many other variables:** budget, climate, regulations, etc.

(See "How to Choose an Ecological Wastewater System" in the Appendix.)

NOTE: Again, beware of designers or authors who provide designs, products, material sources and general guidance for homeowners without requiring an analysis of the soils and consideration of the above variables. (Plus, you can figure out on your own how to attach a nylon stocking to a hose-end and spray your washing machine water over your plants.) You may find your money wasted if a local inspector discovers your system.

In most states, a property owner legally must have plans prepared by a professional engineer or certified wastewater system designer prior to submitting them for state and local approval.

Best Practice: Let Plants Do the Work and Use It Up!

From an ecological point of view, the best approach to managing graywater is to treat it in a planted aerobic system, preferably a plant-based utilization system, not a disposal system. As environmentalist David Rapaport says, grow it away, don't throw it away!

Filter, Not a Septic Tank

Given that graywater does not have excrement solids to be settled out, a smart approach is to avoid the septic tank and simply filter graywater to reduce its solids. For the most part, your composting toilet systems takes the place of the septic tank by diverting and managing the blackwater solids (and does a far better job of treatment!).

Filter out the grease, oils and fats and nonbiodegradable particles, such as fabric lint in graywater. Exclude toxic chemicals that can kill plants.

Test it for pathogens, if graywater is to be used for anything other than subsurface (below-ground) irrigation. If they are present, disinfect it prior to reuse.

Then, drain or pump it directly, immediately and in a fashion that evenly distributes it to the subsurface root zone of plants, where the bioplex of microbes will transform the pollutants into simple soluble constituents that are either used by plants, exhausted to the atmosphere as vapor or harmless gases, or stored as a soil constituent (a little is fine).

Aerobic planted systems provide highly adaptive, flexible and robust environments that can handle extremes—and are far more complex and dynamic than any mechanical treatment system or septic tank.

Mechanical/Disinfection Systems

If you want to treat your graywater to drinking-water quality, you must use advanced treatment, which includes disinfection. Mechanical/disinfection systems using reverse osmosis technology and an ultraviolet light disinfection system are now able to bring effluent to drinking-water (or near-drinking-water) quality. Examples of these systems include Recyclet and a system by AlasCan. The sewage-to-potable-water approach will need a lot of regulator eduacation before you will get a permit. However, these systems are off-the-shelf the components have been thoroughly tested by USEPA-approved analysis and NSF International Standard 53 for drinking water systems.

Below is a technology by AlasCan, sold as an add-on to its composting toilet system. More information can be seen at www.alascanofmn.com/disinfection.html.

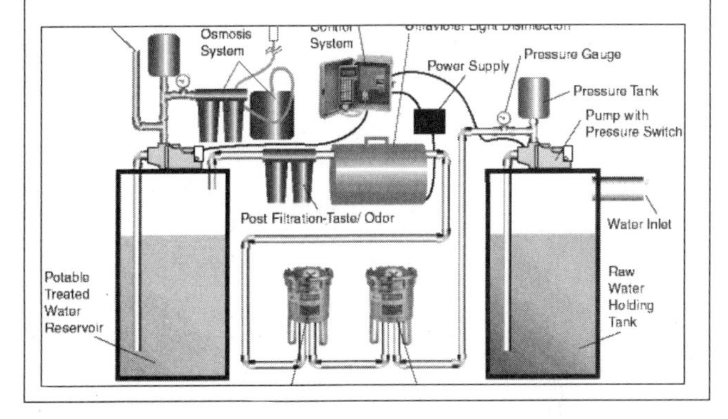

About State Graywater-Reuse Regulations

Many states, including Arizona, California, Colorado, Florida, Hawaii, Idaho, Illinois, Indiana, Nebraska, Nevada, Oregon, South Dakota, Tennessee, Texas, Vermont, Washington, West Virginia, Wyoming and others have adopted progressive wastewater-reuse policies. (See the "Regulations" in the Appendix for contact information your state.) That ranges from permitting household graywater irrigation systems to allowing commercial buildings to use treated graywater to flush toilets.

Although this chapter is about graywater, keep in mind that the treatment and regulatory issues for graywater are, for all practical purposes, the same as for wastewater.

California allows municipalities to recharge groundwater with treated reclaimed water, and it is looking into using treated wastewater in drinking water reservoirs.

Most states require using potable water for flushing toilets and urinals. In these states, your plumbing or health inspector may have you take out any unauthorized graywater-reuse system. However, that is changing. In 1993, the California Plumbing Code and the International Association of Plumbing and Mechanical Officials (IAPMO) adopted standards and regulations allowing the installation of reclaimed water systems to supply water closets, urinals and trap primers for floor drains and floor sinks in nonresidential buildings. Under the provisions of this regulation, tertiary-treated* wastewater can be used for this purpose. This is allowed only in nonresidential buildings, and entails using expensive systems—but it is a step in the right direction.

**Defined as "water which is adequately oxidized, clarified, coagulated, filtered and disinfected, so that the seven-day median number of coliform bacteria in daily samples does not exceed 2.2 per 100 milliliters, and the number of total coliform bacteria does not exceed 23 per 100 milliliters in any sample. The water shall be filtered so that the daily average turbidity does not exceed 2 turbidity units upstream from the disinfection process." (Section R-2 Definitions, Appendix R of the Uniform Plumbing Code)*

Examples of these systems are artificial wetlands, sequenced solar aquatics and Living Machines®, planted rock filters and Washwater Gardens™. They can be indoors, outside or in greenhouses. Some serve double-duty as landscapes and tree groves. They will always provide the necessary and complete treatment at a lower cost than biochemical/mechanical devices. At the same time, they provide habitat, beauty and air purification for the users at no additional cost.

Graywater Disinfection

If graywater is to be used in a manner in which it can come in direct contact with humans or their domestic animals, such as surface drip or spray irrigation, the prudent measure would be to disinfect the effluent first, such as through ozone, ultraviolet light, hydrogen peroxide or halogens (chlorine, bromine iodine, etc.). Disinfection may be the only way a state approving agency would permit its reuse other than subsurface irrigation or disposal. Disinfection would also help in getting graywater reuse approved for edible garden vegetables.

Good Books on the Topic...

The following books cover most, if not all of the technical specifics of wastewater systems and issues:

The best handbook for wastewater utilization is the self-published guide by Dr. Alfred Bernhart, P.Eng., professor *emeritus* University of Toronto, *Evapotranspiration, Nutrient Uptake, Soil Infiltration of Effluent Water* (Toronto, Canada: self-published, 1985). If you can find it, hold on to it. It's the best.

Readers interested in engineered designs of all kinds of black- and graywater treatment systems should consult the excellent technical textbook by Ron Crites and George Tchobanoglous, *Small and Decentralized Wastewater Management Systems* (WCB/McGraw-Hill, 1998). While not specifically targeted to graywater, this text features systems with small design flows, as contrasted with truly large municipal or industrial flows. It features detailed descriptions, calculations and references on constituents in wastewater, the fate of wastewater constituents in the environment, process analysis and design, pretreatment operations and processes, alternate collection systems, biological treatment and nutrient removal, lagoon treatment systems, wetland and aquatic systems, land treatment, intermittent and recirculating filters, effluent repurification and reuse, disposal options, biosolids and sludge management, and system management.

Conventional Systems

Disposal: Septic Tank Treatment

Draining graywater to an existing septic system is often the easiest and most prescribed option, and offers the advantage of requiring no additional plumbing after the composting toilet system has been installed.

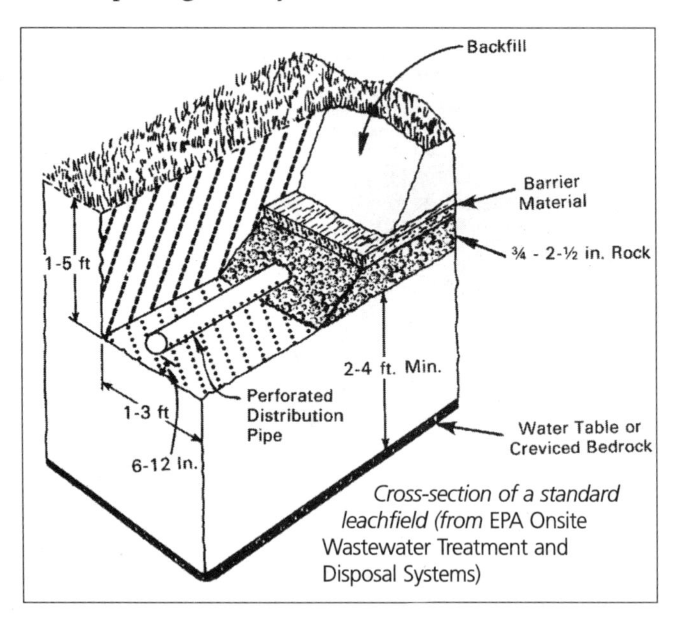

Cross-section of a standard leachfield (from EPA Onsite Wastewater Treatment and Disposal Systems)

In many states, a property owner can install a smaller-sized septic system—usually 35 to 40 percent smaller than a conventional one—if it is used for graywater only. (In Massachusetts, the graywater flow is 66 gallons per bedroom per day versus 110 gallons for combined gray- and blackwater.) In other states, regulators may require installation of a full-sized septic system.

Because septic systems are well known to public health officials, they may be the easiest to get permitted.

Soil Absorption Systems: Leachfields or Mound Systems

Whether the graywater is filtered, flowed through a natural plant-based system, or treated by a septic tank or a packaged treatment technology, the disposal system of choice of most engineers and state regulators is soil infiltration. A soil infiltration or absorption system, usually a leachfield, is mostly a disposal system—a way to get water into the ground.

The most common types are gravity and pressure-dosed (pumped) leachfields. (There are many others, such as sand filters and mound systems.)

Unless it is planted, this is not a utilization

Consider the Septic Tank...

Septic tanks and cesspools settle out solids, which either float to the top or sink to the bottom. Then the effluent drains to a soil absorption system of one kind or another, usually a leachfield.

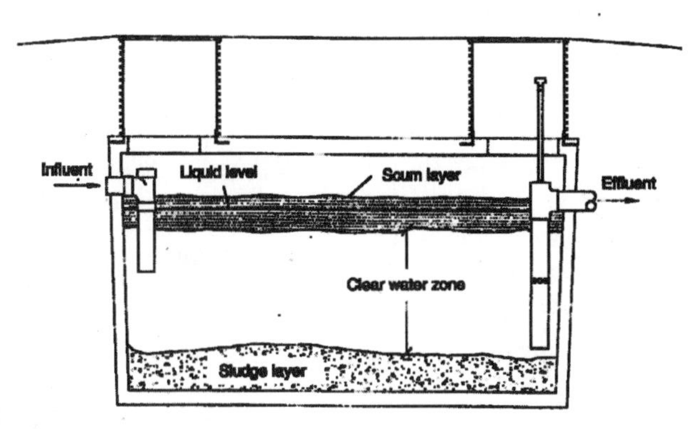

A conventional septic tank

Anaerobic vs. Aerobic Again

A septic tank is a concrete or fiberglass box buried in the ground. No oxygen gets to the process, so the biological process is "anaerobic." It has been reported that only 55 to 75 percent BOD removals can be achieved by an anaerobic septic tank, regardless of the season, after 18 hours of retention. (Crites and Tchobanoblous, 1998.) An "aerobic" tank, such as a composter, provides more efficient biological conversion. (So why not an aerobic "anti-septic" tank, you ask? Supplying a septic tank with compressed air would require some engineering and significant electrical power to maintain an oxygen rate high enough to work efficiently.)

Even Septic Tanks Like It Warm

Biological conversion happens better when it is warm. Most septic system designers do not account for ground temperature in their calculations, because heretofore it was not an important factor to them. But now we know that no microbial growth occurs below 41°F/5°C, and in the northern climates, the winter ground temperature can slow or stop any biological process. Then the tank is only a solids separator. All microbial activity doubles with every 18°F/7.8°C rise, so keeping the tank warmer is better. Some have actually solar-heated their septic tanks for this purpose.

Better Septic Tanks

Septic tanks can do a pretty good job of solids removal by slowing down the flow rate, as Peter Warshall writes in his excellent book, *Septic Tank Practices*. He quotes Dr. Tim Winniberger, who suggested that a good septic tank design should be modeled after a slow meandering river that drops its suspended silt-solids at every bend, with minimal turbulence. The three-chambered tank designed by Winneberger is a better system for that reason. Unfortunately, we do not see many like this, because it costs more to fabricate and many installers are not aware of its advantages.

(Illustration by Peter Aschwanden in The Septic Tank Owner's Guide *(Shelter Publications).*

The Solar Septic Tank by Earthship in New Mexico uses a solar window on the septic tank to enhance microbial treatment. Effluent then flows to two lined constructed wetlands or evapotranspiration beds, depending on the effluent volumes.

approach. However, it can replenish groundwater at sites where the soils are deep, permeable and conducive to transforming pollutants into safe constituents. Otherwise, it can pollute groundwater.

These systems should not be used where

- groundwater levels are close to the surface;
- the hydraulic gradient can carry the untreated effluent into receiving surface waters; or
- bedrock or clay renders the soil impermeable (won't absorb water).

Absorption for Wastewater Disposal

The soil absorption field is the land area where the clarified and partially treated wastewater from the septic tank is spread into the soil.

One of the most common types of soil absorption field uses plastic distribution pipes with drain holes, which extend away from the distribution box in a series of two or more parallel trenches, usually 1½ to 2 feet wide. In conventional, below-ground systems, the trenches are 1½ to 2 feet deep.

Some absorption fields must be shallower due to some limiting soil condition, such as a clay hardpan or a high water table. In some cases, they may be placed in soil brought to the lot from elsewhere.

The distribution pipe is surrounded by gravel that fills the trench to within a foot or so of the ground surface. The gravel is covered by fabric material or building paper to prevent plugging.

Another type of drainfield consists of pipes that extend away from the distribution box, not in trenches, but in a single, gravel-filled bed that has several such distribution pipes in it. As with trenches, the gravel in a bed is covered by fabric or other appropriate material to keep the soil separated from the gravel.

Usually the effluent flows gradually downward into the gravel-filled trenches or bed. In some instances, however, as where the septic tank is lower than the drainfield, a pump must be made a part of the installation in order to lift the effluent up into the drain field.

Whether gravity flow or pumping is used to get wastewater into the drain field, effluent must be evenly distributed throughout the drain field.

It is important to ensure that the drain field is installed with care to keep the distribution pipe level, or at a very gradual downward slope away from the distribution box or pump chamber, according to specifications stipulated by local officials.

Soil beneath the gravel-filled trenches or bed must be permeable so that air can move through it and contact the effluent. Good aeration is necessary to ensure that the proper aerobic chemical and microbiological processes occur in the soil to cleanse the percolating wastewater of contaminants. A well-aerated soil also ensures slow travel and good contact between wastewater and soil.

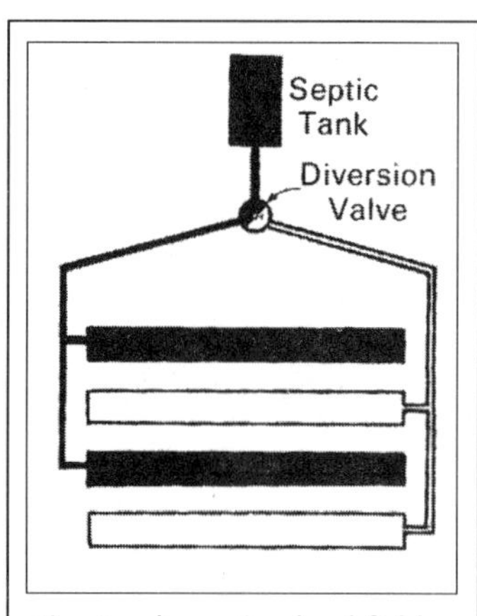

Tip: An alternating leachfield allows fields to rest, allowing better microbial transformation. (from EPA Onsite Wastewater Treatment and Disposal Systems*)*

★ ***This layout is fundamentally the same for subsurface planted irrigation systems.***

One day, leachfields may be more like this graywater bed by Clivus Multrum: a mounded planted bed with pipes just below the surface, in the aerobic zone and in the root zones of plants. (Photo: Clivus New England)

Natural, Ecological and Advanced Systems

Natural or "ecological" systems are usually called that because they duplicate and optimize what occurs in nature, and use up the potential pollutants in wastewater. Many have been effectively used for combined systems, but they can be used for graywater alone, too. Each has to be engineered to a site's specific flow rate, environmental conditions and wastewater characteristics.

Remember, that all systems, whether they are natural or not, have to remove solids. That means filtering, and possibly pretreating and disinfecting, the graywater first.

Graywater Irrigation

California was the first to allow graywater irrigation systems for residences, when, in 1993, the California Plumbing Code and IAPMO Uniform Plumbing Code adopted standards and regulations governing the installation, construction, alteration and repair of equipment for the subsurface use of untreated graywater for irrigation. Many states are following California's lead.

One's natural inclination is to imagine hosing graywater over the lawn. However, surface application of any undisinfected wastewater is not a good idea, as disease organisms can come in contact with people and domestic animals simply walking in the grass or brushing against the surfaces of the leaves. So unless you disinfect graywater, do not spray it. Apply it through pipes beneath the surface. This is usually the only legal way to manage it.

One day, it may be standard to install graywater systems that irrigate the the shrubbery and some flower gardens around homes. Sheltered somewhat from rain by roof overhangs, this application makes a lot of sense.

A wastewater-irrigated landscape (Photo: Bill Wolverton, NASA)

The Washwater Garden™

Although there are many ways to irrigate, one method with which one of the authors has had direct experience is what Del Porto calls the Washwater Garden, for graywater, and the Wastewater Garden, for combined effluents.

Based initially on the work of Dr. Alfred Bernhart at the University of Toronto, the Washwater Garden operates by Aerobic Microbial Respiration and Evapo-Transpiration, or "AMRET." That means that, unlike most leachfields, it is designed to encourage high-rate aerobic processes (faster processing!) and utilization of the nutrients in graywater to feed microorganisms living in the sand, which transform these larger molecules into forms that can be utilized by the plants. In this way, it accomplishes what a leachfield doesn't: faster processing by aerobic (instead of anaerobic) microbes and uptake or utilization of the nutrients by plants.

The treatment occurs in the sand by microbes that are supported in part by the roots of the plants. Together, they create an interdependent system that transforms pollutants into plant food, and assists in the evaporation of the water and the transpiration of it by the plants—which are literally "growing away" the graywater.

The key features of this system are:

- Sand and gravel chosen for their ability to encourage aeration (sand is chosen because graywater has a lot of carbon in it, and so the carbon-rich humus that soil offers is not required)
- Plants chosen for their high needs for water and nutrients
- Shallow depth
- Breezers to vent gasses and vapor and provide oxygen

In environmentally sensitive areas, Washwater Gardens are contained and lined, so no effluent leaves the system.

In addition to a planted bed, the Washwater Garden can take the form of standard landscaping around a building, as well as indoor planters.

In a Wastewater Garden, prefiltered, pretreated wastewater, from all graywater and/or blackwater sources at a site, drains or is pumped into ecologically engineered gardens where aerobic microbes in the planting medium digest the organic molecules and transform them into heat, carbon dioxide, water vapor, and oxidized water-soluble compounds.

Not just any plants work. Plant species are selected for their hardiness, rapid growth rates and high water requirements. These are called phreatophytes and include broad-leafed wetland and rainforest plants, densely leafed evergreens, and fast-growing plants, such as bamboo and many grasses and reeds.

The size of the system and the number of plants are determined by the volume and the constituents of the effluent and the climate.

As in a garden, the plants—or biomass—are harvested and naturally defoliate, thereby removing the nutrients in the wastewater.

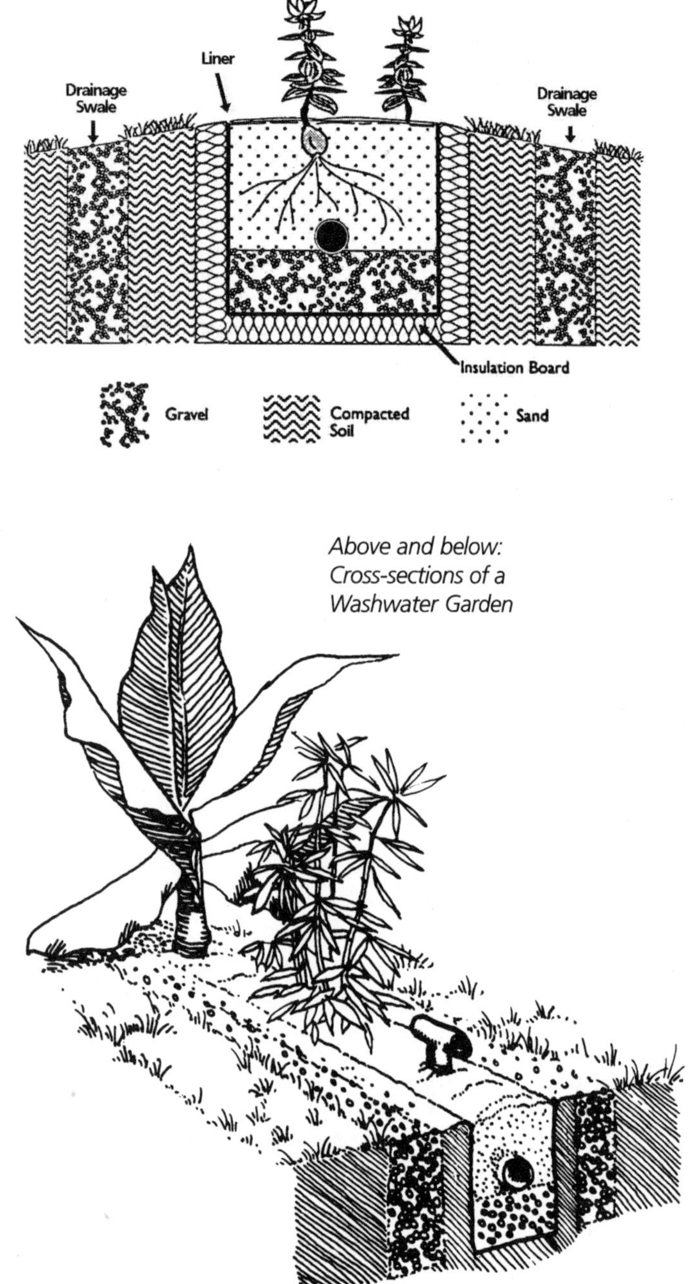

Above and below: Cross-sections of a Washwater Garden

Technical Aspects

Wastewater Gardens are shallow, flat-bottomed lined trenches or beds, typically filled with 6 inches (152 mm) of crushed stone and covered with 12 inches (304 mm) of sharp sand. Perforated pipes on the stone and beneath the sand distribute the filtered wastewater along the length of the trenches. The trenches should be no more than 18 to 24 inches(457-610 mm) deep to maximize the uptake of water and nutrients by plant roots and to allow proper oxygenation. Air is introduced by venting the distribution pipes with riser vents or "breezers" (these may be augmented by blowers to increase aeration, heat the beds and remove carbon dioxide and water vapor).

Top: Washwater Gardens work best covered, such as under roof eaves, or enclosed so that precipitation is kept out. In greenhouses, they can be sized smaller, as heat increases the microbial and evapotranspiration processes. Above: A Washwater Garden in Massachusetts planted with varieties of bamboo and popular shrubs.

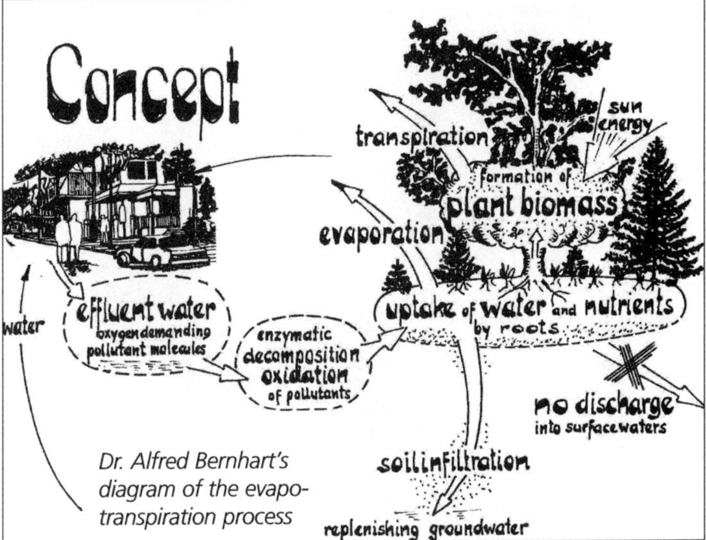

Dr. Alfred Bernhart's diagram of the evapo-transpiration process

Washwater Gardens work best sheltered from rain and snow. Enclosing them in a warm greenhouse powers up the process, which adds capacity.

How AMRET Works

Evaporation occurs as water moves upward through the sand by capillary action and is evaporated at the surface. Evaporation also occurs within the bed as aerobic bacteria release heat as they metabolize the nutrients in the effluent. This heat drives water into vapor, which is forced to the surface. The spaces between the sand particles allow this to happen.

Transpiration occurs as water absorbed by plant roots is drawn up the stalk and stems of plants into the leaves, where it is released as pure water vapor through openings called stomata.

Aeration is critical to support aerobic bacteria and other soil organisms, which transform any remaining pollutant-nutrients in the wastewater at exceptional speeds into plant-available forms.

It is important to vent the system. The resultant gasses, carbon dioxide and water vapor must be moved out of the bed so they don't prevent new oxygen from entering. This is enhanced by the void spaces and surface area of the stone and sand granules in the upper part of the bed and the surface area of the leaves of the plants in the Wastewater Garden.

The microbial respiration also increases the rate of evaporation as the microbes metabolize the nutrients in the effluent.

Heating It

As in composting, increasing the temperature of the system by heating—from the surface and/or from beneath the trenches—has a super-charging effect on the AMRET rate. More heat can double or triple the activity of aerobic microbes. (See the "Heating It" section in Chapter 4 for the technical explanation.)

Warmer temperatures also increase the rate of transport of water from the roots of plants to the surfaces of the leaves, where it is released as a vapor.

Performance

Typical average annual outdoor AMRET rates in the temperate zones of the United States are from 0.05 to 0.2 gallons per day per square foot of surface area of Wastewater Garden. However, this rate can be significantly increased to tropical AMRET rates (0.5 to 3.0 gallons/1.9-11.4 liters per square foot) by enclosing the beds in a greenhouse or a translucent cover.

Washwater Gardens can also be source-specific: The one on the right manages the washwater from a clothes washing machine. Located in Massachusetts, this one is planted with ornamental perennial grasses, evergreen bamboos and shrubs, and hostas. (Photo: David Del Porto)

Indoor Graywater-Irrigation Planter Beds

Some people manage their graywater right inside their homes. Big enclosed beds full of plants can manage part or all of a building's graywater. Be aware that these can add a lot of moisture to the building. Always provide an overflow mechanism, so untreated graywater cannot spill into the home. Also, a diversion valve allows you to divert flow to a holding tank or another system on high water-use days. Public health officials will be reluctant to approve such systems, due to the fear of introducing pathogens in the home.

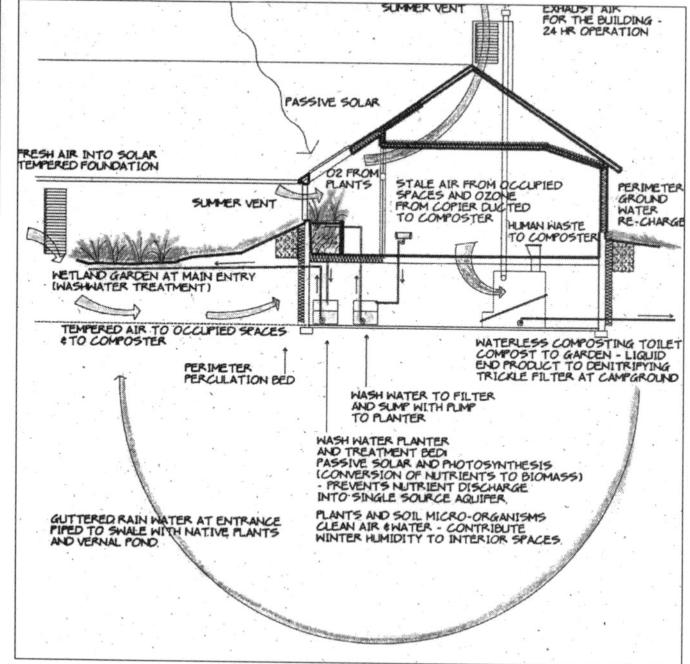

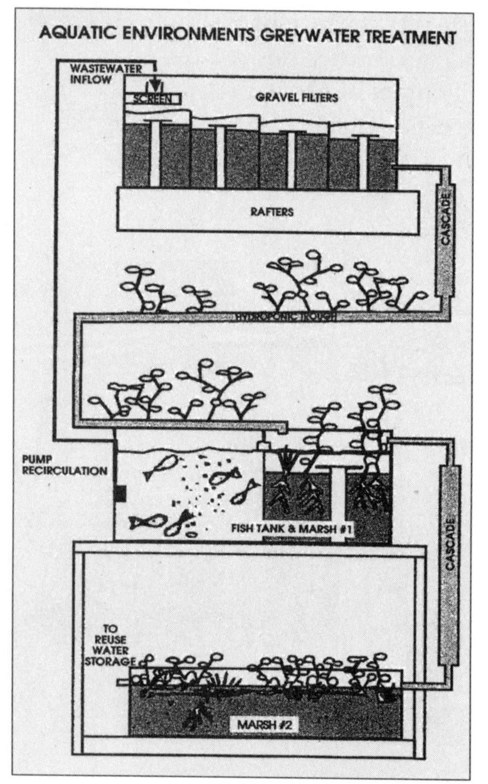

Some examples of indoor graywater planter beds. Clockwise from top left: An indoor graywater planter (Photo: Earthship); a hallway planter manages the washwater from bathroom sinks at an Audubon nature center in Wellfleet, Massachusetts (Photo and diagram: Gerald Ives, architect); a diagram of architect Jorg Ostrowski's graywater system in his Alberta Sustainable House (Graphic: Jorg Ostrowski).

Clockwise from top left: What looks like a simple solar greenhouse and landscape details are doing double-duty as graywater systems for different sources in the home of one of the authors. The greenhouse also helps heat the house. (Photo: David Del Porto); David Bearg in his Concord, Massachusetts graywater greenhouse, which manages graywater from his home's bathrooms. (Photo: Carol Steinfeld); a graywater planter bed thrives in greenhouse conditions in a home in New Mexico. (Photo: Solar Survival Architecture)

Diversion Valves and Graywater Planters

Especially in homes with indoor graywater planters, one should install diversion valves, to allow for the service of the planters or the unexpected (when the football team comes to shower, etc.).

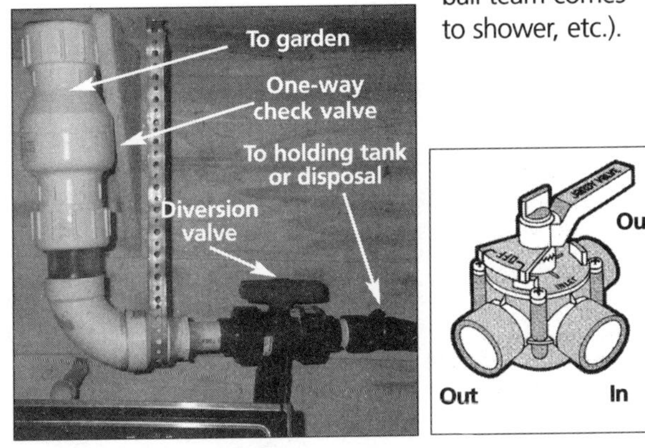

Phreatophytes: Plants for Using Up All of the Water

Many of the hungry, thirsty plants that are key workers in Washwater Gardens are aggressive, invasive plants—not fussy or sensitive plants these. But when properly controlled, they make great machinery for an irrigation system. Remember that unless you add acid to the graywater to neutralize the high pH from soaps and detergents, you should select those plants that tolerate high alkaline conditions, not acid-loving plants.

Remember to avoid using graywater to irrigate food crops, except perhaps orchard crops and fruits that do not touch the ground, because graywater can, and often does, contain pathogens.

Care and Feeding

For some varieties of plants, your graywater will not supply enough nutrients. Remember to fertilize for nitrogen, phosphorous, potassium and micronutrients and maintain the proper soil pH (acid, neutral or alkaline) for the specific plants' requirements.

Check the American Horticultural Society Plant Heat Zone Map and the U.S. Department of Agriculture Plant Hardiness Map for the optimum temperature range for plant selections.

A great all-purpose book on garden maintenance is the American Horticultural Society's *Encyclopedia of Gardening* (New York: Dorling Kindersley, 1993).

Use household chemicals that are good for gardens, such as hydrogen peroxide bleach; soaps and detergents that contain ammonium or potassium or phosphate instead of sodium; and solvents made from citrus oil. The key word is "biodegradable," but use your judgment, because some companies hide dangerous chemicals in biodegradable products. If possible, use potassium-based soaps and detergents; potassium is a fertilizer and does not create the problems associated with sodium-based products. Oxygenating bleaches, such as hydrogen peroxide, are better than chlorine-based products. Liquid soaps are better than powdered soaps, because the latter tends to clog pipes and spaces between soil particles.

Salt

Graywater can be rich in various salts, the most common of which is sodium chloride, or table salt. (Remember that a salt results from the chemical reaction between metals and acids.) Salts cause trouble in a system by binding up the soil and giving plants hypertension.

One of the best sources for this information is from the U.S. Salinity Laboratory. Here is what they say about salt and plants:

Sodium chloride (NaCl) and other salts separate into positive and negative charged particles called ions when they are dissolved into water. Ions that are at lower concentrations outside the plant in the rhizosphere than inside in the root cells move into the plant by a process called "active transport." This transport is catalyzed by enzymes (certain proteins) and consumes energy. Ions at higher concentration outside the plant than in can transport by diffusion. Protein "transporters" assist in this process via pores or channels in the plant. The stems and roots of plants irrigated with salt-laden water or in salty soils will have higher salts levels.

Halophytes

Halophytes, such as the salt cedar, are salt-managing plants commonly found in salty water and soils. They manage toxic ions like Na + and Cl - through a process called "ion exclusion," whereby they limit the influx of ions at the root level or by storing ions in tissue. In some plants, the ions are stored in older leaves and excreted when the leaves are eventually dropped (abscised). Others have salt glands that excrete salt through specialized stomata, usually in the leaf or lenticles.

Study Shows Plants Are Unharmed by Graywater

A two-year study (1996 to 1998) was conducted by the state of California, with the City of Santa Barbara and East Bay Municipal Utilities District in Oakland, to determine the benefits, costs and impacts of irrigating with graywater from single-family homes. Soil and water-quality samples were tested for plant-growth conditions: pH, hardness (dissolved hardness as CaCo3), sodium (Na), specific conductivity (SC), sodium-adsorption ratio (SAR), calcium, magnesium, chloride and boron. The study results indicated that "there were no significant consistent upward trends or spikes in chemical concentrations from the soil samples in any of the constituents. The use of graywater did not appear to have any impact on plant material."

(A. Whitney, R. Bennett, A. Carvajal, M. Prillwitz, "Monitoring Graywater Use: Three Case Studies in California." Sacramento: California Department of Water Resources, 1999.)

How Do Plants Transpire?

How does water move up to and out from the leaves?

"Because water molecules cohere to each other via chemical bonds, called hydrogen bonds, water molecules at the top of the plants are connected to water molecules in the soil much like the cars of a railroad train. When water transpires (a diffusion process) from the leaves, other water molecules are brought closer to the root surface. This waterway is actually called the transpiration stream. Ions move in the transpiration stream much like a non-powered boat floats along a river stream. Thus, the transpiration stream brings ions from the soil water, first to the root where they must cross the plasma membrane barrier, and eventually to the leaf. At the leaf, the water molecules can escape back into the atmosphere through another specialized leaf cell called a stomate. Ions, however, will be left behind."

Fact Sheet SL-59, a series of the Soil and Water Science Department, Florida Cooperative Extension Service, Institute of Food and Agricultural Sciences, University of Florida. Publication date: February, 1993. http://www.ussl.ars.usda.gov/answers/cellroot.htm

Thirsty Plants for Graywater-Utilization Systems

Here is a list of nonedible, thirsty plants (not in any particular order):

For a Summer Garden or Tropical Greenhouse

- *Hibiscus rosa-sinensis* (tropical), *syriacus* (rose of Sharon), *moscheutos* (rose mallow), *cannabinus* (kenaf) and *Rubiaceae* (gardenia)
- The large-order *Zingiberales* are very thirsty and beautiful plants that can be found for tropical or greenhouse applications. The family members include: *Musaceae* (Banana), *Strelitziaceae* (bird of paradise), *Lowiaceae, Heliconiaceae, Zingiberaceae* (ginger), *Cannaceae* (canna lilly) and *Marantaceae* (prayer plant)

  Our Zingiberales favorites:

  - *Zanteceschia aethiopica* (calla or arum lilly)
  - *Heliconia* (standeyi Macbride or psittacorum)
  - *Strezlitzia reginae* (bird of paradise)
  - *Musa paradisiaca* (common banana) or *Musa coccinea* (flowering banana)
  - *Alpina zerumbet* (shell ginger) or *Alpina* purata "Rosea" (rose ginger)
- *Typha sp.* such as *glauca, angustifolia, latifolia, minimus* (bull-rush, cattail)
- *Monstera deliciosa* (monstera)
- *Hosta fortunei* (plantain lilly)
- *Aroids* such as *Alocasia macrorrhiza* (elephant ear) or *Colocasia esculenta* (taro)
- *Melaleuca quinquenervia* (Cajeput tree) or *nesophylla* (western tea myrtle)
- *Cyperus papyrus* (Egyptian paper plant)
- *Eichornia crassipes* or *azurea* (water hyacinth, which will live in wet sand)
- *Rhizophora sp.* (mangrove)
- Bamboo (genus *Arundinaria, Bambusa, Phyllostachys, Sasa* or *Shibataea*)
- *Tamarix ramosissima* (salt cedar, a very thirsty halophyte)
- *Phragmites australis* or *Phragmites communis* (common reed, disturbance weed)
- *Deconia verticillatus* (swamp loosestrife)

Again, remember to adjust your graywater's pH levels for the plants that do not thrive in high-alkaline conditions.

For a Winter Washwater Garden

- *Salix sp.* (All willows are very thirsty. Some are decidous trees, others are evergreen bushes.)
- *Phyllostachys aureosulcata* (evergreen yellow groove bamboo)
- *Phyllostachys bissetti* (a very hardy evergreen bamboo)
- *Phyllostachys nuda* (evergreen snow bamboo)
- *Fargesia murielae* (umbrella bamboo)
- *Fargesia nitida* (fountain bamboo)

  Subspecies include *Ancepts, de Belder, Eisenbach, Ems River, McClure* and *Nymphenburg.*
- *Arundinaria* or *Pleioblastus pygmaea* (evergreen dwarf bamboo) (This can be used for ground cover together with the tall bamboo.)
- *Thuja occidentalis* (emerald or American dark arborvitae)
- *Ilex verticilatus* (swamp holly) or the hybrid, Blue or Meserve
- *Klamia latifolia* (mountain laurel)
- *Taxus canadensis* (American yew)
- *Tsuga canadensis* (American hemlock)

Grasses

Phalaris (canary grass), *Lagarus* (hare's tail), *Briza* (quaking grass), *Pancium* (red switch grass), *Zizania* (wild rice) and *Erianthus saccharoides* (spreading witch grass) are excellent seasonal or warm-weather plants.

Other Flowers

Nasturium, nicotiana, cosmos, snapdragons, salvia, swan river daisy, passiflora and blue splendor will add color to your summer garden.

Vines

Plumeria, ipomoea and dioscorea will add fragrance and drama to a summer garden.

Corn

Yes, corn. To supplement a Washwater Garden during the start-up period, try planting ornamental corn, such as Fiesta, which will take up a lot of water and nutrients. It matures in 60 to 90 days.

You can mix in seasonal with winter-hardy plants and any of your favorite plants alongside the worker plants. Some plants, like bamboo, will be hardy to the point of being invasive if you don't contain them.

Artificial Wetlands

Also known as "constructed wetlands" and "rock filters," artificial wetlands have been used successfully for both on-site systems and large municipal treatment works. Well-publicized municipal systems are in Calcutta, India, and in Arcata, California.

Many design manuals for creating these are available from the EPA the Tennessee Valley Authority.

The advantages are that they provide a high level, or tertiary, treatment of effluents, and can do so with minimal management. Increasingly, they are also used for treating stormwater prior to discharging it into ground and surface waters.

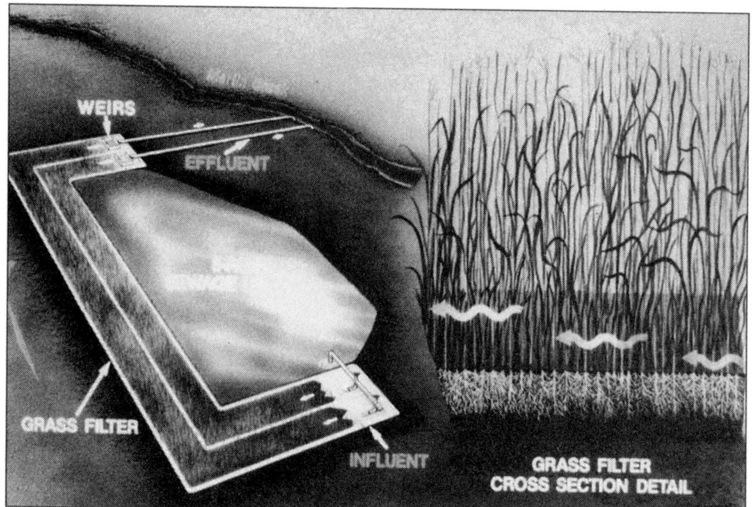

A cutaway of an artificial wetlands system (Graphic: Bill Wolverton, NASA)

As the name implies, an artificial wetland is a constructed ecosystem in which wastewater is fed in at one end, and, thanks to hydraulic gradient or pumps, comes out the other end after being exposed to a bioplex of organisms, both aerobic and anaerobic, that cling to the root systems and planting media (such as gravel). This reduces pathogens, removes nutrients and mineralizes organics. Unlike Washwater Gardens, constructed wetlands are saturated (immersed in water) anaerobic systems with an aerobic zone only at the surface (although this varies with the system).

In Denmark and parts of the United States, such as Arkansas, artificial wetlands are common forms of post-septic tank treatment, often featuring an outfall into groundwater or surface water. In warm summer months, many of these systems become zero-discharge, because the evapotranspiration rates are higher than the in-flow rate. In cool, rainy climates, there are usually outfalls from the system. Artificial wetlands perform better where it is warmer; for that reason, they are most common in warmer and drier climates. Some are enclosed in greenhouses.

An artificial wetlands system in Arizona (Photo: Southwest Wetlands)

In environmentally sensitive areas, they are constructed with impervious geotextile liners. In others, the sides and bottoms are sealed with bentonite clay, or they self-seal as they mature.

Artificial wetlands are typically planted with marsh plants, such as cattails, reeds, sedges and rushes.

Types of Artificial Wetlands

Typically, there are three different kinds of artificial wetlands:

- **Free-water-surface (FWS) and overland flow constructed wetlands:** Plants are periodically submerged or flooded (essentially, allow a grassy area to flood).
- **Subsurface-flow (SF) constructed wetlands:** The effluent is treated as it flows through a porous medium such as crushed rock or gravel. Vegetation is planted in the medium, so you don't necessary see the wastewater flow, which is 18 inches to three feet (0.46-0.9 m) below the surface.
- **Floating aquatic plant systems:** Very shallow one- to three-foot ponds are planted with duckweed, azola, water hyacinth and other fast-growing aquatic plants. Their root systems provide excellent habitat for the microorganisms that transform the constituents of the effluent. Duckweed, a.k.a. "lemna," has a short root system. It sequesters nutrients in the biomass as plant protein and is harvested with special equipment and used for biogas generation, animal feed or composts.
- **Combination systems:** These are usually sequenced systems of the above, so that there may be a series of cells, each providing different levels of treatment, some of which are organized in totally enclosed systems. (Tchobanoglous and Crites, *Small and Decentralized Wastewater Management Systems,* McGraw-Hill, 1998)

In addition to being effective wastewater treatment systems, artificial wetlands provide natural areas and habitat for wildlife.

Possible disadvantages are that they require significant land area due to the shallow nature of the systems and require longer detention times. They can also fill with rain and snow. In the case of floating systems, special equipment is needed to harvest the plants.

They tend to be self-regulating, so they need little management, but their treatment and process rates can vary significantly based on environmental factors such as temperature fluctuations, plant diseases, etc.

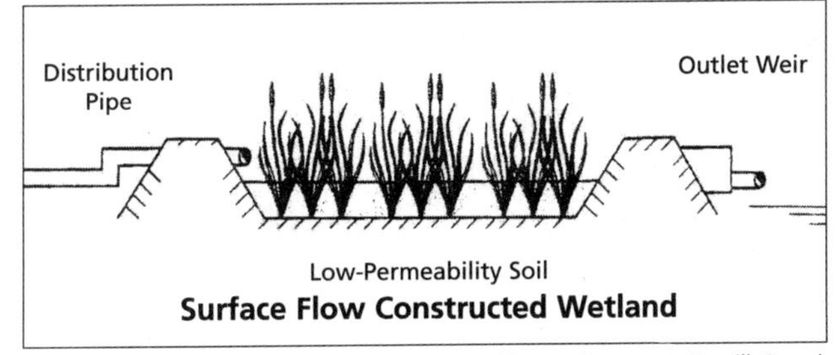

A simple side view of an artificial wetland (Illustration: R.M. Towill Corp.)

The Willows That Sucked Up

In 1991, planners in a Florida town wondered why the water level of an aquifer was dropping. It could not be explained by pumping rates or lake effect evaporation. Then they discovered a large stand of willow trees. Researchers measured the amount of water withdrawn from the aquifer by the trees. They reported in the *Miami Conservancy District Aquifer Update No. 1.1,* that on a hot summer day a mature willow tree was removing 5,000 gallons of water from the aquifer!

Willow trees aren't the only big drinkers: One hectare of an herbaceous plant, such as saltwater cord grass, evapotranspires up to 21,000 gallons of water per day. (Schwab et al., Elementary Soil and Water Engineering, 2d ed. New York: John Wiley & Sons, 1957).

That is the power of plants.

The following is an example of an integrated conventional and natural system.

Living Machines® and Solar Aquatics® Systems

Living Machines and Solar Aquatics systems are enclosed systems usually for combined wastewater. However, they can be scaled down and used as graywater systems.

These were designed by John Todd as a result of his National Science Foundation–funded aquaculture research at the New Alchemy Institute in Falmouth, Massachusetts. Todd was seeking to optimize fish production in transparent solar silos. He found that fish excrement and photosynthesis caused the growth of algae. Tilapia and other fish were introduced to eat the algae and to be later harvested in a backyard fish farm. However, fish urine (uric acid and urea) turned to toxic ammonia levels in the tanks. A colleague, Bill Wolverton of NASA, was doing parallel research on water hyacinth's up-take of nutrients. When they put the two together, they had a combined system that provided a very high level of nutrient and BOD removal.

Todd patented this application for wastewater treatment, and it has been successfully used for domestic and food-production wastewater treatment by for-profit companies, such as Living Technologies in Burlington,

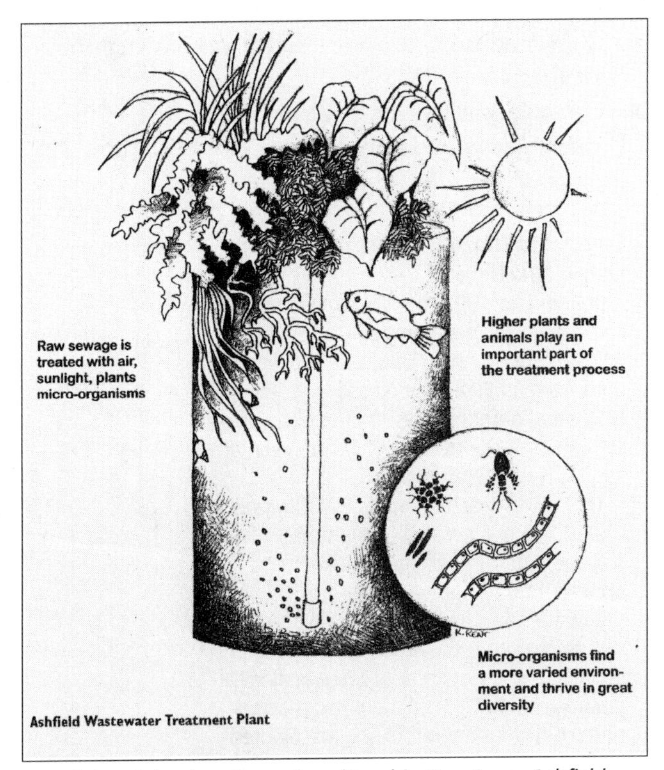

A tank in a Solar Aquatics system (Graphic: Tom Leue, Ashfield, Massachusetts Wastewater Treatment Plant)

Vermont, and Ecological Engineering Associates of Marion, Massachusetts Living Machine–type systems can be found in the United States, Mexico, England and in northern Sweden. Some are managing as much 1 million gallons (3,785,00 liters) a day, and some manage only food-processing residues at food-production factories.

A Living Machine system is also effectively used in a closed-loop zero-discharge system at a Vermont highway rest area. There, combined black- and graywaters are treated in sequenced ecosystems, then disinfected by ultraviolet light and reused as the make-up water for toilet and urinal flushing.

A Living Machines system manages wastewater at a highway rest stop in Vermont.

The idea is to seed an enclosed aerated wetland environment with a large diversity of plants, animals and microbes, and allow them to self-select to optimum levels, following the maxim that "if you build it, they will come." Higher temperatures improve their performance rates.

The key elements of these systems are:

- Large sequenced tanks that contain the process (Solar Aquatics systems feature transparent tanks, a patented feature.)
- Oxygen bubbled into the tanks through air stones
- Plant roots that provide an optimal environment for the aerobic bacteria and other biota
- A greenhouse to provide shelter, warmth and photosynthesis
- Optionally, fish, such as tilapia and catfish, to provide an advanced polyculture. (These fish, however, are not eaten. Although in developing countries, fish production is integrated into the end processes of these systems, after nutrients, chemicals and pathogens have been removed.)

With this system, there is still treated effluent to dispose or utilize.

Regulators may consider the cost of Living Machines high compared with conventional secondary treatment systems. They involve high capital costs for greenhouses and treatment works, power for maintaining high oxygen levels in the water, and regular maintenance. However, compared to conventional tertiary treatment, they are more cost-competitive. The reality is that these are being installed where there is no possibility of using conventional secondary treatment systems.

South Burlington Municipal Living Machine

Built in late 1995, the South Burlington Living Machine was ramped up to full design flow by April 1996. Water chemistry data from an EPA-certified independent laboratory demonstrated the excellent performance of the treatment system even at very cold temperatures.

This Living Machine treats 80,000 gallons (302,800 liters) per day of municipal sewage, an amount typically generated by about 1,600 residential users.

Sewage flows to a greenhouse housing two treatment trains, each with five aerobic reactors, a clarifier and three ecological fluidized beds. The open aerobic reactors have diaphragm aerators and are planted with a variety of aquatic plant species in floating plant racks.

—from Living Machines, Inc.

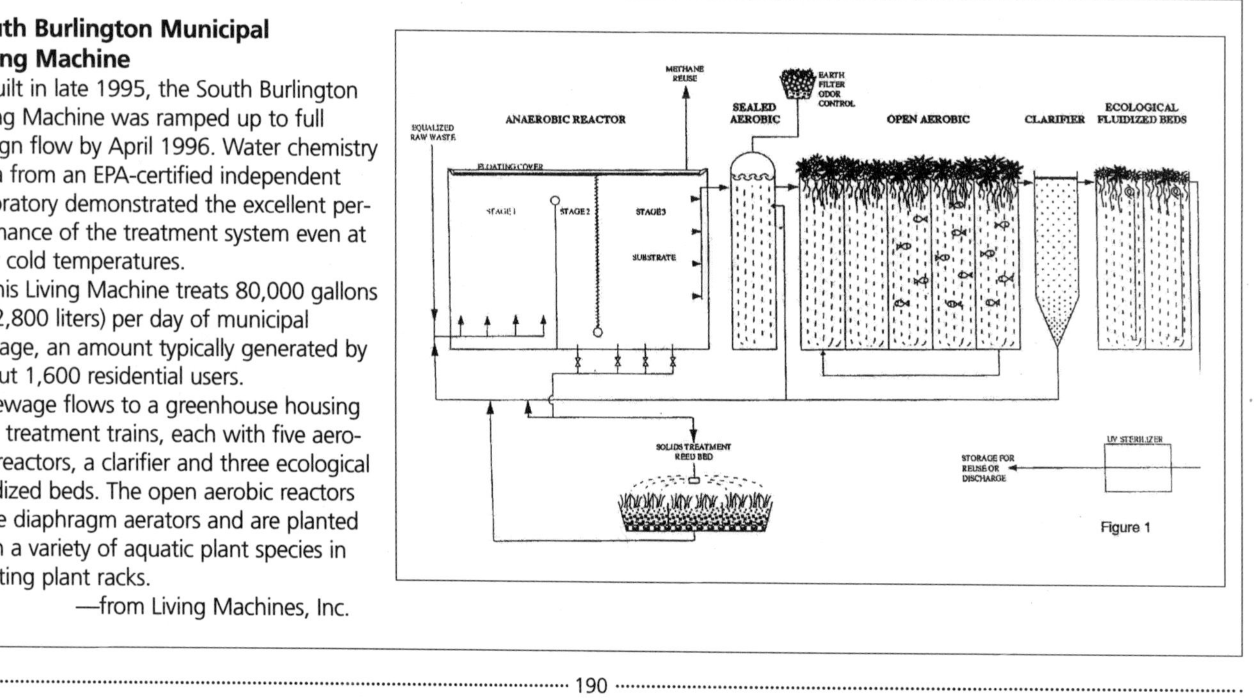

Figure 1

References

Van Nostrand's Scientific Encyclopedia, 5th ed. (New York: Van Nostrand Reinhold and Company, 1976).

American Water Works Association Research Foundation's Residential End Use Study. Available on the World Wide Web: http://www.waterwiser.org/wtruse98/main.html

Carl Lindstrom, "Key Graywater Characteristics" (Cambridge, Massachusetts: privately published, 1997)

Peter Warshall, "Above-Ground Use of Gray water," Bolinas, California

Peter Warshall, *Septic Tank Practices* (Tucson, Arizona: Mesa Press, 1976)

Ron Crites and George Tchobanoglous, *Small and Decentralized Wastewater Management Systems* (New York: McGraw-Hill, 1998)

Soils for the Management of Wastes and Waste Waters (Soil Science Society of America, 1977)

Fact Sheet SL-59; a series of the Soil and Water Science Department, Florida Cooperative Extension Service, Institute of Food and Agricultural Sciences, University of Florida. February, 1993.

Alfred Bernhart, *Evapotranspiration, Nutrient Uptake, Soil Infiltration of Effluent Water* (Toronto, Canada: self-published, 1985)

References for study on page 165:

Gray water Pilot Project: Final Report, City of Los Angeles Office of Water Reclamation November, 1992.

This study referenced the following:

Robert L. Siegrist & William C. Boyle. On-site reclamation of residential gray water.

Joan B. Rose et al. Microbial quality of and persistence of enteric pathogens in gray water from various household sources. Water. resources. Vol. 25 No. 1, PP 37-42, 1992.

CCDEH issue paper: Gray water use in California. California Conference of Directors of Environmental Health. 1992.

W. C. Boyle et al., Treatment of residential gray water with intermittent sand filtration. A. S. Eikum and R. W. Seabloom (eds.). Alternative wastewater treatment, 277-300 Copyright 1982 by D. Reidel Publishing Company.

Kevin M. Sherman, On-site Gray water System Research from a National Perspective. Florida, Journal of Environmental Health, Issue 134, May 1991.

Mark Brandes, Characteristics of Effluents from Gray water Septic Tanks. JWPCF, Vol. 50, No. 11, November, 1978.

Martin M. Karpiscak et al., "Residential water conservation," CASA DEL AGUA, Water Resources Bulletin, Vol. 26, No. 6, December 1990.

Karol M. Enferadi et al., "A Field Investigation of Biological Toilet Systems and Gray Water Treatment." State of California Department of Health Services, Berkeley, California, 1980

Alan T. Ingham, "Residential Gray Water Management in California." State Water Resources Control Board, Division of Planning and Research January, 1980.

Enviro-Management & Research, Inc., "Assessment of On-Site Wastewater Treatment and Recycling Systems." Springfield, Virginia. March, 1992.

Bill Roley, "Home and Community Water Management." Sustainable Cities Concepts and Strategies for Eco-City Development, Eco Home Media 1992.

Trimming the "biomass" (here, a taro plant) is a key task in maintaining a solar aquatics system in Ashland, Massachusetts (Photo: Carol Steinfeld)

Principles of Graywater Use

Here are the four Rs of ecological graywater approaches:

- ☐ **Reduce:** Conserve water to lower the net volume of graywater to be managed. Avoid using or draining toxic chemicals.
- ☐ **Reuse:** Cascade high-quality graywater (condensate and shower "warm-up" water) to other uses in the home (washing machine, toilet flushing) to further reduce the final water component in the effluent to be recycled.
- ☐ **Remove:** Filter out grease, oils, solids and especially nonbiodegradable particles, such as lint from washing machines.
- ☐ **Recycle:** Following appropriate treatment, put it to safe and productive use, such as subsurface irrigation, evaporative cooling or groundwater recharge.

Graywater Resources

Again, remember when looking for graywater information, that the key term for many organizations working in this realm is "water reuse." Here are some sources:

The World Health Organization (WHO) provides publications on water supply and sanitation. Some good titles include:
Analysis of Wastewater for Use in Agriculture, Guide to the Development of On-Site Sanitation, Guidelines for the Safe Use of Wastewater and Excreta in Agriculture and Aquaculture, Operation and Maintenance of Urban Water Supply and Sanitation Systems and *Parasitic Diseases in Water Resources Development.*
World Health Organization,
Distribution and Sales
CH-1211 Geneva 27
Switzerland
41 22 791 24 76; Fax: 41 22 791 48 57
publications@who.ch
In the United States:
WHO Publications Center USA
49 Sheridan Avenue
Albany, NY 12210

USEPA Guidelines for Water Reuse, published by the U.S. Environmental Protection Agency (USEPA is a classic comprehensive manual covering all aspects of reuse, including planning reuse systems and state requirements (USEPA/USAID EPA /625/R-92/004: September, 1992). The book and updates may be available from:
USEPA National Center for Environmental Publications
USEPA-NSCEP
P.O. Box 42419
Cincinnati, OH 45242-2419
513/489-8190
800-490-9198
www.epa.gov/epahome/publications.htm

A wide variety of clearly written resources are available from the **National Small Flows Clearinghouse (NSFC)**. Funded by the USEPA, the NSFC provides information about innovative, low-cost wastewater treatment. Its emphasis is practical, alternative solutions for "small flows" wastewater problems, versus those of large urban treatment plants. The NSFC houses a number of extensive and comprehensive computer databases that contain the latest information about small flows technologies. Searchable by key word, these databases allow technical assistants to access a variety of wastewater information and provide printouts of the searches.
National Small Flows Clearinghouse
P.O. Box 6064
Morgantown, WV 26506-6064
800-624-8301 or 304/293-4191
http://www.estd.wvu.edu/nsfc

Create an Oasis with Greywater by Art Ludwig
One of a very few books solely about graywater, with discussion of cleaners, system options, and more.
The Builder's Greywater Guide features more nuts and bolts information for designers and installers, as well as the complete text of the California graywater law.
$14.95 each from Oasis Design
5 San Marcos Trout Club
Santa Barbara, CA 93105-9726
805 967-9956
http://oasisdesign.net

The water industry has the most to gain when it comes to water reuse. Here are some sources that should prove valuable as time goes on:
WaterWiser, the Water Efficiency Clearinghouse is a cooperative project of the American Water Works Association, the USEPA and the U.S. Bureau of Reclamation. It conducts studies, produces information materials and provides an internet forum for those seeking information.
WaterWiser
6666 West Quincy Avenue
Denver, CO 80235
800-559-9855
http://www.waterwiser.org

Water-reuse conferences: Water-reuse technology and applications, as well as the need for water-reuse expertise, are expanding every day. The proceedings of AWWA water-reuse conferences offer a wide spectrum of water-reuse research and the practical application of new technology and ideas.
American Water Works Association
6666 West Quincy Avenue
Denver, CO 80235
303.794.7711
www.awwa.org

American Desalination Association (ADA) advocates and reports on technologies, including membrane and other processes to remove salts, organic materials, microbial contaminants and other water impurities.
American Desalination Association
915 L Street, Suite 1000
Sacramento, CA 95814
916/442-9285; fax: 916/442-0382
http://www.desalting-ada.org

Some states are on top of water-reuse issues and can be counted on to keep up to date. Texas is one of them.
On-Site Insight, the newsletter of the Texas Water Resources Institute
Texas A&M University
301 Scoates
College Station, TX 77843
409/845-1851

California is another. The booklet, **Using Graywater in Your Home Landscape,** is a good introduction for homeowners about how to irrigate with graywater, under California law. Available from:
California Department of Water Resources
Publications Office
P.O. Box 942836
Sacramento, CA 94236
916/ 653-1097

Standards for the Use of Adequately and Reliably Treated Wastewater in Place of Drinking Water for Non-Potable Uses
The title says it all. Request Publication #97-23 from:
Department of Ecology Publications
Distribution Center
PO Box 47600
Olympia, WA 98504-7600

The Septic System Owner's Manual by Lloyd Kahn, Blair Allen and Julie Jones, Shelter Publications, is an excellent user-friendly book about current septic tank practices and alternative wastewater treatment systems for homeowners. Clear, concise and easily understood, while technically accurate. $14.95 from Shelter Publications.

The **Uniform Plumbing Code** is used by many states and some countries (Palau, for example). It has a detailed section on graywater irrigation and water reuse in Appendix G & H.
International Association of Plumbing and Mechanical Officials (IAPMO)
Publications Department
20001 Walnut Drive South
Walnut, CA 91789
800-854-2766

Education plays a critical role in making shift happen. Even NASA is interested in graywater, and offers an educational program about it for students.

"My Water is Gray Water" is a unit in NASA's STELLAR Program, in which participants utilize Space Life Sciences and Shuttle mission research data to create K-14 classroom activities that engage grade level 5 through 8 students in investigating recycling aboard spacecraft. Students are encouraged to recycle graywater and to find ways to purify it for future use. Activity modules are available on the Web, CD-ROMs, video and in printed form. See the STELLAR web site:
http://stellar.arc.nasa.gov/stellar/Activities/hydro/GrayWater/GrayWater.html

Success Factors for Natural Systems

The effectiveness of natural wastewater systems depends on:

1. The biodiversity of the biological community (the "biota") of both aquatic microorganisms and higher-order animals and plants
2. Temperature (and to a lesser extent, light) and dissolved oxygen
3. Strength and constituents of the wastewater
4. Time, also known as "detention," that the wastewater spends in the system

A big advantage of ecological systems is that they adapt well to changes in environmental factors and wastewater characteristics. That is due to the ability of the microorganisms in them to adapt rapidly.

The plants provide several functions, but mainly a rhizosphere—a conducive setting for microbial action.

- The roots provide surfaces for microbial colonization, as well as some adsorption on their surfaces, and a complex of oxygen and other nutrients that the microorganisms require.
- To the extent that the biomass (leaves, stems, etc.) can be harvested, plants take up and utilize nutrients, and they can remove hyperaccumulated minerals and some toxic compounds, such as heavy metals.

Start-up Time

Remember that all natural wastewater systems, like infants, require time to mature before they are at their maximum efficiency for managing wastewater. It takes time for plants to get acclimated and hearty and for a community of microorganisms to form in the rhizophere. Two years is the minimum in temperate climates.

For a graywater irrigation system planted with new shrubs and bushes, you might also succession plant some ornamental corn the first year to establish a strong rhizophere and some uptake of nutrients as the longer-term plants take hold.

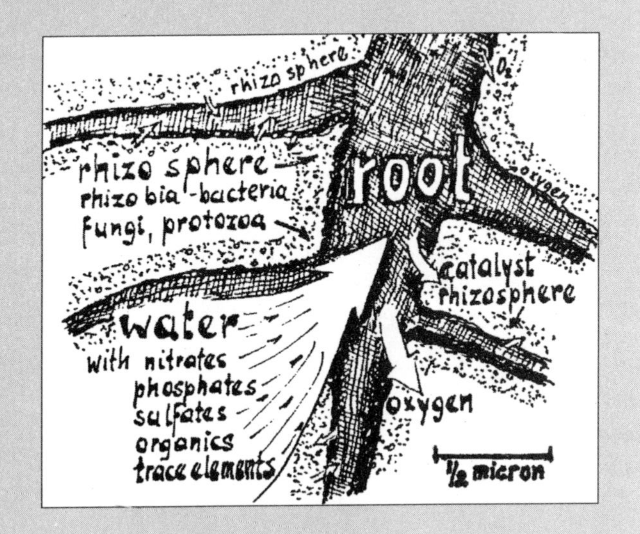

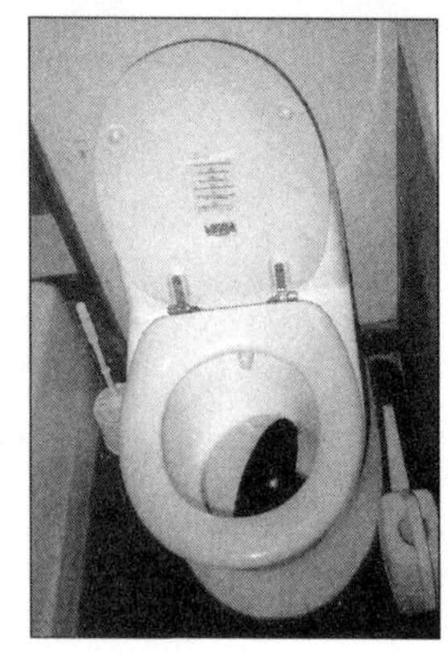

Ecological proved economical for this Massachusetts homeowner, whose home was located in a water-sensitive area (the first clue: its address is "Swamp Road") with high groundwater and the town's wellfields nearby. The local permitting board required the installation of a zero-waste system—no effluent could enter the ground. The home was retrofitted with a Wastewater Garden in a greenhouse (above left), which also provides passive solar heat to the home. Inside, phreatophytes (especially hungry and thirsty plants such as coco palm, bamboo and other tropicals) assist the evapotranspiration process. Above right: one of the authors checks out a bird of paradise plant. The home uses waterless toilets (left) which flow to a composting toilet system.

CHAPTER TEN

The Future of Composting Toilet Systems

As costs continue to rise for cleaning up water, wastewater and water pollution—and as this pollution affects more of us more directly—the sound eco-nomics of ecological wastewater solutions will make them viable and mainstream choices.

In many parts of the world, acute water shortages call for the most strategic and efficient use of water, making waterless and low-water toilet systems a viable—if not overdue—solution.

What you see in this book is certainly not the end of the evolutionary line of composting toilet systems. These systems will evolve into microbial reactors that control all of the composting process variables automatically. They will be a more transparent technology requiring relatively little responsibility from their owner-operators. Periodically, the systems will deposit packets of high-powered designer composting microbes into the composter. In the future, we may refer to composting toilets as "unsaturated aerobic systems," a broader term that better describes this approach to waste management.

Mass production and increased demand will bring down the cost of manufactured systems. At the same time, composters may be built right into building foundations in new construction. Self-contained composting toilets will become more aesthetically acceptable to more people. As in a garbage compactor, the compost/end-product will be removed fully packaged in biodegradeable bags that you either throw away or use on your yard where you need a soil conditioner.

A municipal service person will come around every few months, check your system, and perhaps take a full composter to a central composting facility.

Washwater will no longer be considered wastewater, because it is too valuable a commodity. It will be recycled for irrigation, oftentimes on rooftops of high-rise buildings and hanging gardens of vertical walls, as well as in landscapes.

Micro-flush toilets will advance, and will likely be used with all types of systems.

Ecologically integrated homes, with passive solar features, composting rooms and removal doors, and graywater-irrigated landscapes, will be the norm—and will look completely conventional.

Some key changes are occurring now. In rural and suburban communities, management districts are

The Smart Toilet: James Sullivan, a Harvard University engineering student, demonstrates "The Smart Toilet," which he designed and built. A self-contained composting toilet system, it features solid-state sensors that automatically adjust moisture and temperature levels. The Smart Toilet still has design and cost issues to be worked out, but the electronic-control approach is surely the future for composting toilet systems. (Photo: Carol Steinfeld)

In a small community in Rotterdam in the Netherlands, roll-away composters will be collected and managed by a central authority as a pilot project exploring ecological wastewater solutions. (Illustration: De Twalf Ambachten)

being formed for financing the construction, installation and management of on-site wastewater treatment systems. The goal is to extend the management and financing mechanisms for central systems to on-site systems, which are far more feasible and economical for many communities.

Today, many on-site systems are shared by small clusters of homes. Construction, operation and maintenance are handled through modified condominium agreements.

It is quite possible that our central collection system will be used for commercial and industrial use only.

Some suggest this is harkening back to the past. A more accurate view is that it is a timely converging of common sense, economics, resource management know-how, and technology. ✡

More homes in the future will incorporate garden features that do double-duty as wastewater systems. (Illustration: Matthew Loden/Insite)

Sewage Meters: The Advent of Pay-per-Use Wastewater Treatment?

In the past, when water was cheap and sewage treatment costs were low, the wastewater system of choice was the sewer. Sewers—central collection and treatment—transported away the wastewater management responsibility and afforded cities more opportunities for growth. (It is just this growth and sprawl that now prompts many towns to avoid constructing sewer systems.)

At first, the cost of running the sewer plant was built into property owners' taxes. Then it was included in water bills. In Newton, Massachusetts, for example, wastewater treatment costs account for three-quarters of property owners' water bills. Newton is one of 43 towns and cities in Greater Boston that make up the Massachusetts Water Resources Authority (MWRA) district, which is served by one large central treatment plant.

As costs to maintain the MWRA's sewer system rose, a sewage meter was installed to measure each member town's sewage flow, so they would be charged in proportion to their measured sewage flows. (An aside: Then water bills jumped in some areas, and it was discovered that a huge amount of groundwater was infiltrating the sewer pipes and going to the treatment plant!) However, folks who used water to water lawns and fill pools, etc., didn't want to pay for the cost of treating sewage they didn't create. So, some communities are now considering allowing the installation of sewage meters in buildings. A property owner who shows a difference in water input and sewage output could qualify for an abatement of the sewage portion of the bill.

Ultimately, when the use of such meters is more widespread, property owners will be more aware of just how much sewage they create—and there will be greater incentives to conserve water and not produce sewage, so those who are not creating the problem will no longer have to subsidize those who are.

Central Management of Composting Systems

Central sewer systems work as far as moving waste to a central area for broad treatment. From most users' points of view, these technologies perform reliably and consistently (until they hear of problems). They were developed in response to urban wastewater problems where population densities and pressure for development were high. However, they were not intended for rural and suburban communities, with relatively fewer users and lower flow rates, which make central sewage excessively expensive.

Central systems are capital intensive because the collection component —pipes, pump stations—of a centralized system adds 60 to 80 percent to the total cost. This capital cost, plus operating costs, are divided by the number of users, or the gallons of sewage treated. With fewer users or lower flows, the cost per unit of service is much higher. Water conservation further reduces the flows, driving up the unit cost. This in turn drives more conservation, and so on...

Bringing Central Advantages to On-Site Systems

Increasingly, the United States Environmental Protection Agency and state planners are seeking ways to bring the benefits of central treatment to on-site systems, which are more affordable and feasible for many communities, as they alleviate the need for miles of pipe and infrastructure. The problem is that individual on-site systems are not centrally managed, so they do not offer these benefits. At the same time, most federal funds are for central systems only.

A direction many are taking now is forming management districts: essentially municipal organizations that oversee and manage on-site systems, and have the authority to finance construction. The user would have to be willing to pay for this service, just as they would for sewer service.

According to Richard Otis, a well-respected on-site systems engineer, central management district provide owners of on-site systems the advantages usually associated with central sewer systems, such as: freedom from uncertainty, freedom from responsibility, available public financing and environmental protection.

How Does that Relate to Composting Toilet Systems?

Management districts could pave the way for further installations of composting toilets, graywater systems and other ecological wastewater management solutions. A classic concern about these systems is that they require monitoring and some maintenance responsibilities; management districts would manage that.

Management districts would also allow a community to integrate a diversity of collection and treatment technologies based on what is appropriate for each neighborhood or area.

Cities and towns could turn down proposals for new multi-million-dollar central wastewater treatment plants in favor of less expensive individual on-site systems that are monitored regularly by a municipal management district organization. This way, they both limit growth and assure accountability.

Reference

R. Otis, "Decentralized Wastewater Treatment: A Misnomer," ASAE Conference, Orlando Florida, March, 1998

General Resources

The following are some resources for composting toilet and graywater system information (some of these are listed elsewhere in the book—check the specific sections for others).

Books on Composting Toilets

***Kompost-Toiletten* by Claudia Lorenz-Ladener (151 pages. 1992, Ökobuch Verlag)**

Sprechen sie Deutsche? This is a well-formatted book about composting toilets, written in German, with many good illustrations and photos, including a few nice diagrams of ecologically designed homes. About 40DM postpaid from Ökobuch, Postfach 1126, 79216 Staufen, Germany. (Website: www.okobuch.de)

***Lifting the Lid* by Peter Harper and Louise Halestrap (1999, Centre for Alternative Technology Publications)**

A follow-on to *Fertile Waste,* this book provides an overview of several composting toilet systems in use in Europe, with discussion of the science of composting. Particularly interesting are the photos of a onions grown in a garden irrigated with urine and graywater compared to those that were not (the former are visibly larger). Another group of photos tracks some experiments with composting urine at the Centre for Alternative Technology (CAT). Available in the United States from Jade Mountain: 800-442-1972. Or see CAT's website: www.ccat.org.uk.

***Fertile Waste* by Peter Harper (1998, Centre for Alternative Technology Publications)**

This booklet includes descriptions of a few systems, and general plans for a two-vault system (without an aerator). Available in the United States for $12 from Jade Mountain: 800-442-1972.

***Ecological Sanitation* edited by Uno Winblad, et al (Sida, Stockholm, 1998)**

A clear overview of some composting and dry toilet options for developing countries, with an emphasis on drying toilets and urine diversion. Order on a "print-on-demand" basis from Arkitektkopia, Attn: Lars Lundberg, Box 3436, 103 68, Stockholm, Sweden. Email: LLU@akopia.se

***The Humanure Book* by J.C. Jenkins (2nd ed. 1999, Jenkins Publishing)**

This folksy book contains much good information about why composting it, not flushing it, is a better way, as well as the fertilizer value of composted human waste, and how to use and manage a direct outdoor composting system with a bucket collector. $20.95 from Chelsea Green Publishing: 802/295-6300.

***The Toilet Papers* by Sim Van der Ryn**

Originally published in 1977, *The Toilet Papers* features nice photos and illustrations of some composting toilets, with philosophy, discussions of natural postures for defecating and information about some ecological wastewater systems. Available from Chelsea Green Publishing.

***Future Fertility: Transforming Human Waste Into Human Health* by John Beeby (164 pages, 1995, Ecology Action)**

A well-researched report on composting human excrement for fertilizer in an agricultural setting. This is about manure management not composting toilet systems. Order from Ecology Action, 5798 Ridgewood Road, Willits, CA 95490-9730.

General

***Small and Decentralized Wastewater Management Systems* by Ron Crites and George Tchobanoglous (1998, WCB/McGraw-Hill)**

Not a homeowner book, this technical textbook features in-depth details about engineered designs of all kinds of black- and graywater treatment systems. You'll find detailed descriptions, calculations and references on constituents in wastewater, the fate of wastewater constituents in the environment, process analysis and design, pretreatment operations and processes, alternate collection systems, biological treatment and nutrient removal, lagoon treatment systems, wetland and aquatic systems, land treatment, intermittent and recirculating filters, effluent repurification and reuse, disposal options, biosolids and sludge management, and system management. About $80.

***Sewage Solutions* by Nick Grant, Mark Moodie and Chris Weedon (1996, Centre for Alternative Technology)**

Mostly geared to the United Kingdom, this excellent and reader-friendly book features facts about various lower-impact on-site wastewater solutions (about a page on each), from rainwater collection to better faucets to reed beds. Includes comparison charts and diagrams. Available in the U.S. for $14.95 from Jade Mountain.

Factsheets about on-site systems

The National Small Flows Clearinghouse is a government-sponsored source of information about all kinds of on-site and smaller-system wastewater issues, including compendia of articles and tech sheets on composting toilets and graywater systems (revised by this book's authors). National Small Flows Clearinghouse, West Virginia University, P.O. Box 6064, Morgantown, WV 26506-6064; 800-624-8301.

Managing Organic Wastes

"Recycling Organic Waste: From Urban Pollutant to Farm Resource," Worldwatch Paper 135 by Gary Gardner (1997, Worldwatch)

An excellent report on the effects of breaking the "organic loop"—moving around organic material and soil nutrients from their points of origin, then paying a high cost to buy them back or clean them up, when they are in the wrong places. Highlights the need for better and more strategic resource management.

BioCycle, Journal of Composting & Recycling

This the trade journal for the recycling industry. In addition to municipal and industrial composting issues, the magazine occasionally runs articles on vermicomposting and composting toilets. The same company also publishes *Compost Science and Utilization,* a more technical journal about composting research. The JG Press, 419 State Avenue, Emmaus, PA 18049.

World Wide Web Sites

International Composting Toilet News www.compostingtoilet.org

Produced by Kevin Ison, this is a web site featuring several composting toilet sources and consultants throughout the world, as well as a forum for readers' questions. Donations are accepted.

CEPP's Ecological Sanitation Email Discussion List: http://EcoSan.listbot.com

Exchange ideas and experiences about composting toilet installations on this email discussion list originally established for buyers of the Center for Ecological Pollution Prevention's composting toilet system plans.

Glossary

The following are terms defined as they relate to this book.

Actinomycetes: Unicellular, mostly aerobic micro-organisms found in soils and composters. Originally thought to be fungi, they are now proven to be more closely related to bacteria. Many produce antibiotics such as *Streptomyces.*

Aerobic: Refers to the respiration process of organisms that use free molecular oxygen from air to release the energy from nutrients (see Respiration). Aerobic organisms are 10 to 20 times more efficient than their cousins, *anaerobic* organisms. These live in anoxic (without oxygen) environments, and first must digest oxygen-containing chemicals and compounds, using energy in the process.

Additive: Additive are dry solids added to the composter to provide porosity, absorb urine and water, or supply carbon for the excrement in a composter. Manufacturers and distributors sell or recommend that it be added periodically to the toilet. In municipal composting of wet sewage sludge, the term "bulking agent" is used to describe the additive used to decrease the moisture content so that composting can occur. See the section on Additives for more detail.

Available or **plant-available:** Refers to the ability of a particular molecule or chemical to be digested and metabolized by an organism for energy or growth. Example: The carbon in sugar is "available" to humans, but the carbon in most plastics is not available.

Bacteria: Bacteria are the simplest and most numerous of the living microorganisms. Although small (0.1 to 5 microns), they are comprised of a nucleus containing DNA molecules enclosed in a membrane. This type of organism is called eukaryote. A bacterium is a single cell which gains its energy from nutrients in the surrounding environment (including the intestines of humans) by means of extracellular enzymes which transform complex molecules into simple forms that can pass through the cell membrane to nourish the organism. In most cases, the bacteria provide a beneficial service by assisting in digestion of food in the stomach and intestines for example or composting organic matter. However, some bacteria are pathogenic.

Biofilter: A filter which uses biological processes to decompose that which has been separated. They are typically used for odor management. However a composting toilet is also a biofilter in the broader sense of the word.

Biological treatment: A process that involves living organisms to utilize the nutrient constituents in effluent to transform pollutants into more benign forms. Usually the term refers only to bacteria-based systems as opposed to **"natural"** or **"ecological"** treatment, which employ higher-order plants and animals as well as bacteria.

Biosolids: (Earlier called "sludge" and "septage" which are the remaining stabilized and dried matter from municipal wastewater treatment plants.) This new term was developed by the wastewater treatment industry to obscure the danger of using these materials on farmland for food production. Biosolids can and do contain many of the worst chemicals that institutions, commerce and industry can dispose into a sewer, but are only tested for pathogens and certain "heavy" metals, such as lead and cadmium. See appendix for more details.

Blackwater: A term that originated in the 18th century when excrement was collected in pots and "thunder mugs" and carried out for disposal. Today blackwater is generally toilet and urinal water and not washwater, although some states include kitchen and laundry water, owing to its potentially troublesome constituents.

BOD: Biological Oxygen Demand. One of the tests to measure the efficacy of a wastewater treatment plant by testing the before and after BOD. As EPTA decomposes, it requires or "demands" oxygen. Dissolved oxygen demanded by bacteria from pollution in rivers and lakes is not available to fish and other higher animals. Because fish are more desirable than bacteria, the measurement of the amount of the pollutant, carbon in surface water was very important. By adding oxygen to a known amount of polluted water at 68°F (20°C) over a specific incubation period, then measuring the remaining dissolved oxygen will yield the BOD of the sample in milligrams per liter. The five-day test, or "BOD5," was the incubation period in England, as it took five days for excreta in the Thames River to flow from London to the sea. BODu or L are used to denote the total or "ultimate" carbonaceous oxygen demand or UBOD.

Compost: As it relates to composting toilet systems, this is often a vague term used by some to describe ETPA in a composter that is undergoing composting, but has not completed the process.

Composter: A device, system or machine that provides the necessary environment and process controls to optimize the composting process.

Composting: The controlled biological decomposition of unsaturated but moist organic solids, such as ETPA, by aerobic (requiring free or molecular oxygen found in air) organisms. The desired end-product of composting is an inoffensive, partially oxidized, safe-to-handle material that no longer can support further bacterial decomposition. Often, difficult-to-degrade lignins and celluloses from paper and additives remain, as fungi and actinomycetes that degrade cellulose and lignin take longer to do their work, often requiring temperatures lower than the ideal temperature for aerobic bacteria.

Composting privy or **composting latrine:** A stand-alone building or shelter that houses a composting toilet.

Composting toilet system: A composter that processes human excreta. Also "composting toilet."

C:N ratio: The ratio of dry available carbon to one part of dry available nitrogen. Often expressed as a single number representing the carbon content. Example: The ideal C:N ratio for a microbial nutrient is 25. Combining foods with high C:N (sawdust = 400) and low C:N (urine = 0.8) can, for example, provide sufficient carbon for microbes to utilize all of the nitrogen in urine.

Disposal: The disposition of wastewater into a place that has no consequences or penalties that can be identified and/or attributed to the disposer.

Dry toilet: A toilet that does not require water, such as a composting toilet. Drying toilets are those systems which intentionally desiccate and dehydrate excrement with no biological process desired. (See "Composting Toilets versus Drying Toilets" in Chapter Three for more details.)

E. coli: Echerichia coliform (E. coli) is a common bacteria found only in the intestine and is therefore used as an indicator organism for testing for the presence of fecal bacteria in the finished product of composting toilets. Most coliform bacteria are harmless, aiding in digestion, but some, such as E. coli O157:H7 enterohemorrhagic (ECEH), have been responsible for several serious recent food-borne and waterborne disease outbreaks.

Effluent: A general term to describe a mixture, solution or suspension of water and anything else that is drained, flushed, pumped or vacuumed away from the place where it was generated. It has a neutral connotation. We use this term because effluent has only constituents and characteristics that need to be thoroughly understood in order to prepare it for other purposes or use. Effluent that is not "wasted" is not wastewater.

ETPA: The authors' acronym for Excreta (E), Toilet Paper (TP) and Additive(A), the usual materials entering a composter. This term is used because more than just excrement is put into composting toilets.

End-product: The material that is removed from a composter after all processing has taken place. This material may or may not be completely processed.

Excreta: That which is excreted by humans. Feces and urine are its primary constituents, however a broader definition should include vomit.

Fungi: Also called molds, these are a large group of micro- and macro- organisms that are primarily involved in the decomposition of cellulose and lignin. Their fruiting structures and mycelium give them a plant-like appearance. Like bacteria and actinomycetes, extracellular enzymes are excreted to break down larger molecules that are absorbed over the entire mycelium cell surface. Many produce antibiotics, such as *penicillium* or common bread mold.

Graywater: Generally, all washwater in a home or office. It is more often classified by what it is not, i.e., it is not toilet or urinal effluent. Some states exclude kitchen and laundry water from this classification, owing to their potentially troublesome constituents.

Humus (or end-product): The desired end-product of composting. Humic substances (humic and fulvic acids) may constitute 90 percent of the soil organic fraction. These represent the near complete mineralization of carbon to CO2. (This term also describes the third layer of a forest floor; the first is the "litter" layer and the second the "duff" layer.)

Leachate: The liquid filtrate that has settled by gravity to the lowest point in the composter. It does contain a wide variety of beneficial nutrients (nitrogen being the significant constituent) as well as table salt (sodium chloride). It may contain significant pathogens from dissolved fecal particles carried by urine and water added to the compost toilet. Some manufacturers use the term "compost tea" to avoid any negative connotations, implying that it is safe and ready to use.

Nutrient, or **biostimulant:** Any substance, element or chemical that is available to living organisms for growth, cell maintenance and energy. The essential or macro-nutrients are carbon, nitrogen, phosporous and potassium. Oxygen and hydrogen are required by organisms to utilize nutrients. Micro-nutrients or trace elements, such as sulfur, calcium, metals and salts, are required for specific life functions or structure formation.

Pathogens: Organisms that cause infectious disease, such as bacteria, virus and parasites.

Respiration: The metabolic process by which living organisms produce needed energy by the controlled oxidation of nutrients, such as carbohydrates. In this enzyme-controlled process, sequenced reactions slowly release the energy from nutrients. In most cases, oxygen, either molecular or that bound in organic compounds, is utilized, and carbon dioxide (CO_2), water (H_2O) and energy are produced. In the case of the sugar, glucose, the following formula typifies the reaction: $C_6H1_2O_6 + 6O_2 \rightarrow 6CO_2 = 6H_2O$ + Energy + Dissipated Heat.

Saturated: A state in which all of the void spaces in ETPA in the composting toilet are filled with liquid (urine or water). Composting, by definition, cannot occur in a saturated environment.

Solid: A material in a non-amorphous form, often containing a significant percentage of moisture that is not a liquid or a gas.

Toilet: A device with collects excreta and transports it elsewhere for composting, storage, treatment or disposal.

Toilet paper: Manufactured cellulose tissue paper that Westerners use to clean themselves after urinating and defecating. Moslems use little or no toilet paper, because Koranic law requires water to be used for the cleansing function. Some populations in tropical climates and rainforests use certain leaves. Women can use 200 to 400 percent more toilet paper than men, as women use toilet paper after urinating and men do not.

Toilet stool: That upon which one sits when excreting into a toilet.

Virus: The smallest of all biological entities (0.02 to 0.03 microns), it can only develop within the cells of host organisms. Viruses are parasites that are neither cells nor organisms. In fact they may not be "alive" by some definitions, although they do reproduce. As the smallest pathogen, they are composed of a single nucleic acid for containing genetic information, surrounded by an outer shell of protein. Although viruses are themselves inert, they possess a highly evolved survival mechanism that parisitizes a host cell's metabolic processes by invading and using the cell itself to replicate themselves hundreds of times. Viruses then leave the host cell to penetrate other cells and start the process again. Outside the host, viruses can survive for hours, weeks or months in the proper environment until another host cell arises.

Vectors: Any organism that conveys disease to another organism. Flies and mosquitoes are common vectors, but there are others.

Ventilation, or **Vent Pipe:** A misnomer used by manufacturers to describe the exhaust pipe that conveys odors and gaseous composting by-products out of the building and up above the roof where they are not likely to offend. True ventilation pipes would provide fresh air to the user or to the composter. A better term is **exhaust pipe** or **exhaust chimney.**

Wastewater: Effluent that has been devalued following an analysis that determines that costs of disposal are less than the benefits of reclamation.

Zero-discharge: A treatment system that does not discharge any treated or untreated effluent into the soil, ground or surface water environment.

State Regulations

Use this state-by-state regulatory information for general guidelines only. Sometimes your local health agent may call the shots. If your health agent shrugs his or her shoulders or gives you a firm negative, consider checking with the state folks, particularly anyone who has a reputation for being progressive, supportive and knowledgeable of innovative and alternative systems. Many composting toilets are approved this way. Note, too, that regulations change often—shift happens!

To compile this section, a questionnaire was sent to every state, and followed up with several phone calls. Some states were forthcoming and some were not, and the information may reflect that. To get more information, call your state department of health or environment protection agency.

For more information about regulations:

The USEPA-sponsored National Small Flows Clearinghouse offers a free list of state contacts for on-site systems, as well as a regulations repository (for a fee, you can receive a state's on-site system approval regulation, although you will have to find relevant requirements on your own). Call 800-624-8301.

As Small Flows reports, "homeowners and developers may have a hard time getting approval for some systems because of inflexible regulations or because health officials are unaware of certain alternative system designs or have questions concerning their performance, operation or maintenance." Small Flows offers many technical bulletins and publications about on-site and small community systems. NSFC, PO Box 6064, Morgantown, WV 26505-6064, World Wide Web site: http://www.nsfc.wvu.edu

NSF Listing

NSF International, Inc., is an independent, not-for-profit organization that develops standards for a wide range of public health-related technologies, ranging from wastewater treatment systems and drinking water filters to food-handling equipment and water pipe. It works with the American National Standards Institute (ANSI) to develop performance standards, and are internationally recognized by regulators, who will usually approve a product that has been listed and certified by NSF. Composting toilet systems are tested against ANSI/NSF 41-1998 Non-Liquid Saturated Treatment Systems. The standards include everything from the materials to the design and construction, labeling, O&M manuals and odor. But most important is pathogen testing. The presence of colonies of Escherichia coli are analyzed in the end-product from composting toilets and may not be greater than 200 mpn (most probable number) per gram. Moisture content of the solid end-product must not exceed 65 percent by weight, and shall not produce objectionable odors immediately upon removal from the systems. Liquid end-products shall not produce an objectionable odor immediately following removal from the system and shall not contain E. coli more than 200 CFU per 100 ml.

Listing by NSF almost always guarantee that a state or local regulator will approve the technology because NSF's testing standards are highly reputable. NSF testing requirements and listing maintenance requirements are costly to a manufacturer—in the thousands of dollars annually—so smaller manufacturers may seek approval at the state level.

NSF International
3475 Plymouth Road
P.O. Box 130140
Ann Arbor, MI 48113-0140
734/769-8010
info@nsf.org
Web: http://www.nsf.org

The green Nordic environmental label with the flying swan graphic or Eco-label, helps consumers to identify European products that cause the least damage to the environment. This multinational certification closely cooperates with the EC eco-labeling system. Products from Norway, Sweden, Finland and Denmark should have this label.

Look for CTs tested against the Closed Toilet Systems Criteria document Version 2.0 (Which is also a good read for system designers - ed.) www.interface.no/ecolabel/english/criteria.html

Eco-labeling Norway
Kristian Augusts gate 5, N-0164 Oslo, Norway
47 22 36 07 109
info@ecolabel.no

CSA Listing

The Canadian Standards Association (CSA) has adopted NSF testing protocols, and some manufacturers have their systems tested under CSA..

CSA International
178 Rexdale Boulevard
Etobicoke, Ontario M9W IR3
1-800-463-6727; Web: www.csa.ca

The High Cost of NSF

Performance standards for composting toilets and other wastewater treatment systems are important. NSF International is a highly respected developer of performance standards. NSF also tests products against those standards, and publishes a list of those products that have been tested and certified by NSF. These certifications are the basis technology approval in many states.

The cost for testing a composting toilet can run from $10,000 to $20,000 per model submitted. Every five years and every time the performance standard is revised, the unit must be retested. As of 2000, the annual administration fees run from $5,000 to $7,000, including the costs to have an NSF representative visit the company at a present rate of $120 per hour.

It is the authors' opinion that this high cost stifles innovation, because small companies cannot afford the fees necessary to meet NSF listing requirements. Many composting toilet companies are opting not to seek or maintain NSF listing due to these financial hardships, and that means it becomes more difficult for a state or local approving authority to make a judgment on a product.

NSF and other testing laboratories should have a program for small companies whereby fees are based on annual sales of the product, thus allowing a company that is not well funded to benefit by this service.

Local Certifying Agencies

Several states, such as Massachusetts, have developed their own testing facilities and offer their own state approvals in recognition that most wastewater treatment systems are affected by regional environmental factors, especially climate.

Call your town and county health offices for pertinent regulatory information first. Usually they must follow state regulations; however, in states such as California and Ohio, the counties have their own onsite system regulations. With composting toilet systems and graywater systems, you may have to do some educating; always bring as much literature and helpful information as you can.

(CT=Composting Toilet)

> **Note:** Washington state is now developing evaluation protocols for owner constructed composing toilets and is the first in the country to do so.

Alabama

Alabama Dept. of Public Health
Division of Community Environmental Protection
The RSA Tower, Suite 1250
Montgomery, AL 36130
334/206-5373
http://www.alapubhealth.org

State regs do not specifically address composting toilets, however, one can apply for a permit for a system. Provide supporting literature, examples of approvals in other states, a maintenance plan, warranty information, a plan for disposing/utilizing the end-product and any test data on performance, etc.

Alaska

State of Alaska Dept. of Environmental Conservation
35390 Kalifornsky Beach Road, Suite 11
Soldotna, AK 99669
907/262-5210

Alaska (18AAC72) requires engineered plans for all "alternative"-type domestic wastewater treatment and disposal systems. Regulations do not differentiate between black water and graywater, so they must be treated and disposed of in similar manner. Engineered plans are required for all non-conventional-type wastewater systems. "Conventional systems" are defined as a STSAS serving a single-family or duplex dwelling. Conventional on-site systems for single family, duplex and commercial systems with less than 500 gpd do not need plan approval prior to construction. Complete engineered plans are required, including a report on site soil conditions, detailed systems plans, detailed site plans that address all required separation distance requirements, information on proposed treatment system and in some cases verification that the treatment systems proposed are NSF listed.

(Lots of CTs have been sold in Alaska over the years—whether or not they are permitted is another issue. —Ed.)

Arizona

AZ Dept. of Environmental Quality
3033 North Central Avenue
Phoenix, AZ 85012
602/207-4335

Engineering Bulletin #12 speaks to graywater reuse in such technologies as evapotranspiration, irrigation and other subsurface systems. Basically, composting toilets have to have NSF certification or approved equal. Graywater plans need to be designed and certified by a registered professional engineer (refer to Article 7: Min. Requirements for the Design & Installation of Septic Tank Systems and Alternate On-Site Disposal Systems). Disposal of composting residues are subject to ADEQ special waste rules.

Arkansas

Arkansas Dept. of Health
State Health Building
4815 West Markham
Little Rock, AR 72201

Composting toilets must be NSF listed. There is a locally issued alternative systems permit. With a composter, you get a 35% reduction in leachfield size. Standard information required for all permits. Refer to AR Alternative Systems Manual. (Artificial wetlands are widely used here.)

California

California is unique in that, officially, the state provides no guidance about alternatives to septic systems of any kind. Each of the nine counties and water resource districts make their own decisions with guidance from the state Department of Health. Composting toilets are okay in some counties, especially in northern California, such as Humbolt and Calaveras. You really must contact your county department of health for specific guidance. In the late 1970s, owner-built and early commercial CT systems were evaluated by the State Department of Health and, as reported earlier in this book, they did not hold up to critical examination. Understandably, this may have colored state officials' attitudes about these systems. The good news is that a new research and training facility, is now established to learn about and evaluate alternatives and recommend them to the regional offices. Contact its director, Tibor Banathy, for updated information at
Northern California Wastewater Training & Research Center
California State University at Chico
311 Nicholas C. Schouten Lane
Chico, CA 95928-0270
tbanathy@csuchico.edu
530/898-6027

In California, some key players in graywater research are the California Department of Water Resources in Sacramento; the City of Santa Barbara; East Bay Municipal District in Oakland; and the U.S. Bureau of Reclamation in Sacramento.

Colorado

CO Dept. of Public Health & Env.
Water Quality Control Division
4300 Cherry Creek Drive South
Denver, CO 80246-1530
303/692-3525
http://www.cdphe.state.co.us/environ.html

Permits come from the county health departments, which vary in their enthusiasm for composting toilets. No leachfield size reduction is allowed with composting toilets. Graywater regs are peculiar here—you have to put water back into the ground. No second use, such as irrigation, is presently permitted (that will probably change). (However, there are local health agents who are very supportive of alternative systems. —Ed.)

Connecticut

CT Dept. of Environmental Protection,
Permits & Enforcement
State Office Building
165 Capitol Ave.
Hartford, CT 26115
860/240-9277

Local and state health departments have been designated by DEP to permit on-site systems. CTs can be installed only if the soils are capable of supporting a full-size conventional STSAS system. No credit for graywater-only leachfield, according to state guy, however we know of at least one case in which a reduction was granted. No permitting of alternative systems for individual households or sewage flows under 5,000 gpd. Plans must be certified by a professional engineer. (Not particularly friendly to any alternative systems, however it can be done. —Ed.)

Delaware
Department of Natural Resources
Division of Water Resources
89 Kings Highway
P.O. Box 1401
Georgetown, DE 19901
302/856-4561
Bruce E. Patrick, P.E.

Regulations do not specifically address composting toilets. They would likely be considered alternative systems, which are approved on a case-by-case when a conventional system cannot be installed. NSF certification or other performance data required, and engineer's seal required. Here's an opportunity: Applications for community systems in accordance with standard engineering practices shall not be considered alternative (state review). Graywater is not addressed.

Composting toilets would be permitted as alternative systems in accordance with section 6.12000 of the Regulations Governing the Design, Installation and Operation of On-Site Wastewater Treatment and Disposal Systems.

District of Columbia

"What is a composting toilet?" DC is served by a sewer system, and, as of 1999, no one has yet applied for a permit for a composting toilet.

Florida
Environmental Specialist III
Bureau of Water & Onsite Sewage Programs
1317 Winewood Boulevard
Tallahassee, FL 32399-0700
850-488-4070
holcomb@nettally.com

Permits available through Florida's 67 County Health Departments of Water and Onsite Sewage Programs. Apply for Routine Onsite sewage treatment and Disposal system Construction Permit. Composting Toilets must be NSF approved. Required: site plan, site evaluation, soil profile and establishment flow data—same information as is required for any on-site system.

There is a proposal to reduce the size of the system which treats the remaining black water and graywater when a composting toilet is used. If it passes, it will probably appear in the rule in 1999 or so. Graywater systems allow a reduction in size for the remaining blackwater system.

Georgia
Georgia Department of Human Resources
Environmental Health Section
2 Peachtree St. NW
Atlanta, Georgia 30303-3186
404/657 6534

Rule for On-site Sewage Management Systems, Chapter 290-5-20-.09 paragraph 6-8, pages G-12 through G-15 covers composters and graywater. This rule allows a 35% reduction in the leach field (design subject to soils and perc test) if flow restriction devices such as 3 gpm shower heads are installed. Composting toilets must be NSF or equal certified. Solid end-product must be buried by covering with at least six inches of soil and should not be used as fertilizer for root or leaf crops which may be eaten raw.

Graywater must be managed in accordance with rules for sewage disposal and the rules as currently written have no provision for graywater irrigation or reuse. Although it its the authors belief that a responsible plan certified by an engineer would be approved as an experimental system.

Hawaii
Wastewater Branch
Department of Health
P.O. Box 3378
Honolulu, HI 96801
808/586-4294

Apply to the county. Composting toilets are only generally addressed in Administrative Rule (HAR), Chapter 11-62, which requires the homeowner to submit plans for approval. Graywater design should be based on a minimum of 150 gpd per bedroom (quite high!), and assume that two people are in a bedroom. Graywater to be used for irrigation must be disinfected first. Regs include specs for designing evapotranspiration beds.

Idaho
Barry Burnell
Idaho Division of Environmental Quality
1410 North Hilton
Boise, ID 83706
208-373-0502

Composting toilets require a permit from the district health department. The composting system must have its own ventilation system. Currently, the plumbing code prohibits the use of composting toilets without the permission of the Health Department. Portable systems are allowed for temporary use.

Graywater systems are considered experimental, and would require an experimental permit again from the district health department. See pages 104 to 106 of the Technical Guidance Manual. A variance is required for this use. The system must be provided by a professional engineer licensed in Idaho. Graywater does not include kitchen sinks, dishwashers or laundry water from soiled diapers. It can be used for irrigation of non-edible landscapes.

Illinois
Illinois Environmental Protection Agency
1021 North Grand Avenue East
P.O. Box 19276
Springfield, IL 62794-9276
217/782-5830

Composting toilets must be NSF listed and are approved in Section 905.190 of the Private Sewage Licensing Act and Code. Material from a composting toilet is considered "septage" and must be disposed as such (Section 905.170). With a composter, you get a 25% leachfield reduction. Innovative graywater systems are addressed for experimental use in Section 905.20. "We encourage and alternative solutions," says Thomas McSwiggin, manager of the permit section.

Indiana
Indiana State Dept. of Health
Division of Sanitary Engineering
2 North Meridian, Section 5-E
Indianapolis, IN 46204
317/233-7179
http://www.state.in.us/isdh

Apply to local county health department for a residential sewage permit 410 IAC 6-8.1. Mostly parks have applied here. Regulators want to know what you are going to do with the graywater. No leachfield size reduction is offered with a composter, but a state contact suggested that county ordinances may be less stringent than state requirements.

Iowa
Water Supply Section
Iowa Department of Natural Resources
502 E. 9th St.
Des Moines, IA 50319
515/281-7814

The Iowa Department of Natural Resources deals with regulations and permits for alternative systems. However, each local county is the actual permitting authority.

Composting toilets do not require a construction permit. Graywater systems do. Any discharge of human waste or water contaminated by human living activity need a system construction permit. Iowa regulations do not mention disposal of compost from a toilet, but septage disposal rules may apply. So, if your CT discharges leachate, it must be treated and disposed by an approved system. Quantity and quality of waste is important, as well as type of treatment system proposed and site conditions.

Kansas

Administrator, Onsite Wastewater Program
Kansas Department of Health and Environment (KDHE)

Alternative systems, including composting toilets, are reviewed on a case-by-case basis by the local administrative agency, usually the health department. NSF or other preferred.

"Some counties still consider sand filters to be alternative, others consider them to be conventional. In counties that have no sanitary code, KDHE serves in an advisory capacity but still does not issue permits. KDHE may approve alternative systems in these counties.

"Kansas Administrative Regulations state that all household (domestic) waste must be discharged to the treatment/disposal system. No surface discharge, even of graywater, is allowed without an NPDES permit, which is not issued for private systems. (New regs are in the works.)

Kentucky

Kentucky Cabinet for Human Resources
Environmental. Sanitation Branch
275 East Main St.
Frankfurt, KY 40621-0001
(502) 564-4856

Permit required is the same as for any on-site system. Must apply to state, then the county, which is reimbursed any time and expense incurred in reviewing the applications.

Louisiana

Office of Public Health
Department of Health and Hospitals
206 East Third Street
Thibodaux, LA 70301
504/449-5007

Chapter XIII Sewage Disposal 13-020-1 Non-waterborne Systems references composting toilets but states that the state health officer must determine that is it is "impractical or undesirable to connect to a sewage system or construct a conventional (system)." No special provision for graywater—probably same requirements for all wastewater.

Maine

Wastewater and Plumbing Control Program
Division of Health Engineering
207/287-5695

Composting toilets are approved per se in Maine, as well as on an individual design basis when we are approached by manufacturers or distributors. We have even approved a site-built solar composting toilet based upon a design published in the Mother Earth News, back in the late 1980s. NSF certification helps, but is not mandatory in Maine.

The Department of Human Services, Bureau of Health, Division of Health Engineering, Wastewater and Plumbing Control Program (DHS) promulgates the Maine State Plumbing Code, Subsurface Wastewater Disposal Rules for on-site sewage disposal in Maine. The rules include provisions for use of alternative toilets, such as composting toilets, pit privies and incinerating toilets. A Plumbing Permit, obtained from the Local Plumbing Inspector of a community (or from this office if none is appointed) is required for any residential on-site sewage disposal system, regardless of configuration. A site evaluation and soil test (HHE-200 Form) is required prior to issuance of any Plumbing Permit for an on-site system. This document is prepared by a licensed Site Evaluator, who in turn is licensed by the DHS but is hired by the property owner/applicant.

(Maine approves CTs by brand and by model number on a case-by-case basis. NSF listing is typically required, but some non-NSF systems have been approved. Essentially, the dry CTs are approved. —Ed.)

Maryland

MD Dept. of Env.
Water Management Administration
2500 Broening Highway
Baltimore, MD 21224
410-631-3780
http://www.mde.state.md.us

CTs are approved if they are NSF listed. A 36% design-flow reduction is offered for the size of the constructed graywater system, but the required minimal sewage disposal area must be available. That is, you must have the good soils for a full-sized system, but you need only contruct 64% for graywater only. See COMAR 26.04.02.02B, 26.04.02.04F and 26.04.03.03A(1) for details. For new construction, you may be required to demonstrate that you have a minimum 10,000 square feet of acceptable disposal and recovery area. Innovative greywater systems are currently allowed on a case-by-case basis under a Innovative and Alternative program.

Massachusetts

Executive Office of Environmental Affairs
Department of Environmental Protection
One Winter Street
Boston, MA 02108
617/292-5500
http://www.state.ma.us/dep/dephome.htm

Composting toilets are generally approved (see current list), and are even approved under the plumbing code as plumbing fixtures in lieu of a regular flush toilet. MA has approved composting toilets in 310 CMR 15.289(3)(a) of the State Environmental Code and 240 CMR 2.02.(6)(b) Basic Principles of the Uniform State Plumbing Code. In the former, they are approved by name and in the latter as an "Alternative Toilet Technology Toilet System" that can be built or installed in place of a water closet which is required by the code for occupancy.

Graywater systems are regularly approved if submitted to the state by a P.E. or a Registered Sanitarian under a repair (with Composting Toilets) or as a Pilot project. Mass. is piloting a variety of graywater-utilization plans. All applications and plans must be submitted by a registered engineer or sanitarian.

Michigan

MI Dept. of Natural Resources
Jackson State Office Building
301 E. Louis Glick Hwy.
Jackson, MI 49201
517-335-8286

Composting toilets are not regulated by the state, but are approved by counties. Some counties approve them, some do not. Graywater is the primary concern—no acceptable graywater systems have been approved in Michigan.

Minnesota

MN Pollution Control Agency
Water Quality Division
520 Lafayette Road
St. Paul, MN 55155-4194
612/296-5856

One county sanitarian, David Abasz, says composting toilets are indeed allowed under regs. (Note that some state folks in Minn. aren't quite as sure.) There is no special permit requirement right now. Material removed from a CT is considered sewage and have a pumper remove the material that needs to be removed and disposed of rules (503) or taken to a wastewater treatment plant. Future regs may require NSF listing. With composter, a 40% reduction in the leachfield size is allowed.

(Many CTs—perhaps thousands—have been sold in Minn. to cottage owners. –Ed.)

Mississippi

MS State Dept. of Health
U-226 Div. of Sanitation
P.O. Box 1700
Jackson, MS 39215-1700
601/576-7690
http://www.msdh.state.ms.us

CTs are approved pursuant to MSDH 300, 2.0 Regulation Governing Individual Onsite Wastewater Disposal and Design Standard XIII "Non-waterborne Wastewater Systems" Check the list of approved products. If your selection is NSF listed, chances are that it is on the list. Variations to published design standards (Subsurface drip irrigation, Constructed wetlands and Overland discharge and more all have design standards) are all are considered "Performance-based" and submitted by an engineer with supporting documentation to the Division of Sanitation for review and approval.

Missouri

Stanley R. Cowan, R.S.
Missouri Department of Health
P. O. Box 570
Jefferson City MO 65102-0570
573/751-6095 Fax: 573/526-7377
http://www.dnr.state.mo.us/homednr.htm

Missouri's regulations defined alternative systems include: LPPs, lagoons, elevated sand mounds, holding tanks, sand filters, drip irrigation systems, wetlands, and privies. Composting toilets may be considered under "Other Systems." Graywater must be treated and disposed of the same as black water. The Department of Health (DoH) has jurisdiction for on-site sewage systems with domestic wastewater flows of 3,000 gpd or less. The Department of Natural Resources (DNR) has jurisdiction for any discharge-type system (except single-family residence lagoons), for any system with domestic wastewater flows greater than 3,000 gpd, and for any system receiving other than domestic wastewater. DoH does not implement the state law where a county on-site sewage ordinance is in place. Currently, 55 of 114 counties have on-site sewage ordinances. Permits required: Depending on jurisdiction, either DoH, DNR or county agencies will require a construction permit. Information required: Design of composting toilet and planned disposal of end-product and treatment and disposal of graywater. A septic tank and soil absorption system (STSAS) is required. Also require soils information, absorption field/lagoon design, setback distances, etc. A lagoon or drip irrigation system may substitute for a soil absorption system. (Missouri regs offer a nice, complete constructed wetlands design that even suggests some plant varieties. —Ed.)

Montana

Montana Dept. of Environmental Quality
Permitting and Compliance Division
Metcalf Building
P.O. Box 200901
Helena, MT 59620-0901
406/444-3926 (or 5344)
http://www.deq.state.mt.us

Nebraska

Nebraska Department of Environmental Quality
1200 "N" Street, Suite 400
PO Pox 98922
Lincoln, Nebraska 68509
402/471-2186

Approval is on a case-by-case basis. Plans must be submitted by an engineer or approved systems designer. No credit is currently given for graywater-only systems.

Nevada

State of Nevada Dept. of Human Resources
Health Division
Bureau of Health Protection Services
1179 Fairview Drive, Suite 101
Carson, City 89710-5405
702/687-6353
www.state.nv.us/health/bhps

NV Administrative Code Chpt. 444 Regulations for individual sewage systems essentially mandate that one must have a conventional septic system in place. Systems must be approved by the health division or approved by NSF International. Graywater (disposal field), however, is addressed, and does not include waste from the kitchen sink. (Nevada is likely open to possibilities.)

New Hampshire

NH Dept. of Environmental Services
Bureau of Wastewater Treatment
6 Hazen Drive
Concord, NH 03301
603/271-3711

N.H. approves composting toilets. Graywater systems are approved on a case-by-case basis. With no pressurized water, no problem. If there is pressurized water, one must have a full-sized septic tank and leachfield, or a system approved at the state level.

New Jersey

NJ Dept. of Environmental Protection
Division of Water Quality
Bureau of Nonpoint Pollution Control
P.O. Box 29
Trenton, NJ 08625-0029
609/292-0407

Apply at the county level. Composting toilets are subject to Chap. 199 of the NJ code for individual on-site systems. CTs do required building code approval and approval by local health departments. The system has to be in accordance with the Uniform Plumbing Code. A graywater system would also have to be in accordance with the UPC. To vary from that, one would have to do measured flows (they might accept a margin of safety of 50%, if it's under 2,000 gals. per day). Presently, everything must go to the a sewer or STSAS, unless approved by the county and state boards of health. No changes expected. (Mark Miller, Geologist)

New Mexico

Bob Chacey, Water Resource Specialist
N.M. Environment Department
PO Box 26110
525 Camino de los Marquez, Ste. 4
Santa Fe, NM 87502
505/827-7536

Ah, New Mexico. Bob Chacey wants to see alternatives work, he says. "If somebody's got a good plan, and addresses potential failure issues, we'll look at it—if it doesn't pose a health hazard." However, you still must plumb the bathroom as though a conventional system would one day be installed. For composting toilets and graywater systems, one must apply for a Liquid Waste Disposal permit at District Field Offices, per Regulations .20 NMAC 7.3, as one would for an STSAS. Subpart 306 addresses alternative systems.

NM has an extensive graywater use code; graywater includes water from kitchen sinks and laundry water. However, garbage disposal effluent and diaper washing water is considered blackwater. They do allow evapo-transpiration systems. Above-ground use of graywater is illegal.

Composting toilets are not directly referred to in the regs (Title 20, Chap. 7, Part III "Liquid Waste Disposal"), however 7.3aa speaks to a "enclosed system." (Very positive, pro-active state. —Ed.)

New York
NY State Dept. of Health
Bureau of Community Sanitation and Food Protection
2 University Place, Room 404
Albany, NY 12203-3300
518/458-6706

Composting toilets must be NSF listed and must have a 5+ year warranty. 30% credit on leachfield size with composter. Plans must be designed and submitted by a "design professional." NY does approve evapotranspiration beds for graywater. Graywater systems must be designed for a flow of 75 gallons per day, per bedroom. (Regulation: Section 201(1)(I) of the Public Health Law, Appendix 75-A, 1990+). NY is approving the installation of 100+ composters for a lakeside community, so this may be a very amenable state for CT installations.

North Carolina
Environmental Permit Information Center
919/715-3271

Composting toilets must comply with state standards for conventional treatment. Regulators rely on data from the manufacturers and are particularly interested in pathogen kill and vector control. Material removed from a CT can be land applied. "There have been a few permitted where we saw plans and the manufacturer's specifications," says John T. Seymour. Graywater is addressed by the Division of Water Quality. Graywater must be disposed subsurface (although alternatives have been approved).

North Dakota
North Dakota Department of Health
600 East Boulevard Avenue
Bismarck, ND 58505-0200
701/328-2372
For your county public health contact:
www.health.state.nd.us/localhd

Composting toilets, as with all on-site wastewater technology, must comply with the plumbing code, Chapter 62 of the ND Administrative Code. Individual county agencies administer that code and are open to separated systems but will require NSF listing (unless you can convince them otherwise). Custer District, for example, has approved graywater irrigation and allows a leachfield reduction with a composting toilets.

Ohio
Bureau of Local Services
OH Dept. of Health
246 North High Street
Columbus, OH 443266-0588
614/466-1390, 614-644-3505
http://www.odh.state.oh.us

Like California, individual counties make the rules, although that is changing as the state onsite code is getting updated. Call your county health department. Indoor composting toilets are addressed by the plumbing code.

Oregon
Dept. of Consumer & Business Services
Building Codes Division
1535 Edgewater Street NW
P.O. Box 14470
Salem, OR 97309
503/378-2322

CTs are approved under Sections 447.115 to 447.124 of Title 36 of the Plumbing Code. However, graywater drainage systems are regulated by the Department of Environmental Quality (503/373-7488).

Pennsylvania
Department of Environmental Resources
Division of Certification, Licensing and Bonding
Market Street State Office Bldg., 1st floor
400 Market Street
Harrisburg, PA 17101-2301
717/787-6045
MSection 73.1 (V) of the Pennsylvania Code, Title 25 addresses composting toilets, including site-built types, such as Bio-Sun systems. It does not mention graywater, although by Pennsylvania definitions, graywater may not be considered sewage.

Rhode Island
Division of Groundwater and ISDS
ISDS Section
291 Promenade Street
Providence, RI 02908

Composting toilets are approved in RI in Rules and Regulaitons Establishing Minimum Standards Relating to Location, Design, Construction and Maintenanc of Individual Sewage (SD 14.05). RI provides a 20% leachfield reduction with a CT but you must have 100% available.

South Carolina
SCDHEC Division of Onsite Wastewater Management
2600 Bull Street
Columbia, SC 29201
803/935-7825

Composting toilets are approved for use in SC under Regulation 61-56, which, however, does not approve a specific graywater treatment method. SC does not differentiate beween graywater and blackwater. No leachfield size reduction is offered. However, SC seems open to possibilities (talk to them), and these regs are being revised.

South Dakota
Richard Hanson
Dept. of Environmental and Natural Resources
523 E. Capitol Ave.
Pierre, SD 57501
605/773-3351

Alternative systems are considered on a case-by-case basis. Applicants present their plans to Hanson, who looks at the product brochures and specifics such as where the systems discharge. (Graywater systems and composting toilets are considered separately.) In SD, one can discharge into the ground; he recommends discharging at the roots of trees. He and the staff engineer decide if they are in compliance with state standards. Other than this, the process is the same as it would be for a septic system approval. He sees very few applications for these systems, but is open to seeing more.

Tennessee
TN Dept. of Env. & Conservation
Div. of Groundwater Protection
10th Floor, L7C Tower
401 Church Street
Nashville, TN 37243-1540
615/532-0762
http://www.state.tn.us/environment/gwp/index.html

According to "Regs to Govern Subsurface Disposal Systems," composting toilets must be certified by NSF Standard 41. TN offers up to a 40% reduction in standard leachfield size. A nonstandard graywater system would be applied for as an experimental system.

Texas

TNRCC
Office of Regulatory Services
P.O. Box 13087
Austin, TX 78711-3087
512/239-2428
http://www.tnrcc.state.tx.us/oprd/rules/index.html

The relevant regulation is Title 30. Environmental quality, Part I. Texas Natural Resource Conservation Commission, Chapter 285. "On-site sewage facilities," Subchapter d. Planning, construction and installation standards for OSSFS states: "§ 285.34 Other Requirements Composting toilets will be approved by the executive director provided the system has been tested and certified under NSF International Standard 41.

County regs may be more specific, for example, Austin requires: "Ordinances #880310-H and #880310-I address composting toilets and are a part of Chapters 6-10 of the Austin City Code. The Austin-Travis County Health Department must issue a permit to install a composting toilet. The permit cost is $200 for an NSF-approved unit, or $300 for an engineered unit. The composting toilet is considered an alternative system and must be inspected annually by the Health Department. The inspection fee is $30. When the composting toilet is outside of the City, there is not an annual inspection requirement. The licensing procedure outlined in the Greywater Irrigation Section is required for composting toilets (percolation tests do not apply if a sewer is present)."

Utah

UT Dept. of Env. Quality
288 North 1460 West
P.O.Box 144870
Salt Lake City, UT 84114-4870
801/538-6080
http://www.deq.state.ut.us/

Vermont

VT Agency of Natural Resources &
Dept. of Env. Conservation, WW Mgt. Division
103 South Main Street
Sewing Building
Waterbury, VT 05671-0405
802/241-3027

Under Vermont regulations there are three categories for single-family homes: VT is mostly concerned about developments. (Permits for remote and single-family homes appear relatively easy to get.For lots less than 10 acres in size and subdivided after 1969, a state subdivision permit is required. It would be acceptable to use a composting toilet, but the lot must be capable of supporting conventional toilet facilities. For other properties, about half of the towns in the state have local regulations.

Virginia

VA Office of Env. Health Services
Main Street Station, Suite 117
P.O. Box 2448, Rm. 119
Richmond, VA 23218-2448
http://www.vdh.state.va.us
DALEXANDER@vdh.state.va.us

Any composting toilet that is certified as meeting NSF Standard 41 (without a drain, or with a drain and an approved place to dispose of the leachate) is

approved for use in Virginia whereever a privy can be used. The regulations can be found on the state's web site.

Washington

Washington State Department of Health
1500 West 4th Ave., Suite 305
Spokane WA 99204
509/ 456-2490
http://www.doh.wa.gov

WA state Department of Health (DoH) approves composting toilets and graywater systems listed in its current "List of Approved Systems and Products." Local boards of health can approve composting toilets and wastewater systems listed in Chapter 246-272 WAC On-Site Systems; all others must be approved by the state. Any system not in the list must be reviewed by the DoH Technical Review Committee. Local health depts. can only approve systems approved by the state. The permit is issued by local health officer.

As of March 1997, "Interim Standards/Guidelines for Municipal Greywater Reuse/Conservation Pilot Projects" apply. Graywater is only approved for subsurface landscape irrigation. The regulations document for this is an excellent locating, design, operation and maintenance manual.

Alternative designs can only be submitted by engineers qualified (by DoH) designers and soil scientists to perform soil and site evaluations. There may be some kind of monitoring arrangements—it depends on jurisdiction

(WA is one of the friendliest and most progressive states for alternative systems. Nice literature and thorough, educational and clear regulations—a model for other states. —Ed.)

West Virginia

Environmental Health Services
Public Health Sanitation Div.
815 Quarrier Street, Suite 418
Charleston, WV 25301

Graywater and composting toilets are addressed in WV Interpretive Rules (BoH) was updated by Title 64, Series IX, apply to local boards of health. Require design data sheet and plan

This state's space requirements for leaching fields are already quite high. Standards may be set aside for alternative systems. Will consider alternative systems for lots of two acres and larger.

Wisconsin

WI Dept. of Natural Resources
Water Resources Management Bureau
P.O. Box 7921
Madison, WI 53707
608/266-8631
http://badger.state.wi.us/agencies/commerce

There is a specific provision to allow composting toilets, however they are approved on an experimental basis.

Wyoming

Dept. of Environmental Quality
Herschler Building
Cheyenne, WY 82002
307/777-7075
http://www.deq.state.wy.us/wqd1.htm

CTs are permitted (on a case-by-case pilot basis) under Section 5 of Chapter 11 of the Rules of the Wyoming Department of Environmental Quality, Water Quality Division, adopted in accordance with 16-3-101 through 16-3-115. They allow a 33% reduction for waterless toilets. They do not specifically address composting toilets, only chemical toilets and privies. (Note that WY's specifications for evapotranspiration beds are too deep.)

Appendix

- A history of water-carried effluent
- How wastewater disposal systems pollute
- The trouble with sludge (a.k.a "biosolids")
- System planning and permitting requirements
- A sample maintenance agreement
- Beneficial intestinal bacteria
- Aerobic/Anaerobic summary
- How we process food
- Conservation and drainage fixture units
- Pathogens detailed
- Disease severity scale
- Effect of antibiotics on composting toilets
- Urine and sugar connection
- Sodium and fertilizer effects in plant cells
- E-coli in leachate
- The rhizosphere: where it all happens
- Composting urine
- A new model for managing wastewater

Water and Wastewater: The Origins of the Piped Society

By Joel A. Tarr, Carnegie Mellon University

Science knows now that the most fertilizing and effective manure is the human manure.... Do you know what these piles of ordure are, those carts of mud carried off at night from the streets, the frightful barrels of the nightman, and the fetid streams of subterranean mud which the pavement conceals from - you? All this is a flowering field it is green grass, it is the mint and thyme and sage, it is game, it is cattle, it is the satisfied lowing of heavy kine, it is perfumed hay, it is gilded wheat, it is bread on your table, it is warm blood in your veins.

Victor Hugo, 1862

For Victor Hugo "town dung [was] gold." He believed that if this "gold was returned to the land instead of being thrown into the sea," it would "nourish the world." Hugo was one of many writers, including agriculturists, engineers and sanitarians, who in the 19th century maintained that the most sanitary and logical way to dispose of urban wastes was to return them to the soil where they would be reincorporated into nature's cycle. While the use of sewage (human wastes in water carriage) in agriculture dates back to about 1800, the application of human wastes directly to the land has a much longer history. The Romans used human wastes as fertilizer, and in the Middle Ages, farmers in Flanders purchased "night soil" in the cities to use on the land. Edicts dating back to the 17th century in Flanders compelled settlers in the peat-marsh colonies to manure their fields with urban wastes.

Such waste recycling continued in many European nations throughout the 19th century. In Paris, for instance, the existing sewers were intended only for storm and wastewater. Human excrement was disposed of in cesspits—large underground tanks usually built of some impermeable material. In 1842 there were approximately 50,000 cesspits in Paris serving about 900,000 people. A crew of between 200 and 250 "vidangeurs" cleaned the cesspits by hand and pump and deposited the wastes in a dump on the outskirts of the city. There the liquids and solids were separated and the solid matter dried. Once dried, it was transported to rural areas to be used as fertilizer.

In England the "pail system" was extensively used as a pre-water-carriage-type of waste removal. In Rochdale (population about 70,000) and Manchester (population about 400,000) in the latter half of the 19th century, health department regulations required households to each construct a "pail closet." This pail closet was constructed on a raised flagstone platform and consisted of a hinged seat set over a removable pail or tub. At the side of the closet was an ash sifter which deposited a layer of ashes on the human wastes. The pails or tubs were collected once a week and taken to a depot where their contents were mixed with more ashes, coal dust, and gypsum. After drying, the mixture was sold as fertilizer. In Rochdale in 1873, the sale of the fertilizer supposedly paid for 80 percent of the cost of collecting and preparing it.

The greatest amount of recycling of human wastes has historically been in China, Japan and Korea. Writing in 1850, an admiring American agriculturist noted that the Chinese were "the most admirable gardeners and trainers of plants," and attributed their success in agriculture to their careful use of human urine and night soil from the cities. Urban wastes were collected daily and either used immediately or baked with clay to form dry bricks called *tafeu*. At the beginning of the 20th century, 182 million tons of human wastes were annually applied to the land in these Asian nations; the figures for Japan alone were nearly 24 million tons in 1908.

In America during the 18th and most of the 19th centuries, as in Asia and Europe, wastes from cities were often disposed of on the land for fertilizing purposes. As late as 1910, 62 percent of the population lived in places without water-carriage removal of human wastes—that is, without sanitary sewers. Some American cities, such as Boston and Philadelphia, had sewers as early as the 18th century, but these were for the removal of surface waters only. Their main function was drainage not waste removal. In fact, municipalities often had laws forbidding the depositing of human excrement in the sewers. Urbanites placed their wastes in cesspools or privy vaults (often with removable "ordure tubs"), while household slops were thrown into a cesspool in the yard or into the street. Cesspools and vaults were used until they had filled up and then new ones dug. Some cities required that waste receptacles be regularly emptied and cleaned by scavengers, usually at night. The scavengers disposed of the night soil by dumping it in local rivers or lakes, burying it in the fields, selling it to processing plants to be made into fertilizer, or selling it to farmers. In some towns and cities, farmers paid for the privilege of cleaning cesspools as well as the streets, where they collected valuable horse manure.

American farmers used urban wastes extensively. In nearly half of the 222 cities surveyed by a leading sanitary specialist of the time, George E. Waring, Jr., scavengers or farmers collected the night soil and either deposited it directly on the land or composted it with earth and other materials, or sold it to processing plants to be manufactured into fertilizer. Cities in New England and the middle-Atlantic and upper southern states made the most extensive agricultural use of urban wastes. Farmers utilized the wastes of 43 out of 55 New England cities, 31 out of 49 Middle Atlantic cities, and seven out of the eight Upper South cities.

Eight cities, including New York, Baltimore, Cleveland, and Washington, D.C., sold their night soil to processors who made it into fertilizer.

Scavengers or farmers usually emptied the cesspools by hand receptacles and buckets, although 11 cities reported using "odorless evacuators." Boston described its odorless evacuator in the following manner: "These machines, which are airtight, have the air pumped out of them and suck the contents of the vault for any distance up to 125 feet (38 m) through strong hose. A small charcoal furnace, connected with the air pump, destroys any gases as they are pumped out." In some cities, however, like New York in 1872, the organized scavengers opposed the introduction of this new technology due to their fear that it would drive them out of their jobs.

The Night Soil Fertilizer Industry

Eight cities, including New York, Baltimore, Cleveland and Washington, D.C., sold their night soil to processors who made it into fertilizer. The most prominent firm in 1840 was the Lodi Manufacturing Company, which used New York City night soil and mixed it with "vegetable fibrous substances and chemical compounds" to make a "New and Improved Poudrette." As a fertilizer, Poudrette had an advantage over night soil because of ease of conveyance and handling.

Most night soil, however, was applied directly to the land. It is difficult to estimate just how much, but it was significant. In 1880, Brooklyn reported that 20,000 cubic feet (556 m^3) of night soil was taken each year from the city's 25,000 privy vaults and applied to "farms and gardens outside the city." In the same year, Philadelphia estimated that about 22,000 tons of fluid matter per year were "largely used by farmers and market gardeners of the vicinity...." [It is important to note that this night soil usually had far less toxic chemical content compared to today's sewage. —Ed.]

In Baltimore, the practice of applying night soil to crops continued into the beginning of the 20th century. The city was the last major municipality to construct a system of sewers, and the population depended on 70,000 cesspools and privy vaults to dispose of its excrement. "Night soil men," using either odorless evacuators or dippers and buckets, emptied the vaults. In the 1870s, the city itself manufactured a fertilizer from its night soil, but by the turn of the century, this practice seems to have ended. An 1899 article relates that the scavengers sold night soil to a contractor for 25 cents per 200-gallon (757-liter) load, and then shipped

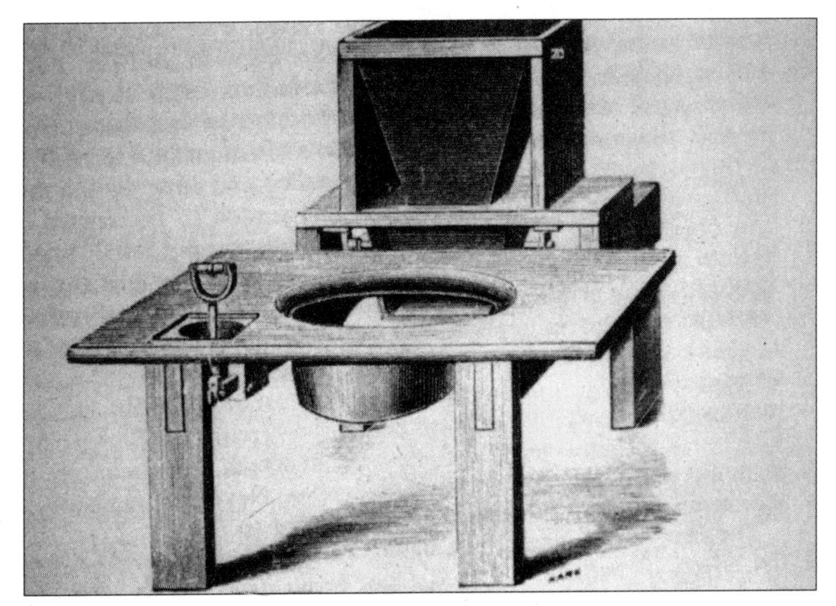

An Earth Closet (Photos supplied by Dr. Joel Tarr)

the wastes by barge eight or 10 miles (16 k) below the city to be sold to farmers. Farmers purchased more than 12 million gallons (45.4 million liters) per year, and used it to grow crops such as cabbage, kale, spinach, potatoes and tomatoes. According to one reference, "little smell" arose from either the pits where the fertilizing material was stored or the lands to which it was applied.

It's the Smell, not the Substance

Nineteenth century medical science posed no barrier to the use of human wastes directly on the soil. For most of the century, doctors believed that either the corrupted state of the atmosphere (miasmatic theory) or specific contagia stemming from decayed animal or vegetable matter, caused infectious diseases. Sanitary reformers about the time of the Civil War were insisting on the removal of filth from towns and cities because they believed that these wastes either generated epidemic disease or threw off "exhalations" that promoted disease. Actually, this belief went back for centuries. The 12th century *Regimen Sanitatis Salernitanum* warned that "evill sents" could breed infection.

When applied to the land "in the open country," however, as the Massachusetts Sanitary Commission of 1850 argued, the wastes were "diluted, scattered by winds, oxidized in the sun: vegetation incorporates its elements." Miasmas and contagions would be dealt with by nature.

In most cities, however, as populations grew, the scavenger system and the existing sewers could not deal with sanitary requirements. Many privies were never cleaned or they were cleaned imperfectly. Although the law required that night soil carts have tight lids, the carts usually left a trail of wastes. The wastes, if not sold to farmers, were often dumped into a local waterway, causing serious pollution.

The Advent of Sanitary Reform

Americans, especially in the growing middle class, became increasingly unwilling to deal with such intolerable conditions. American values concerning public health and sanitation and cleanliness were changing, and the society sought ways to move towards goals of health and science. By the middle of the 19th century the so-called "filth theory" of disease had strong support among sanitarians.

Regardless of the correctness of the filth theory, it had a beneficial impact on the public health by insisting that sanitation be improved. Altered beliefs about the causes of epidemics, such as cholera and yellow fever, reinforced the demand for cleanliness. While many Americans accepted the cholera epidemics of 1832 and 1849 as a product of God's will and a punishment for sin, by the epidemic of 1866, society was more prone to regard cholera as a social disease that could be cured by sanitary reform. And, sanitary reform meant dealing with the removal of human wastes in a more efficient and expeditious manner.

Introducing "The Water Closet"

The technical solution that became most popular in the United States involved the so-called "water closet." The water closet had a long history. Types of closets had been used by early civilizations, such as the Minoan and the Roman, and Sir John Harrington had proposed a water closet in his *Metamorphosis of Ajax*, published in 1596. The first English patent for the water closet, however, was not issued until 1775, and the first American patent not until 1833. In 1842, Andrew Jackson Downing, one of the most popular architects in America, wrote that "no dwelling can be considered complete which has not a water closet under its roof." By 1860, the 16 largest cities of the nation, as well as a number of smaller ones, were served by water works, and many affluent residents of these cities installed water closets in their homes.

These early water closet installations, however, were not without problems. Some of these water closets were connected to sewers, but most flowed into cesspools or privy vaults. Vaults and cesspools had often overflowed before the introduction of the water closet; now the closets caused constant danger of soil saturation and the leaking out of wastes. These problems generated concern among many sanitarians of "sewer gas," which rose from decomposing organic matter; they feared that this would cause epidemic disease. In addition, cesspools required more frequent emptying, creating an added cost.

One solution to the inadequacies of the early water closets was the earth closet, invented by Henry Moule, an English vicar, a version of which was used in various English and European cities. [For more on the earth closet, see Chapter One. —Ed.] The earth closet consisted of a wooden seat with a bucket beneath and a hopper containing dry earth, charcoal or ashes at the back. Pulling a handle released a shower of earth into the bucket. The earth closet was simple to empty, and thus less likely to overflow as the privy vaults did. In addition, the mixture of dirt and wastes made an excellent fertilizer. But while the earth closet was widely adopted in Britain and Europe, it never became popular in America. As one of its chief American advocates, George E. Waring, Jr., explained, the earth closet was "distasteful to a mass of persons in the United States whose necessities demanded immediate relief." But there was an obvious solution for the problems caused by the water closet and the increased amounts of household wastewater—build sewers connected with private homes to carry off the wastes by water carriage. These systems were known as "sewerage systems." The opening of the Croton Aqueduct in New York City in 1842 increased water consumption throughout the city, and caused the number of water closets in use to increase sharply. The law then forbade the connecting of closets to the sewers which were for surface water. In 1844, however, an alderman proposed that householders be permitted to connect closets to the sewers. The connection of households and water closets to the existing sewers, however, overburdened them, requiring the city to accelerate its sewer-building activities. The process of sewering New York for both household wastes and surface water had begun, although the goal of a fully sewered city lay far in the future.

The water closets caused constant danger of soil saturation and the leaking out of wastes.

The Sewer Overfloweth

This process of change initiated in New York by the introduction of running water and the adoption of the water closet was duplicated in other cities. An increasingly health- and sanitation-conscious group of citizens would demand better water supplies and sanitation. A city would introduce running water without constructing a sewerage system; householders would connect their water closets and sinks to cesspools which would then overflow; sanitary problems and general nuisance would result; and the costs for emptying cesspools would accelerate. Citizens demanded relief from these nuisances, and for improved sanitary conditions, stimulating debates with professionals and government officials about solutions. These discussions resulted in cities throughout the nation deciding to construct water carriage systems or sewerage.

While some cities like Chicago and Brooklyn constructed sewerage systems that provided for the removal of a combination of stormwater and household wastes ("combined systems") in the 1850s, it was not until 1880 that the first system providing for household wastes alone was built. Designed by George E. Waring, Jr. for Memphis, Tennessee, this system resulted from the impacts of the yellow fever epidemics that devastated the city in 1878 and 1879. The virulence of the fever was blamed on the filthy condition of the city, and in 1880, the city authorities and the National Board of Health recommended that a sewerage plan proposed by Waring be adopted. It was based upon Waring's theories about the dangers of sewer gas and the need to speedily remove human wastes from the home.

Waring's Memphis system, as well as the writings of other sanitarians, stimulated great public interest in sewer building in the late 1870s and the 1880s. Popular journals carried articles on sanitation. In 1877, the first sanitary engineering journal was published in the United States.

In an article and an editorial published in 1873, *Scientific American* summed up the various sewage disposal options suggested by the British reports. The journal maintained that the profitable disposal of sewage was subordinate to the problem of disposing it. It recognized, however, that sewage had potential value and calculated that the worth of the "annual voiding of an average individual" ranged from $1.64 to $2.01. During the 19th century in England, hundreds of patents were issued that described the production of a usable manure by combining sewage with various materials. The most popular method was known as the "ABC process," since it involved mixing sewage with alum, blood and clay; other processes combined sewage with chemicals such as lime, sulphate of iron or mineral phosphates. While these so-called chemical methods of sewage disposal had some popularity in England, the most widely used approach was that of sewage farming.

Sewage farming was actually an old practice. From about 1800, several towns in Devonshire, as well as Edinburgh, Scotland, had irrigated neighboring agricultural lands with their sewage. The leading English promoter of sewage farming was Edwin Chadwick, who in 1842 advocated the use of untreated sewage as field manure, claiming that the sale of urban sewage to farmers would pay for the cost of maintaining urban sewage systems. By 1880, 19 English cities disposed of their sewage on agricultural land. In Europe, Antwerp, Berlin, Brussels, Paris and Milan all had sewage farms.

In America during the 1870s, several New England institutions began using their sewage to grow crops. The first municipality to use a sewage farm was Lenox, Massachusetts, where Colonel Waring built a system in

1876. The next system was in Pullman, Illinois, followed by Pasadena, Colorado Springs and Salt Lake City, as well as smaller towns. In 1899, a study listed 24 municipal sewage farms. Among the crops grown on sewage farms were potatoes, wheat, oats, barley, carrots and other vegetables and fruits, while Italian rye grass was very common.

A 19th century sewage hauler.

Sewage Treatment Is Born

Sewage farming was appealing because it suggested that wastes could be disposed of in a profitable manner. However, during the 1890s, several new technologies developed that appeared to be as efficient in sewage disposal as sewage farms without the land requirements and the disagreeable (and possibly unhealthful) aspects of the farms. Intermittent filtration, developed by the Lawrence Experiment Station of the Massachusetts State Board of Health, was the most important technology. In addition, sanitary engineers and scientists formulated other chemical, mechanical and bacterial processes. By the turn of the century, more cities were using intermittent filtration and other methods, such as septic tanks, than sewage farming. During the early 20th century, methods using bacterial processes, such as activated sludge, became the most popular forms of sewage treatment in the United States.

By the turn of the century, a flood of public and scientific attention focused upon the problems of sewers and sewage disposal. The affirmation of the germ theory of disease and the belief of sanitarians that "the excreta of man and other animals are the principal original vehicles of infection and contagion" stimulated efforts toward sewering towns and cities. Between 1890 and 1909, the miles of sewers in the United States increased from about 8,000 to more than 25,000 (12,800 to 40,000 k), and the population served by sewers from about 16 million to more than 34 million. While the building of sewers was a health improvement over the use of privy vaults and cesspools, it also created immense problems of sewage disposal. The biggest problem was that most cities, believing that "running water purifies itself," dumped their raw sewage into rivers and lakes without treatment. The most serious aspect of this practice, aside from the aesthetic problems, was that many cities drew their drinking supplies from the very rivers into which other cities disposed of their raw sewage. In those cities downstream from the sewage outlets of upstream cities, such as Pittsburgh or Newark, typhoid death rates soared to high levels. Especially important in drawing attention to the dangers of river pollution was William T. Sedgwick of the Massachusetts Board of Health, who demonstrated the water-borne nature of typhoid fever germs.

Under the prodding of public health authorities, legislatures in states such as New York, Ohio and Pennsylvania passed laws to stop cities from disposing of raw sewage in rivers. However, this legislation was limited to new sewerage systems or extensions to existing ones. Most municipalities continued to dump raw sewage into waterways. Sewage treatment technology was too expensive or uncertain to persuade municipalities to invest large sums of money for the benefit of downstream cities. At the same time, advancements were made in water filtration. Much of the experimental work in this area was done at the Lawrence Experiment Station of the Massachusetts Board of Health. By 1912, over 28 percent of the urban population (10,806,000 people) were drinking water filtered by either sand or mechanical filtration methods, while the water supply of many other urbanites was being chlorinated. In those cities with filtered water, typhoid fever deaths, as well as those from other water-borne diseases, such as diphtheria or infant diarrhea, dropped precipitously.

Increasingly it became clear that, even though cities did not clean up their sewage, death rates from waterborne disease could be reduced.

Cleaning the Water, Not the Sewage

The question as to whether municipalities should purify their sewage, filter their water, or do both became a crucial issue in the second decade of the 20th century. Public health authorities on state boards of health tried to compel cities to purify their sewage before disposal, but sanitary engineers maintained that, given the primary goal of reducing typhoid death rates, water filtration was sufficient. Sewage purification, they held, was costly and reasonably ineffective in reducing pollution. In his book *Clean Water and How to Get It,* Allen Hazen, one of the nation's pioneers in the water-quality movement, expressed the point of view of most of the country's sanitary engineers in the following: "It is . . . both cheaper and more effective to purify the water, and to allow the sewage to be discharged, without treatment, so far as there are not other reasons for keeping it out of the rivers." Sewage treatment, therefore, might pay dividends in "comfort and decency," but it would be costly and not necessarily effective in saving lives.

Many of the issues concerning methods of sewering and sewage disposal that arose in the late 19th century have again become the focus of public and scientific attention

Cities, therefore, made much more rapid progress in developing water-filtration facilities than they did in building sewage treatment plants. As late as 1940, more than 66 percent of the people living in communities with sewers discharged raw sewage into water courses with little more than fine screening.

In the period after World War II, the rate of both sewer building and sewage treatment expanded fairly rapidly, with the greatest increase occurring after the provision of federal construction subsidies in 1957. In 1970, the U.S. Council on Environmental Quality reported that the sewage of about 110 million people—61 percent of the population—was treated by municipal wastewater systems. By 1988, the sewage of about 176 million people—72 percent—was treated at a central facility.

Today's Wastewater Management

But while these developments in the construction of sewage treatment plants sound encouraging when gauged by the standards of 1900, the problem today is far more complicated. Much of the treated sewage effluent from even the more advanced treatment plants are rich in inorganic materials such as carbon dioxide, nitrate and phosphate, which support the growth of algae. The algae bloom and die, creating eutrophication (choking off of life through lack of oxygen) in lakes and rivers into which the sewage effluent is disposed. There are also problems with the major by-product of the treatment processes: sewage sludge. Sludge is created in enormous volume, and is either incinerated or disposed of on the land or at sea. All of these methods of sludge disposal, however, have undesirable results: incineration creates air pollution, land disposal is expensive and creates nuisances, and ocean disposal has fouled many beaches. And finally there is the issue of synthesized organic compounds that cannot be removed by ordinary methods of treatment—and the environmental effects of which engineers and scientists have little or no information.

These problems of sewage disposal have caused some ecologists to recommend abandoning the water-carriage system of waste removal in favor of alternatives, such as waterless and low-flush composting toilets. However, those who expect the water-carriage system to continue recommend *central* solutions that utilize these effluents—almost a return to the sewage farming methods of the 19th century but with greatly improved technology.

The search for an ecologically sound and economical method of sewage disposal goes on, and billions of dollars will continue to be spent combating water pollution in the future. Many of the issues concerning methods of sewering and sewage disposal that arose in the late 19th century have again become the focus of public and scientific attention. Ironically, however, it was the very manner in which these early problems were dealt with that has helped create our present predicament. ☼

How Sewers and Septic Systems Pollute

We can thank central sewers and advanced on-site wastewater systems for allowing us to live in relative cleanliness and for lowering the risk of disease in high-population areas.

But their success is a matter of degree; they still fall short in their charge to protect public and environmental health.

The introduction of the flush toilet and water-carriage plumbing resulted in a flush-and-forget mentality that distanced us from witnessing firsthand the pollution and potential diseases this system can cause.

The problems associated with the common wastewater management approach are many and broad. Simply put, they are:

- It uses too much water.
- It pollutes.
- It disposes of a potential resource.
- It is expensive.

The water-carriage wastewater disposal system has been the wastewater system of choice for more than a century in the developed world, and now in many developing countries. And yet it is far from ideal. Using a lot of water to move a little waste causes a lot of problems.

Think about it: Every day, most of us use an average of 3.35 (range: 1.6 to 5.0) gallons (12.7, 6-19 liters) of drinking-quality (potable) water to flush a toilet just once. At 5.2 flushes per day, the average American uses 7,300 gallons (276,306 liters) of drinking water each year to flush away 1,300 pounds (590 kg) of excrement.

Essentially, we are using a valuable resource—water cleaned at great expense—to dilute and transport another potentially valuable resource—human excrement—and adding significant amounts of industrial and household chemicals, then paying a high cost in the attempt to separate all of those elements and clean them up. And are they cleaned to drinking-water quality standards? No. That would be outrageously expensive. Instead, they are filtered and treated to a degree mandated by federal law, and disposed back into the environment.

Every year we are learning more about the trouble that our wastes cause, and our regulations get tighter. The answer is that we cannot afford to create sewage anymore.

One way to consider some of the issues with the current approach to wastewater management is to consider what happens at the beginning and the end of the wastewater discharge pipe.

At the Beginning of Pipe...

It uses a lot of water.

About 19.5 trillion gallons (73.8 million liters) of water are used annually in the United States to flush toilets.

Sometimes that water is moving waste 20 feet to a septic tank. Sometimes it is moving it to a sewer main half a mile away, which conducts it to a sewer system that transports it perhaps two, five, 10 miles (16 k) to a treatment plant.

That water, mixed with excrement and household and industrial chemicals, must be transported and treated. And more water must be found and cleaned—so we can flush again.

The water needed for this activity taxes the safe yield of our ground and surface water supplies. If too much water is withdrawn, lakes and rivers are taxed and, in some places, simply dry up (witness the Colorado River, which no longer reaches the sea). In coastal areas, withdrawing groundwater sources can result in pulling in salt waters from the ocean. This salt water intrusion contaminates fresh groundwater supplies with sodium ions, rendering it unfit for drinking..

And every year, drinking water standards are raised by regulators, bringing up the cost of supplying and treating water.

We can no longer continue the practice of using dwindling drinking water supplies to flush away our wastes.

At the End of Pipe...

It pollutes our aquifers, water we use for wells and which replenishes our surface waters. It pollutes our lakes, streams and seacoasts. It produces toxic waste.

Generally, septic systems and wastewater treatment plants can produce three forms of pollution:

- **Nutrients, such as nitrogen**
  In surface waters, this results in overgrowth of algae

and aquatic plants, which can use much of the oxygen in the water, thereby killing off other life. This contributes to "eutrophication," the dying of aquatic life (such as the lake that turns into a swamp). This is a classic occurrence in lake and river communities, where an increasing number of cottages result in dying lakes and waterways and inedible fish.

- **Fecal organisms that are potentially pathogens**
- **Toxic chemicals and heavy metals**

How Septic Systems Pollute

The primary function of septic tank and soil absorption systems (STSAS) is to settle out solids and larger particles, and then drain the remaining effluent into the ground.

Only a minor amount of anaerobic decomposition takes place in the tank, as temperatures are usually far too cool for significant biological activity.

Septic tanks cannot remove nutrients and toxic chemicals, and they have a limited effect on disease-causing pathogens.

Nutrients, primarily phosphorus and nitrogen, move through the septic process and into the receiving ground via the soil absorption system (leachfield). The term, "soil absorption system," suggests the ground acts as a sponge. But what really occurs is "adsorption," whereby various particles and chemicals adhere to soil particles, where biological and chemical processes can break them down.

We used to think this was a safe thing to do—and in some cases it is. The issues are, what is the quality of the effluent, what kinds of soils exist and in what direction does your groundwater flow? These are unknowns for many property owners.

Leachfield pipes are typically placed two to 10 feet below the soil surface. This drains septic tank effluent too far underneath the surface, well below where plants and microbes can reach it, and it may move with little or no transformation along a hydraulic gradient and into waters. This same principle applies to toxic chemicals. They, too, pass through the septic tank and percolate through the soils. So, in some ways, septic tanks are just a convenient way to get wastewater into the ground, after settling out the solids.

A case can be made for shallow leaching systems, where effluent is drained to the active aerobic zone that exists just one to two feet below the surface. That's where intricate relationships between the roots of plants, bacteria and protozoa can interact to transform the pol-

A General Description of Central Sewer Systems

Here are some very, very general typical steps of wastewater treatment plants. The particulars vary *widely.*

- ☐ Grates remove large solids. Big pieces are cut into small pieces by comminuters (imagine rotary knives inside a tank).
- ☐ Solids are screened, floated and settled, and a polymer or alum is added to glom the particles together, so they sink and can be removed.
- ☐ Biological processes such as trickling filters digest the digestible carbon component, reducing the BOD but producing a lot of bacteria and concentrated solids called "sludge." The more advanced the process, the more sludge is created. Sludge is 5 to 15 percent solids, the rest is effluent with troublesome constituents and characteristics.
- ☐ Then there is secondary settling: More chemicals are added to make things stick together and sink (this is called "precipitation").
- ☐ Fats, oils and greases (FOG) and other floatables float to the top and are skimmed off.
- ☐ Sludge goes to a processing stage where it is dewatered by filter-pressing or centrifuge activity.
- ☐ Now it's a sludge cake solid. This is now referred to as "biosolids."
- ☐ That is landfilled, incinerated (very expensive but thorough), land-applied or composted for agricultural use.
- ☐ The liquid sludge continues to be filtered, and is sometimes sent back to re-inoculate the entire process with beneficial bacteria (in that case, it is called "activated sludge").
- ☐ The effluent separated from the sludge goes to a disinfection stage. Usually that means it is chlorinated (creating carcinogenic chlorine compounds, such as trihalomethanes and dioxin). Better plants use ozone or ultraviolet light to disinfect.
- ☐ Some plants then dechlorinate, often by adding sulfur dioxide and filtering with carbon adsorption.
- ☐ At this point, the effluent must be aerated to a dissolved oxygen level that is equal to that of the receiving water.
- ☐ Finally, the effluent is disposed into a river, lake or ocean (or injection wells, if there are no surface waters).

lutants into forms that can be utilized by plants.

When you think about your wastewater as irrigation, you change your whole mindset. You think about what you are putting in the water, whether it is good for gardens, and what's the best way to get the nutrients into the rhizomes so they can be transformed. The answer is shallow subsurface irrigation by such means as gravity, pressure dosing and drip irrigation.

The sludge in your septic tank, or "septage," is usually either taken to central treatment plants, incinerated, composted (more about that later) or illegally dumped in landfills.

How Central Wastewater Treatment Plants Pollute

Once we have added every imaginable solid, liquid and even gas as unwanted residues to our waste stream, we now relegate the separation and management of it to what are euphemistically referred to as "treatment" plants. Treatment plants are handed this chemical stew, and through a series of physical, chemical and biological processes, attempt to return the water portion back to its surface and groundwater source in an innocuous form. The hope is that the expensive drinking water that we have contaminated can be returned to its source in a form that harms neither the environment at large nor the living inhabitants of it. But guess what? The end-products aren't radically different from what went into it.

Central wastewater treatment plants produce two byproducts: Solids and liquids (and some gasses). These are usually disposed of.

Liquids are treated, disinfected and sent into receiving waters: lakes, rivers, oceans, and rarely, the ground. Very occasionally, it is introduced to the groundwater recharge through a variety of means.

As with septic systems, nutrients and toxic chemicals essentially are discharged from the plant untreated into the water. These plants are charged with reducing pathogens, certain bacteria, solids and BOD (or, biological oxygen demand, which is carbon).

The nutrient portion is mostly nitrogen compounds—80 to 90 percent from urine, the rest from protein-based compounds—and phosphorus compounds, which leave the plant often converted via nitrification to nitrate or ammonia. Those simply are dissolved in the water and go right into the receiving water or leach bed.

For information about conventional wastewater treatment systems, as well as a technical explanation of the contents of sludge, see *Wastewater Engineering, Treatment, Disposal and Reuse* by George Tchobanoglous and Franklin L. Burton (New York: Irwin-McGraw Hill, 1991).

And Generally...

It wastes nutrients that could be used to improve the soil, for which we often import expensive fertilizers.

The recycling of nutrients (phosphorus, nitrogen and potassium) in excrement offers an opportunity to produce food without an excessive use of fertilizers. The recycling of this carbon and nitrogen to agriculture helps prevent further degradation of agricultural soils, especially in arid and semi-arid countries. Instead, we flush it down the toilet, and then buy expensive fertilizers to produce our food crops. That's disrupting the nutrient cycle. It is not only wasteful, it produces pollution. And that makes it expensive, too.

In all of human history, humans have been using excrement-derived fertilizer (EDF) on food crops. However, we do not recommend the traditional method of putting fresh excrement on fields. Proper composting and long-term containment ensures the demise of human disease organisms prior to use and makes the nutrients available to plants.

Today, the regulatory community is wisely playing it safe when it comes to using composted human excrement from a composting toilet system. They know that burying it six inches below the surface removes it from the range of people and animals and allows the nutrients to be utilized by the root zones of plants. And to further ensure safety, they often exclude edible plants. In the future, composted human waste (only!) from a monitored and controlled composting operation will likely be acceptable for this purpose.

For an excellent report on this topic, see "Recycling Human Waste: Fertile Ground or Toxic Legacy?" by Gary Gardiner (WorldWatch, January/February 1998).

It is expensive to install and operate.

Costs for water and sewage treatment and waste disposal in general are skyrocketing due to pressures from development, increasing public and environmental health concerns and growing realization of the true costs of water reclamation.

Throughout the developed world, conventional wastewater treatment and disposal systems are failing to meet current public heath, safety and environmental protection standards, and regulations for disposal are tightening. The costs of remediating failed systems and constructing new ones are skyrocketing. Certainly with the high capital outlay and operating expense of sewage disposal, wastewater is now an economic commodity to be prudently managed like other commodities such as heating oil, gasoline or electric power.

Billions (at least!) are spent in North America alone to treat wastewater. It is hard to pinpoint an average cost for sewer systems and central wastewater treatment plants, as the variables are so many. In New England, a new treatment plant for 200 homes and businesses was estimated to cost $10,000 per property to build, and at least $2,000 each annually to maintain. And, as the cost of central treatment plants grow, so do the pumping costs for septic tanks. In southeastern Massachusetts, for example, pumping costs in 1998 were as high as 19 cents a gallon. Consider the cost of pumping 1,000 to 2,000 gallons every two or three years. That adds up.

And wastewater treatment plants need updating. In 1998, upgrading Boston's old primary wastewater treatment system to 1972 standards cost $5 billion—and required installing a disposal pipe that runs nine and one-half miles out to sea!

Just getting sewage to a treatment plant by pipe or by truck is a resource-intensive process involving miles of pipes and/or trucks and fuel.

Millions of homeowners in this country and others are finding they can no longer install simple septic tank systems or foot the costs of expensive central sewage treatment options. In Massachusetts, for example, on-site wastewater regulations were tightened so much in 1995 that an estimated two-thirds of homeowners will have to upgrade their septic systems, use holding tanks or pay thousands per household to build or expand a central sewer system.

Due to the high cost of central sewer systems, more homes are using on-site wastewater systems (usually septic systems). According to USEPA figures, about 30 million dwellings will be served by individual or non-sewered systems by the year 2000 in the U.S.

Regulators are no longer recommending maintenance-free on-site technology. That's because "maintenance free" historically equaled "out of sight, out of mind," which provided a false sense of security about the polluting effects of wastewater. In a report to Congress, the USEPA recommended greater emphasis on on-site and decentralized wastewater systems. It also recommended establishing management districts to monitor and help manage these systems.

Whether you are concerned about the environment or whether your more immediate worry is your household costs, it is obvious that a better way is needed.

Composted Excrement versus Composted Sewage: There *Is* a Difference

Of course, there is much more than simply excrement and washwater in sewage. From the toilet to the wastewater treatment plant, excrement is joined by toxic chemicals and heavy metals poured down drains by both households and industry. Thousands of chemicals and metals—including PCBs, pesticides, dioxins, heavy metals, asbestos, petroleum products and industrial solvents—can be part of the sewage cocktail. The 1972 Clean Water Act standards allow the toxics and nutrients to remain because remediating them completely would be expensive, if not impossible.

Solids left over from central wastewater treatment are called "sludge." This is a viscous mixture of fecal and other organic matter, toxic metals, synthetic organic chemicals and settled solids removed from domestic and industrial wastewater.

Sludge is either landfilled, incinerated, land applied or composted and used for fertilizer after composting. All of these processes have their problems, but composting sludge for agricultural use has been the most publicized in recent years. In 1997, new U.S. standards for organically grown crops allowed the application of composted sludge on certified organic crops. This added fuel to the fire of the long controversy about using sludge for its nutrient value on crops.

Consider: According to the book, *Toxic Sludge Is Good for You*, "Over 60,000 toxic substances and chemical compounds can be found in sewage sludge, and scientists are developing 700 to 1,000 new chemicals per year. Stephen Lester of the Citizens Clearinghouse for Hazardous Wastes has compiled information from researchers at Cornell University and the American Society of Civil Engineers showing that sludge typically contains a long list of toxins. (John Stauber and Sheldon Rampton, *Toxic Sludge Is Good for You! Lies Damn Lies and the Public Relations Industry.* Common Courage Press)

Some of these materials, reports Worldwatch Institute researcher Gary Gardiner, "degrade quickly with no harm to the environment, while others persist for decades or even centuries; some soils in Italy, for example, still contain lead leached from the pipes of ancient Rome. Natural cycling tends to keep potentially polluting materials spread thin. Human economies, however, often distill and concentrate harmful pollutants in our waste streams. Depending on the levels of concentration, returning these wastes to farmland can be more dangerous than beneficial." (WorldWatch Paper 135: Recycling Human Waste: Fertile Ground or Toxic Legacy?, Gary Gardiner, WorldWatch Institute, January 1997.)

The only sound way to make constructive use of the nutrient value of excrement and protect public health, is to compost only human excrement and kitchen wastes. Don't create problems—divert and utilize!

How to Choose an Ecological Wastewater System

Be prepared! The following are some sample criteria and required components you will need when planning a wastewater system. You might create a "Design and Implementation Matrix" with comparisons of the different systems, costs, maintenance ratings, site considerations, regulatory ratings

Steps in the Design and Construction Process

1. What is your budget?
2. Usage: How much effluent do you create? Know its quality, make-up and peak flows.
3. Decide what conservation measures you will implement to reduce volumes.
4. Calculate effluent design flow based on your commitment to #3 above.
5. Do you want to divide the design flow (blackwater and graywater)?
6. Recalculate design flow based on the above variables.
7. Conduct a soils analysis; determine the percolation rate and subsurface hydrogeological conditions. Survey the landscape position, drainage patterns and topography.
8. If you might choose a plant-based system, know your micro-climate for plants (what grows where). A greenhouse may be required in cold and rainy climates.
9. What's your plant hardiness/growing zone?
10. Consider your lifestyle and how much maintenance and involvement you are willing to commit.
11. Find out what can be permitted in your state and requirements for application for innovative/alternative systems.
12. Choose a system (include composting toilet system, if that's your choice).
13. Select a systems designer with credentials acceptable to your regulatory agencies and who has designed and permitted alternative/innovative systems you have selected.
14. Determine the data required by state and local regulatory agencies on the engineered plan to be submitted for approval.
15. Prepare the plan, with any variances from the existing regulations in accordance with #14 above.
16. Submit the plan to the appropriate regulatory agencies in accordance with their submittal requirements.
17. When the plan is approved, submit the plan to several qualified contractors for bidding and construction.
18. After construction, prepare an "as-built" plan to identify field changes that occurred.
19. Contract for or arrange to perform the requisite operation and maintenance functions.

Getting Them Approved (What Regulators Want to See)

The following should be on the plan you submit for approval for new or existing construction:

1. A plan of the system on the property that includes:
   - Address, lot size, lot lines, floor plan of dwelling, number of bedrooms, , page and plot numbers of your property from town or county deed records
   - Location of street, building, garage, outbuildings, driveways
   - Accurate perpendicular distance from street and both sidelines to building
   - Elevation of the top of foundation, cellar floor, garage floor
   - Existing or proposed street and driveway centerline grade
   - Location and logs of all soil observation holes and percolation tests ever performed on the lot, even if not used for waste treatment system purposes. Soils should be identified and classified by type and layer.
   - Elevations of ground surface at test pits, bottom of test pits and any rock formations or other impervious strata and observed groundwater encountered. Ledge outcrop location, if any; elevations and bounds
   - Profile of the existing wells and sewage disposal system showing invert elevations at building drain inlet and outlet from the septic tank; inlet and outlet from distribution box; invert of leach line; and bottom of soil absorption system
   - Cross section of proposed blackwater and graywater treatment system showing all construction details
   - Existing and proposed grading at building corners, from the lot corners, soil absorption system, and at such other areas where the existing ground contours are being changed
   - Location of all existing and proposed water or injection wells, wetlands, water courses, rivers, streams, brooks, ponds, lakes, swamps, marshes, vernal pools, culverts, pipes, swales, etc. Flood plain and any mean high tidal water within 100 feet of the lot line as shown on an accepted town map. Any seasonally wet area, swamp marsh area or area of temporary or permanent, ponded water within 100 feet of any lot line of the proposed system must be shown on the plan.
   - Location of existing municipal, county or state water resource districts, i.e. well fields and recharge areas, drainage areas from watersheds, sole-source aquifers or designated impoundments, lakes, rivers or ponds from which public drinking water is drawn.
   - Septic facilities and wells on immediately adjacent lots must be indicated on the plan
   - The plan shall by contours and narrative explanation demonstrate that drainage patterns will not be detrimental to the proposed lot/dwelling and executed in two-foot gradients. Conditions which will result in prolonged standing water are unacceptable. Conditions which will significantly increase drainage onto a driveway or street/roadway are unacceptable.
   - Underground heating oil tanks, water, electric, gas and other utility lines shall be shown and their size designated
   - Required inspections shall be noted. If required, they consist of excavation inspection and inspection of the fill material to include a sieve test of soil and trench bottom, final construction inspection before backfilling, final grade inspection
   - Variances requested from local and state codes, regulations
2. Brochures on products or technology specified, to include: Performance evaluations by independent testing organizations, such as ANSI/NSF International, Academic institutions or state laboratories, UL, CSA; warranty; maintenance manual; descriptive literature with parts list, electrical diagrams, etc.
3. Maintenance contracts or clear maintenance instructions for user implementation
4. Properly fill out special state, county or local permit applications and provide approval letters by regulatory agencies.

Adapted from MA 310 CMR 15.00 (Title 5 regulations for on-site systems)

A Sample Maintenance Agreement for a Composting Toilet And a Graywater System

SCOPE OF SERVICES

Tasks to be performed by a licensed plumber or skilled service contractor approved and/or licensed by the local Board of Health

- Discuss with Client issues of routine care and elicit reaction, comments and observations.
- Ensure that Client has all technical manuals, drawings, plans and contact phone numbers.
- Inspect and service oil/grease trap (ask Client for service intervals and volumes).
- Inspect and service all lint and particle filter (ask Client for service intervals).
- Inspect all pipe connections for leaks.
- Inspect and test pump:

Disconnect line and pump water into a 5-gallon (16 liters) bucket. Measure volume per 15-second intervals and multiply by a factor of 4 to get gallons per minute at zero head and compare to pump specifications. Clean filter screen.

- Inspect and test electrical components and pipe heaters and insulation (freeze protection).
- Inspect and test check-valve(s), back-flow prevention valves and ball valves.
- Inspect overflow buffer tank and high-water alarm components for general integrity.
- Secure from Client the pump-out frequency of buffer tank (note volume of tank below alarm level).
- Inspect and service composting toilet system (if applicable):

Ensure that Client has all manufacturers' operating instructions and contact phone numbers.

Determine frequency, volume and fate (ultimate destination, use or disposal) of humus end-product removed from unit. Remove humus end-product if required.

Elicit from Clients their general reaction and comments.

Check for signs of leachate leakage. Remove accumulated leachate if required.

Check for odor at unit and in toilet room or elsewhere.

Check electrical components (heater, fan motor, thermostats, fuses, etc.).

Check mechanical systems (mixer, inspection doors, trays, fasteners, etc.).

- Inspect and test alarm system:

  Level setting
  Audible and visual alarm
  Autodialer

Tasks to be performed by a horticulturist, arborist or landscape contractor

- Discuss with Client issues of routine care and elicit reaction, comments and observations.
- Ensure that Clients have all manuals, drawings, planting plans and contact telephone numbers.
- Recommend plant additions to enhance performance (nutrient and water utilization) and aesthetics.
- Identify and record the number and genus of plants in the garden.
- Check perimeter of garden for evidence of bamboo rhizomes escaping the containment barrier. Prune rhizomes to control if required.
- Inspect the integrity of covers or greenhouse (glazing, environmental controls, etc.).

- Inspect health of the plants for pests, disease or disorders to include but not be limited to:
  1. Die-back
  2. Unspecified plants (weeds, Client substitutes, etc.)
  3. Evidence of insect or nematode invasion
  4. Chlorosis (loss of green color) brought on by a lack of nitrogen and other nutrients
  5. Wilted, moldy, mildewed, spotted, dried, deformed, holes, discolored or curled leaves
  6. Dead branches & stalks
  7. Root health (rot, under-developed, blackened or peeling indicates water logging)
  8. Discoloration
  9. Stunted growth

- Apply nutrients, integrated pest management (IPM) and disease mitigation measures as required.
- Prune stalks (roots as well), train, harvest and replace plants as required
- Core sample of gravel & sand column nearest the inlet to bed

1. Check for mottling indicative of anaerobiosis, check pH with litmus paper

2. Clogging of gravel and sand particles due to lint or grease

- Inspect sand surface for crowning and texture to shed rain and snow melt
- Odors at the inspection pipe and breather vents (musty or lake-y is good - hydrogen sulfide "rotten egg smell") may indicate standing water in sand column.

Any subcontractors and outside associates or Contractors required by the Contractor in connection with the services covered by the Agreement will be limited to such individuals or firms specifically identified and approved by the Client during the performance of this Agreement.

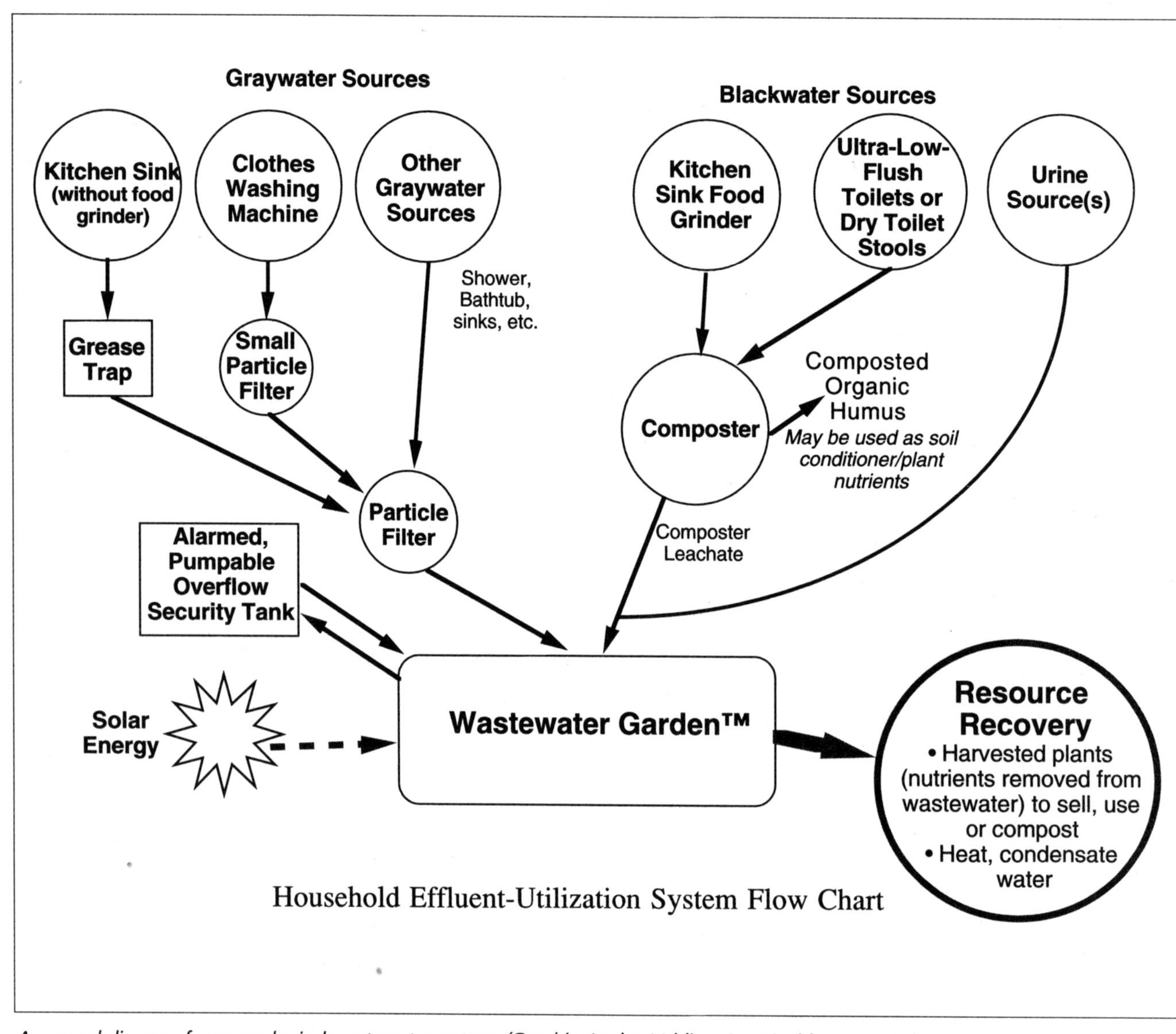

A general diagram for an ecological wastewater system. (Graphic: Jordan Valdina, Sustainable Strategies)

Changing Plumbing Codes to Reflect Water Conservation?

The Stevens Institute of Technology, with the support of the USEPA studied the impact of emerging water conservation technologies on water supplies and drainage systems within buildings. The study provides plumbing code developers the data which supports the revision of the water supply fixture unit (WSFU) method of sizing drain pipe size. This report recognizes the impact of more efficient fixtures on outdated plumbing codes and the continued adherence to outdated data. The recommendation is to use the Hunter method and expand the types of buildings addressed, calculates new fixture unit values and reduces the required pipe diameters for any given fixture or bathroom group. For example a 1.6-gpf (6 lpf) gravity toilet can now have a WSFU of 2.5 versus 4 WSFU in the old codes—a reduction of 37.5 percent.

Source: Tom Konan, Stevens Institute of Technology, "Impact of Water Conservation on Interior Plumbing," Proceedings of the American Society of Plumbing Engineers Convention, 1994.

Facts about the Microbial Inhabitants of the Large Intestine and the Gasses They Produce

- Human intestines are hosts to many billions of microbes. Each gram (about a thimble-full) of undigested food and excrement from the large intestine contains up to ten trillion (10,000,000,000,000) microbes!
- Microbes provide nutrients to animals.
- Microbes in the intestines perform many essential tasks. For example, microbes make several vitamins in the intestines. One of these is vitamin K, a vitamin that is otherwise lacking from human diets.
- Matter passes through intestines in approximately 24 hours.
- Some of the different microbes in the intestine are:

  - Bacteroides
  This is one of the most abundant microbes in the intestine, and is found in higher numbers in people who eat meat than in vegetarians.

  - Lactobacillus acidophillus
  This is a normal, helpful inhabitant of intestines. The billions of Lactobacilli in your intestines crowd out harmful microbes so they cannot grow. This is why many travelers take capsules filled with Lactobacillus, so they don't get traveler's diarrhea.

  - E. coli (Eschericia Coliform)
  This is the most well-studied organism. It is a normal resident of your intestine and provides you with vitamin K and some of the B vitamins.

  - Klebsiella
  It is speculated that these bacteria can fix nitrogen into protein in the intestines of people who have too little protein in their diet.

  The intestines of newborn, milk-fed infants contain almost a pure culture of this bacterium.

  - Methanobacterium smithii
  This bacterium produces methane in human guts. Normal humans fart about 10 to 15 times per day, passing a few hundred milliliters of gas per day. Boys and girls fart about the same amount. Though more than 50 percent of this gas is nitrogen, some of the gas is microbially produced, including the gas methane, which is flammable and is also a powerful greenhouse gas. Carbohydrates that cannot be digested by humans, but can be digested by microbes, are the leading cause of farts. Methane is an odorless gas: smells come from dimethyl sulfide and methanethiol. Other gases include oxygen, hydrogen, carbon dioxide, and nitrogen.

Source:
The Microbe Zoo, The Digital Learning Center for Microbial Ecology [http://commtechlab.msu.edu/sites/dlc-me/].

More on: Aerobic Bacteria vs. Anaerobic Bacteria

Aerobic microbial organisms require an oxygenated environment for their metabolic process. This requires the processing system to be aerated. Why use aerobic bacteria which require this specific environment? These micro-organisms provide essential benefits to the system as a whole. Anaerobic bacteria do not. And aerobic bacteria breaks down material much, much faster than their anaerobic counterparts.

Free Energy Production

During metabolism, aerobic bacteria produce about 800 Kcal/g Mol of energy. This energy is used for nitrogen fixation, the breaking of N - N bonds to form free nitrogen ions, which can then recombine with free oxygen to form nitrates, an essential plant food. This process requires 160 Kcal of energy to break one N - N bond. Anaerobic bacteria only produce about 50 Kcal/g Mol of energy, because they utilize available energy while breaking oxygen atoms from the molecules they feed on.

Oxygenated By-Products and No Toxic Build-Up

The by-products of the aerobic metabolic processes are oxygenated compounds and plant food, which are readily absorbed by plant roots in the system for their nutritive value.

Anaerobic metabolism results in non-oxygenated compounds—except for phosphates and water—such as methane, ammonia and hydrogen sulfide. These by-products are not plant food, and instead serve as a system contaminant. (See above diagram below.)

Natural Population Control and Avoiding Clogging of Voids in the System

Due to the aerobic environment, larger micro-organisms, such as protozoa, ciliata and amoebae, are able to exist. Protozoa (50 to 100 microns in size) prey upon the smaller aerobic microbial organisms (1 to 5 microns in size) of the system, devouring daily about one million bacteria, thus, naturally controlling their ever-increasing numbers. This predator-prey relationship prevents the clogging of the system with excessive bacteria, which would lead to inhibition of the capillary rising of nutrients and water to the plant root systems. These protozoa cannot exist in an anaerobic environment.

Climate Adaptability

Optimal conditions for aerobic microbial metabolic processes are a temperature between 59 to 168°F, and a D.O. (dissolved oxygen) concentration of >3.0 mg/L. Aerobic bacteria are adaptable. When placed in unfavorable climatic conditions, their metabolic processes greatly reduce, and the organism goes into a state of dormancy, which can last up to several years. Once favorable conditions return, the organism "wakes up," and its metabolism rises to system-efficient levels. Anaerobic bacteria do not have such adaptability.

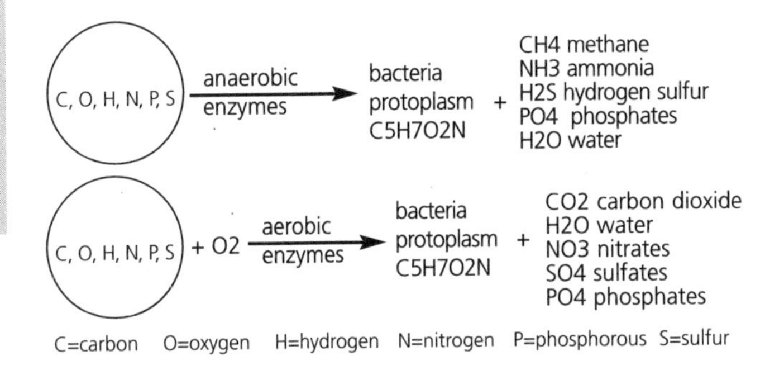

More on: Requirements and Processes for Life

Food

Among other activities, life requires a continuous supply of certain chemicals and energy. The primary source for the energy is sunlight and the chemicals are recycled from those found in the earth's crust. The program to manage the life process is contained in information and instructions learned over time and maintained in the genetic structures of each living organism. The chemicals must be taken in by the living organisms and processed for use in life maintenance. These chemicals are bound up in different forms ("food" and "water") of matter such as minerals and organic compounds held together by weak and strong bonds that must be broken in order to release the chemicals for life.

Green plants can photosynthesize the chemicals required for growth and energy from sunlight, soil chemicals, water and the carbon dioxide in the atmosphere and are called autotrophic organisms. The rest of us heterotrophic organisms in the Animalia kingdom must ingest organic plant and animal tissue composed of carbon, hydrogen and oxygen and other chemicals in order to form the chemical building blocks of life and to provide the energy to do so. In the process of living, heat is released and the by-products of metabolism are excreted.

Digestion and Metabolism

The breaking of the bonds that hold the chemicals in food and water together is called "digestion" Living organisms, digest and metabolize the food with specific enzymes tailored to transform food molecules into energy and to extract the chemicals they require for metabolism, growth and cellular maintenance. Enzymes (protein catalysts) produced by cells are the chemical machinery that do the work of digestion and metabolism and control all the chemical processes in living organisms.

Respiration

To accomplish this activity, cells that make up living organisms require significant and continuous amounts of oxygen. Chemicals and specifically carbon, in food are oxidized to carbon dioxide, water and energy. This is called respiration. Certain organisms use free atmospheric oxygen for respiration and they are classified as aerobes. Other organisms must first extract the oxygen from the materials they are digesting. These are anaerobes.

Excretion

Plants excrete gases such as oxygen, chemicals and water from openings in the leaf and stalk called stomata. They also excrete unhealthy fiberous material, bioconcentrated metals, minerals and other nutrients during leaf-drop. Animals excrete water, carbon dioxide and other chemicals and undigested plant and animal tissue. In addition to the by-products of metabolism, excreta contains chemicals and organisms that are not required by or are toxic to the excreting organism.

The excreta of plants and animals contain the building blocks of life itself and are used as food for other organisms. In animals, the excreta also contains undigested food and large volumes of bacteria that grow in the intestine, so it is rich in nutrients that require additional digestion to be released. In an ecosystem, the materials produced by one class of organisms are used by others in a continuous cycle driven by the energy from the sun.

Decomposition

When enzymes digest non-living compounds such as minerals and dead plant and animal tissue its called "decomposition". The decomposition activity is a necessary phase of the life, death and renewal cycle that releases the chemicals and energy bound up in the earth's crust, dead plant and animal tissue so they can be recycled for use in the life processes of other organisms.

Stabilization

When the supply available nutrients is consumed, the microorganisms begin to consume their own protoplasm to obtain energy for cell maintenance. When this occurs the microorganisms are said to be in the endogenous phase. When these organisms die, their protoplasm and cellular matter is digested by other organisms. Eventually, if no new food sources are presented, all the energy will be released and the matter fully oxidized. The end of this phase results in an end-product that is very stable and safe. This material contains the chemicals, now in the form that can be utilized by the plant kingdom and the cycle begins anew.

More About Drying Toilets

That drying toilets can work well was shown by a 1999 study.

In 1999, a team of Vietnamese and Swedish public health researchers studied 60 single- and double-vault toilets with slight differences in construction details. After six months of use, 12 toilets were tested for the survival of microbial indicators, Salmonella typymurium phages 28B and Ascaris suum eggs. In both cases, the pathogen indicators were placed into the middle of the fecal material pile in the chamber or bucket. Weekly samples were taken to monitor die-off of Sal. phages, and samples were taken of the Ascaris eggs every two weeks.

Findings: "The survival time of the A. suum eggs was generally longer than that of Sal. typymurium phages. Their shortest survival time was 51 days and the longest was 154 days."

Because wood ashes were added to all of the toilets, the researchers surmised, "pH seemed to be the most considerable factor in governing the die-off of the organisms."

The study concluded, "It takes six months of retention of fecal materials in the tested toilets to become absolutely safe."

The researchers did not test well-aerated composting toilet systems, which would have measured pathogen die-off without high-pH alkaline additives (ashes).

Nguyen Huy Nga (Viet Nam Ministry of Health); Bui Trong Chien, Duong Trong Phi, Bui Chi Chung (Nha Trang Pasteur Institute, Viet Nam); Thor-Axel Stenström, Anneli Carlander, Therese Westrall (Swedish Institute for Infectious Disease Control); Uno Winblad (WKAB consulting group), "Testing Eco-San Toilets in Central Vietnam," presented at the Stockholm Water Conference, 1999.

More on: Pathogens

Pathogens Are a Moving Target

Pathogens are an ever-changing challenge to the public health community and the designers and manufacturers of all waste treatment technologies. Not only do pathogens co-evolve with disinfectants to evolve into new, resistant forms, but we continue to discover new threats that were heretofore never known. Cryptosporidium parvum, for example, is a single-celled aquatic protozoan animal, which is also an obligate intracellular parasite. The sporocysts of this organism are resistant to most chemical disinfectants, such as chlorine. Fortunately cryptosporidium are susceptible to predation and antagonism by other aerobic organisms, as well as drying and the ultraviolet light from the sun.

Witness the recent discovery of prions—a protein that is neither dead nor alive, but nonetheless a pathogen—which have become famous because recent research has confirmed they are the infectious particles in a disease group found in humans and animals called transmissible spongiform encephalopathies (TSEs). These include Creutzfeldt-Jacob disease (CJD) and Mad Cow Disease. The discovery that proteins alone can transmit an infectious disease has come as a considerable surprise to the scientific community, and the mechanisms underlying the propagation of the infectious proteins and their pathogenic effects remain a matter of hot debate.

References:

Monari, L., Chen, S. G., Brown, P., Parchi, P., Petersen, R. B., Mikol, J., Gray, F., Cortelli, P., Montagna, P., Ghetti, B., et al. (1994). Fatal familial insomnia and familial Creutzfeldt-Jakob disease: different prion proteins determined by a DNA polymorphism. Proc. Natl. Acad. Sci. USA 91, 2839-42.

Tateishi, J., Kitamoto, T., Doh, u. K., Sakaki, Y., Steinmetz, G., Tranchant, C., Warter, J. M., and Heldt, N. (1990).

Types of Wastewater-Borne Pathogens

Pathogenic Bacteria

Bacteria are the simplest of the living microorganisms. Although small (about 0.02 to 0.1 micrometers), they are comprised of a nucleus containing DNA molecules enclosed in a membrane. This type of organism is called eukaryote. A bacterium is a single cell which gains its energy from nutrients in the surrounding environment (including the intestines of humans) by means of extracellular enzymes which transform complex molecules into simple forms that can pass through the cell membrane to nourish the organism. In most cases, the bacteria provide a beneficial service, for example, by assisting in digestion of food in the stomach and intestines, or composting organic matter. However, some bacteria are pathogenic. Echerichia coliform (E. coli) is clearly identified by intestinal diseases. E. coli O157:H7 enterohemorrhagic (EHEC) has been responsible for several serious recent food-borne and waterborne outbreaks. Antibiotics are the treatment of choice.

Some bacteria that could be found in domestic wastewater:

- Salmonella spp.
- Staphylococcus aureus
- Campylobacter jejuni
- Yersinia enterocolitica and Yersinia pseudotuberculosis
- Vibrio cholerae O1
- Vibrio cholerae non-O1
- Vibrio parahaemolyticus and other vibrios
- Vibrio vulnificus
- Clostridium perfringens
- Aeromonas hydrophila and other spp.
- Plesiomonas shigelloides
- Shigella spp.
- Miscellaneous enterics
- Streptococcus

Enterovirulent Escherichia coli Group (EEC Group)

- Escherichia coli - enterotoxigenic (ETEC)
- Escherichia coli - enteropathogenic (EPEC)
- Escherichia coli O157:H7 enterohemorrhagic (EHEC)
- Escherichia coli - enteroinvasive (EIEC)

Viruses

Viruses are parasites that are neither cells nor organisms. In fact they may not be "alive" by some definitions, although they do reproduce. As the smallest pathogen (0.02 to 0.03 microns), they are composed of a single nucleic acid for containing genetic information, surrounded by an outer shell of protein. Although viruses are themselves inert, they possess a highly evolved survival mechanism that parisitizes a host cell's metabolic processes by invading and using the cell itself to replicate themselves hundreds of times. Viruses then leave the host cell to penetrate other cells and start the process again. Outside the host, viruses can survive for hours, weeks or months in the proper environment until another host cell arises. Usually the host cell cannot survive the invasion and dies. Vaccines can prevent some infections by adding additional antibodies to the body's natural immune system, but once the invasion is under way, vaccines can do little to cure the infection. Antibiotics are of little help with viruses.

Some viruses that could be found in domestic wastewater:

- Hepatitis A virus
- Hepatitis E virus
- Rotavirus
- Norwalk virus group
- Other viral agents

Parasitic Protozoa and Worms

Parasites are viruses or living organisms that live at the expense of other organisms. Most are animals, but some are flowering plants and fungi. They are a diverse population, and include microscopic non-cellular or unicellular (not divided into multiple cells) animals, such as protozoa and larger animals, which include nematodes or helminths (worms) that invade the human body and use it as a nutrient-providing host while they multiply. Like bacteria, most are beneficial (protozoa eat millions of bacteria daily in a compost or soil system and earthworms loosen and aerate the soil) but some are pathogens that are excreted in feces and can infect others who ingest them.

Some parasites and worms that could be found in domestic wastewater:

- Giardia lamblia
- Entamoeba histolytica
- Cryptosporidium parvum
- Cyclospora cayetanensis
- Anisakis sp. and related worms
- Diphyllobothrium spp.
- Nanophyetus spp.
- Eustrongylides sp.
- Acanthamoeba and other free-living amoebae
- Ascaris lumbricoides and Trichuris trichiura

Other

There are other pathogens of major concern in moist environments such as the ever-present threat of aflatoxin, produced by fungi of the aspergillis genus. Aspergillis are significant to this book because they are present in composting of any organic matter. Fortunately, relatively few are pathogens. Only those that have been characterized as aflatoxins are of concern.

About Ascaris lumbricoides (roundworm)

This round worm must be about the

most ubiquitous parasite of humans with an estimated 1 billion people infected world wide. In some communities infections rates reach 100%. We give it special attention here, because of its reputation for long-term survival in dry or anoxic environments. However, moist composting is a hostile environment for parasites out-of-host due to nutritional, physical, chemical and thermal conditions falling out of the narrow range for long-term survival and the antagonism with other . better adapted organisms that will kill foreign organisms. One report identified viable eggs after years after being painted into the floor of an abattoir (slaughter house). Ascaris can be the "old man" of pathogens when excreted into a composting toilet. But if you have Ascaris in your composter, you probably have it in you, as well as your pets, yard, animal pens and garden too!

Ascaris is the second most common nematode infection (pin worm is first). It is a large intestinal round worm found in warmer climates world-wide. Ascaris is the largest nematode found in humans (20-35 cm or 12 inches). It is the most difficult to control because its eggs can survive for years in soil and has a diversity of human and non human hosts. Important epidemiological factors include the fact that female worms may produce as many as 1 to 2 million eggs per day and these eggs are highly resistant to drying because of the thick egg shell. It is believed that these eggs can survive for as long as 5 years in a dry environment.

Life Cycle and Morphology

The life cycle, like that of all the other nematodes, has two distinct phases, a preparasitic phase which is direct and a parasitic phase which is migratory. The preparasitic phase begins with the passage of eggs in the feces of infected animals. One important feature of the early life cycle is that the eggs do not hatch, per se. The larval stages (L1 and L2) develop inside the egg. Thus the infective stage for a susceptible host is an egg containing a second stage (L2) larva. Two other phases (L3 and L4) will be followed by the egg-laying adult.

Transport hosts, such as earthworms and beetles may also ingest these infective eggs, in which cases, the eggs molt and the L2 migrate to the tissues of these transport hosts. When an infected transport host is ingested by a definitive host, a human, pig or other warm-blooded animal, these encysted L2's are released and begin their migration. A human or warm-blooded animal host ingests embryonated eggs from soil or transport or by ingesting infected transport hosts. These eggs do hatch in the stomach and small intestine, where worms penetrates the intestinal walls and enter the portal blood stream and migrate to the liver within 24 hours after ingestion. There they molt to L3's and via the blood stream into the lungs. From the lungs they travel through the trachea and pharynx and are swallowed (they can also migrate out the mouth or anus or nose). After being swallowed, they mature (L4) in the gut and produce eggs which are excreted in the infected hosts excrement. This cycle takes from 6 to 12 weeks and produces approximately 27,000,000 eggs.

Ascaris causes diseases by the physical presence of large worms in the body. Worm migration and intestinal blockage; entering the bile and pancreatic ducts, liver, lungs (Ascaris pneumonitis) and peritoneal cavity: As with all parasitic worms, nutritional deficiencies in the host appear due the nutrient requirements of adult worms (The more worms the greater the nutritional deficiency).

References

The Bad Bug Book, U.S. Food & Drug Administration Center for Food Safety & Applied Nutrition

Foodborne Pathogenic Microorganisms and Natural Toxins Handbook, Center for Food Safety and Applied Nutrition; available on the World Wide Web: http://vm.cfsan.fda.gov/~mow/badbug.zip.

"Staff Summary: Pathogens of concern and their control through removal and or disinfection," Massachusetts Water Resources Authority , May 6, 1998

Karol Enferadi, Robert Cooper, Storm Goranson, Adam Oliveiri, John Poorbaugh, Malcom Walker, Barbara Wilson, "Field Investigation of Biological Toilet Systems and Grey Water Systems" (State of California Health Department, Department of Health Services and the Water Engineering Laboratory, Office of Research and Development, U.S. Environmental Protection Agency. EPA/600/2-86/069, 1980).

Severity Scale 1 - 100

Problem	Mean Severity
Infectious hepatitis	83
Encephalitis	69
Salmonella	68
Shigella	64
Typhoid	64
Enteric viruses	62
Entmeboeba histolytica	59
Polio virus	55
Giardia	48
Ascaris	43
Hookworm	40
Cestodes	37
Nitrate (in drinking water)	36
Rodents	35
Mosquitoes	34
Flies and gnats	32
Animal parasites	26
Venomous arthropods	24
Odor	21
Beetles	15

This chart is very helpful for composting toilet system designers and regulators, because the ability of any one technology to manage all the problems at an affordable cost to the users is a function of relative risk of infection. For example, using the chart above, it is not a priority of a composter to eliminate beetles, although they may be a nuisance.

(Enferadi et al., 1980)

The Effect of Antibiotics on Composting Toilets

There has been considerable discussion regarding the use of anti-infective drugs, such as penicillins, tetracyclines and sulfonamides by users of composting toilets and their effect on the composting process. The following study found that the addition of antibiotics through excrement had no detribmental effect on the microbial activity in the composting process!

In 1979, the Department of Microbiology at the Agricultural University of Norway conducted tests with human body wastes and toilet paper (200-gram portions) collected from bucket privies in Olso by the waste collection department. These tests were conducted in controlled batch composting systems equipped with respirometers to measure the air uptake and the carbon dioxide output on an hourly basis. They also counted viable bacteria using standard methods of sampling, culturing on sterile beef extract-peptone-yeast extract agar plates which were incubated aerobically for 48 hours each.

A series of 200-gram portions of waste material was used for the compost experiments. To each portion was added an innoculum of 2 grams of old compost and a 10-mililiter solution of the anti-infective drug to be tested. The material was thoroughly mixed and 30 grams of unfertilized peat was added as a moisture absorbant. The final dry matter content was approximately 16 percent.

The drugs potassium phenoxymethlypenicillin (Apocillin A.L.), tetracycline chloride (Tetracycline A.L.) and sulfamethizole "NAF" were added directly to the excrement in amounts equal to conventional oral doses (0.1, 1, 10 grams per day) The excrement was added in amounts according to Gotaas (1956), 1.5 liters (1.25 liters urine and 0.25 liters feces). Three parallel experiments were run with each drug and a control with no drugs. The duration was 15 days.

The investigators noted that the doses were applied directly to the excrement which would equal massive doses, because only 20 percent of an oral dose of penicillin is excreted in the excreta (primarily in the urine). Twenty to 60 percent of an intravenous dose of 0.5 g of tetracycline is excreted in the urine during the first 24 hours. Generally, the excretion in the feces is one tenth of these amounts.

Discussion:

Generally the results on respiration rate, effect on total CO2 evolution and the effect on the number of viable bacteria were the same for each antibiotic with only slight differences.

In all parallel experiments (which is typical for all composting toilets), the first phase (day 1 and day 2) was characterized by a sharp rise in the number of viable organisms and respiration rate as the sugars, starch and to a certain extent lipids are utilized very quickly. Thirty percent of the total CO2 was released during this first phase. The second phase began on day 3 through day 15. In this phase, the more difficult cellulose and lignin compounds are decomposed. The number of viable bacteria stayed constant until the 10th day, later slowly diminishing.

The results:

"The results indicate that 50 percent inhibition of the respiration is the maximum and that was by sulfamethizole and tetracycline, and the number of the organisms seems to be maintained relatively well. After 13 days the respiration rate recovered to that of the control series"

"Ten days after the addition of 10 daily doses of tetracycline the number of viable organisms was even higher than the control series."

Penicillin had no effect on respiration or the number of viable organisms after the second day. In fact the respiratory rate on the third and fourth day was 50 percent higher than the control.

Conclusion:

The rates of respiration were generally depressed only for the first one to three days and even with continued inoculation with the antibiotics, all recovered to a rate equal to the control.

"In practice, only a fraction of the given anti-infective drug is excreted. In view of the dilution which takes place in biological toilets, this makes it unlikely that any of the tested drugs will to any practical extent affect the composting process that is vital to the functioning of such toilets"

References:

"Effect of some Anti-Infective Drugs on Composting of Human Body Wastes," Ove Molland and Tor Arve Pedersen. Department of Microbiology, Agricultural School of Norway, N-1432 As-NLH Norway. Acta Agriculturae Scandinavica 29 (1970)

As the Compost Turns...

Perhaps you can turn too much. "Turning a compost pile tends to reduce the degree of pathogen removal achieved by the process, because the parts of the pile that have been hot enough to kill microorganisms can then be recontaminated by contact with those that have not. Turning also increases the cost and the amount of space required. Avoiding the need for turning can therefore be an advantage."

—*Environmental Health Engineering in the Tropics*, Feachem and Cairncross

How Thermal Transfer Takes Place

Thermal transfer and storage is accomplished by various heat transfer methods:

Conduction, where the heat must diffuse through solid materials or stagnant fluids; by convection where the heat is carried by actual movement of the hot materials (usually liquids or air); and by radiation, where the heat is transferred by means of radiant wave energy. The mass and thermal conductivity of different materials will determine whether it is a good storage material. Some, like aluminum and copper, lose energy too fast for storage but are good conductors. Others such as glass and plastic are better insulators. Hot water is one of the easiest forms of storage, but others, such as stone and phase-change salts, are common.

The Urine Connection and the Nitrogen Cycle

How Nitrogen Is Transformed Into Plant Protein

Nitrogen is an odorless, invisible gaseous element that comprises about 20% of our atmosphere. In its molecular or atmospheric state, it is chemically inactive, but can combine with oxygen, hydrogen and sulfur to form hundreds or thousands of organic and inorganic compounds.

Inorganic nitrogen compounds, such as nitrates (NO3), are synthesized into organic nitrogen, such as amino acids and plant protein, by autotrophic plants, which get their energy from sunlight. When these plants are eaten by animals to produce animal protein from amino acids, the unused or residual nitrogen compounds are excreted as urea, creatine, ammonia and uric acid. These compounds are then consumed and used by living soil organisms (also found in composters), which in turn produce oxidized forms which are again used by plants in a neverending cycle.

The conversion of molecular or atmospheric nitrogen into inorganic nitrogen compounds is called "nitrogen fixation." In soils, this process is carried out by algae and bacteria. Nitrogen-fixing bacteria, such as Azobacter and Clostridium pasteurianum, are free-living. Others, such as rhizobium, are found in the nodes of roots living symbiotically with fungi as micorrhizae. The so-called blue-green algae cyanophyceae and myxophyceae are unicellular, colonial or filamentous algae that contain chlorophyll. Using light as an energy source, they fix nitrogen and oxygen into inorganic compounds.

• *Nitrification and De-Nitrification*

The nitrogen in urine, primarily in the form of urea, is quickly transformed to NH3 and NH4 in a two-step chemeo-autotrophic* processes. In an aerobic soil system or composter, nitrosomonas bacteria transform urea and ammonia (NH3) to nitrite (NO2), and nitrobacter utilize the nitrite and transform it to nitrate (NO3), which are fully oxidized nitrates. That means they are now available as plant food. This process is called "nitrification" in the wastewater treatment industry. Composting toilets are hosts to these organisms, too.

* The organisms produce their own organic constituents from inorganic compounds, using energy from oxidation.

Denitrification is the conversion of nitrites (NO2) and nitrates (NO3) into gaseous forms of nitrogen, such as ammonia (NH4), nitrous oxide (NO) or atmospheric/molecular nitrogen (N2). Once converted to a gas, nitrogen is lost as a fertilizer and wasted. For that reason, we do not want to denitrify urine if we are going to use it as a fertilizer. In wastewater systems that do not utilize urine, denitrification is accomplished by adding more carbon to an anoxic digester to grow anaerobic bacteria, which convert the nitrogen to gasses. Denitrification in natural anoxic soils is usually carried out by anaerobes, such as clostridium, echerichia and staphylococcus. It is also carried out by the actinomycetes myocbacterium, nocardia and streptoyces

Sugar as a Carbon Source

Glucose, a sugar, is also normally present in healthy human urine to the extent of about 0.1%. The C:N ratio of urine is low (0.8:1) because of the relatively large amount of nitrogen in relation to the carbon in the form of glucose. This would require about 31.25 (25/0.8) grams of carbon to achieve a C:N ratio of 25:1, if one wanted to provide an optimum diet to microbially oxidize all of the nitrogen in urine.

• *Sugar as an Easy Carbon Source*

We are not advocating buying sugar for this purpose, but suggesting that there are other ways to prepare urine for microbial processing without buying or transporting a room full of wood chips. The logic is as follows:

• *Some Carbon Is Better than Others*

With the exception of hydrogen, the element carbon (C) forms the largest number of known molecules (hundreds of thousands) among the chemical elements. The molecules of carbon we are interested in are those forms that are "available" as a nutrient that can be readily digested by microbial enzymes as opposed to carbon bound up in plastics, for example, which is "unavailable." Sugars are the most "available" of all the carbohydrates, and can provide a lot of carbon with very little volume compared to straw, leaves and woodchips. These are primarily difficult-to-digest celluloses and lignin. In a composter, cellulose and lignin are decomposed by fungi and actinomycetes, not by bacteria.

Sugar, however, is a simple carbohydrate. Cane or beet sugar, or sucrose, is a disaccharide carbohydrate with the formula $C_{12}H_{22}O_{11}$. One sugar crystal contains millions of sucrose molecules. When it is in the presence of the enzyme, invertase, it yields the sugars, glucose and fructose, which are simple monosaccharide carbohydrates. Carbohydrates are polyhydroxy aldehydes or ketones that are the primary nutrients that provide cellular growth and energy. It is the C12 in sucrose that interests us in this formula. That means there are 12 carbon atoms bound up in the sugar molecule.

• *How Much Sugar?*

About a third of a cup of sugar per person per day should do it. (Here's the scientific rationale: Urine is 98% water, and a gallon weighs about 8.6 pounds or 3,882.8 grams, which is slightly higher than pure water because urine has a specific gravity of 1.024. Of the total, only 60 grams are dry solids. Of the dry solids, only 11 grams are nitrogen. An ounce of sugar weighs 28.3 grams. Of that, 42% is carbon. So, one ounce of sugar contains 11.9 grams of carbon. One cup of sugar is 8 ounces or 95.3 grams of carbon. To obtain a 25:1 ratio for this new mixture, a cup of sugar should be added to a gallon of urine, or a third of a cup per person per day.)

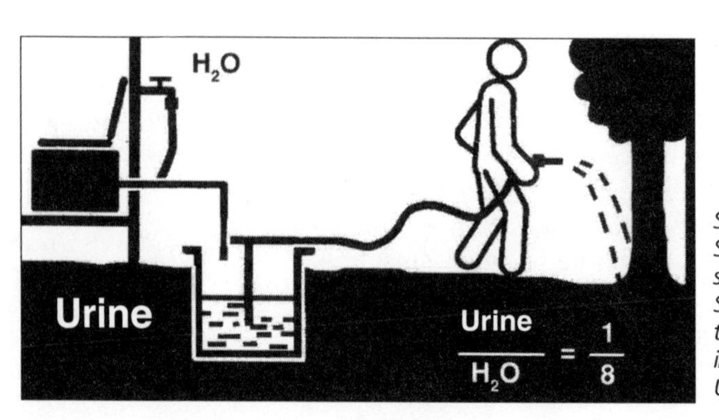

Servator-Separet of Sweden suggests this set-up for using urine. Surface irrigation like this would be illegal in much of the United States.

More on: The Effects of Salt on Plants

Salt and Plants

Because urine has a lot of nitrogen and table salt in it, technical discussion of how plants manage it is interesting.

According to Dr. Alex L. Shigo, a plant physiologist:

"Elements or molecules made up of a few to many different atoms enter the Roots as ions. An ion is a charged atom or molecule. Ions with a Positive charge are cations, and those with a negative charge are anions. Each particle or granule of fertilizer is a salt made up of a lattice of anions and cations, just as ordinary table salt is made up of a grand lattice of connected sodium cations and chloride anions. When salt as sodium chloride dry granules is poured into water, the sodium And chloride ions separate. When they separate, they carry electrical charges and are called the sodium ion and the chloride ion. When a cation enters a root, another cation must exit. This is very important, As we will see. When nitrogen enters a root as nitrate anion, an anion of bicarbonate ion from carbonic acid exits. The bicarbonate ion is probably the second most important compound in nature, next to water, because it drives the absorption process. When a bicarbonate ion exits Into the rhizosphere, the pH increases.

When urea is used in fertilizers as the nitrogen source, the pH in the rhizosphere could increase to 2 or more pH units. The chemistry behind This is complex, but here I present only the conclusion, because a common problem with trees in some high pH soils is chlorosis. There is no easy field method for measuring the pH of the one millimeter wide Rhizosphere. The rhizosphere could be pH 8, and the bulk soil would measure pH 6. As pH increases, the availability of elements such as iron and manganese decreases. In soils, it is one thing to have an element present and another to have it in a form available to the plant as an Ion. As pH increases, iron and manganese element, form molecules that precipitate in water rather than ionize. If they are not available as ions, they will not be absorbed. And, if they are not absorbed, several of the enzymes essential for chlorophyll formation and photosynthesis will not form.

When the energy flow from the top of the pump is blocked, then the bottom does not get enough energy for growth and defense. The pathogens invade, and the plant declines. This scenario does not mean that every time you use urea, plants will decline from chlorosis. But the use of urea could be a contributing factor where plants with genetic codes for growth on low pH soils are planted in high pH soils. If fertilization is a desired treatment, then a fertilizer that has nitrogen in a positive charged ion, such as an ammonium ion, would help to reduce the rhizosphere pH. When the ammonium ion enters the root, a proton of positive charge will exit. The protons in rhizosphere water will bring about more acidic conditions, so there is a way out." (Dr. Alex Shigo in *Plant Care Industry*)

The Problem with Sodium and Clay Particles in Leachfields

It has been said that the sodium equivalent of 56 pounds of table salt are discharged into a septic tank leachfield each year from a household of three to four.

The high sodium content of cleaning products for laundry, kitchen and bathrooms are a primary source of leachfield failures. Water softener wastes and high-sodium tap water also contribute to the problem.

For instance, when chemical (cationic) exchanges occur from sodium, fines or clay particles may bond into a waterproof barrier, which in turn causes the physical flooding, blockage of soil passages and biological death of air-dependent cleaning organisms in the soil. In agriculture, this chemical change causes physical or structural changes in the soil which ultimately leads to loss of biological uptake of plant nutrients.

In the septic system drainfield, problems begin when a thin impermeable layer of bonded fines develops directly under the leach lines or on the trench floor or walls. This layer grows in density over time and soon a "waterproof" barrier prevents access to the absorptive active soil surfaces needed for maximum organism contact and cleanup of wastes.

Let microbes and soil bacteria act as your intermediary between excrement and plants.

More on: Composting Toilet Leachate Characteristics

(More information about the study noted on page 40)

Nepon foam-flush toilet (200 ml water/flush) and a Sealand vacuum toilet (500 ml/flush) microflush toilet. Ground kitchen scraps were also flushed into the composter. According to the report:

"The compost leachate is characterized by a relatively high organic loading (COD and TOC), high nutrient loading (TKN and TP), and the presence of fecal coliform. The leachate is colored by organics but is not especially turbid. However, direct observation of this system and measurements at other AlasCan installations indicate the volume is generally less than 10 liters per day. The compost leachate volume is estimated to be less than 3 percent of total graywater effluent volume based upon comparison of the leachate mass loadings for individual constituents to the total graywater effluent loadings for those same constituents. As a comparison, leachate fecal coliform levels are in the 103 to 104 range while reported values for septic tank effluent range from 105 to 108 (EPA).

The BOD5 values of "zero" are not thought to be representative due to lack of appropriate sample seeding at the laboratory. BOD is primarily used as a process control variable—test results are dependent upon the biological process from which the seed source is derived. TOC and to a lesser extent COD are less subject to this limitation. Ratios of BOD/COD and BOD/TOC in the leachate would be expected to be lower than in typical raw domestic wastewater since the compost process is aerobic, and much of the soluble organic matter available to the organisms involved in the BOD5 test has already been converted to residual organics or carbon dioxide."

The complete report can be seen on the AlasCan web page:
http://cloudnet.com/~alascan/testdata.html

More on: The Value of The Rhizosphere

(This text was originally published in the October 1996 edition of *Plant Care Industry.* Adapted and reproduced with permission of Dr. Alex Shigo, Shigo and Plants Associates, PO Box 769, Durham, NH 03824. Dr. Alex Shigo is a noted authority in the field of modern arboriculture.

Rhizosphere Wars

The rhizosphere is the absorbing root-soil interface. It is the zone, about one millimeter in width, surrounding the epidermis of living root hairs and the boundary cells of mycorrhizae as well as hyphae growing out from some mycorrhizae.

The rhizoplane is the boundary where soil elements in water are absorbed into the plant. Under an electron microscope, the rhizoplane appears as a jelly where microorganisms and plant cells mix, making it impossible to tell which side is plant and which is soil.

A constantly changing mix of organisms inhabit the rhizosphere and surrounding soil. Bacteria, actinomycetes, fungi, protozoa, slime molds, algae, nematodes, enchytraeid worms, earthworms, millipedes, centipedes, insects, mites, snails, small animals and soil viruses compete constantly for water, food and space.

The rhizosphere is a battleground and the wars are continuous. Amoebae are eating bacteria. Some bacteria are poisoning other bacteria. Fungi are killing other fungi. Nematodes are spearing roots. Fungi are trapping nematodes. Earthworms are eating anything they can find. Sometimes the victors benefit the plant, and sometimes they do not.

....Soils and wood share a common problem: They are thought of as dead substances. This has come about because wood-products research gained an early lead over research on wood in living plants. With soils, many texts still define soils as "loose material of weathered rock and other minerals, and also partly decayed organic matter that covers large parts of the land surface on Earth."

Sapwood in living plants has many more living cells than dead cells. In upper layers where most absorbing roots of plants grow, soils have more soil organisms than grains of weathered rock. In great disrespect, most people still refer to soil as dirt! When researchers first discovered the great value of soil microorganisms for human antibiotics and profit, the living nature of the soil began to emerge.

A more correct definition of soil should be that it is a substance made up of sands, silts, clays, decaying organic matter, air, water and an enormous number of living organisms. Survival of all living systems depends greatly on synergy and efficiency to optimize the functioning of all processes and to keep waste as low as possible. When synergy and efficiency begin to wane, declines follow.

When there are troubles in the rhizosphere, there will be troubles with the plant.

Energy and Root Exudates

Microorganisms compete in the rhizosphere, an area rich in exudates from the plant. The exudates contain carbohydrates, organic acids, vitamins and many other substances essential for life. From 5 to 40 percent of the total dry matter production of organic carbon from photosynthesis may be released as exudates! When plants begin to decline, the amount of organic carbon released as exudates increases.

Or, on the basis of the mass-energy ratio law, as some plants on a site get bigger, many smaller suppressed plants will die. As the suppressed plants decline, they contribute a higher percentage of their soluble carbohydrates to the rhizosphere.

A Closer Look at Roots

Woody roots have cells with walls of cellulose, hemicellulose and lignin. Lignin is that natural "cementing" substance that gives wood its unique characteristic for strength. Woody roots also have an outer bark or periderm made up of three layers: the phellogen, phelloderm and phellem. The phellogen is the bark cambium. The phelloderm is a thin layer of cells on the inner side of the phellogen. The phellem is the outer corky layer. Phellem cells are impregnated with a substance called suberin, which is a fatty substance that prevents water absorption.

Some characteristics of woody roots are:

- They do not absorb water.
- They have no pith.
- Their conducting elements are usually wider than those in the trunk.
- They have a greater proportion of parenchyma cells than is usual for trunks. The living parenchyma store energy reserves, usually as starch.

A soft cortex without chlorophyll may be in the bark. In some plant species that thrive in wet soils or have deep roots, the cortex may have many open spaces that act as channels for air to reach the living cells in the roots. It is important to remember that the parenchyma in the woody roots store energy reserves, and root defense is dependent on energy reserves. When reserves are low, defense is low. When defense is low, weak or opportunistic pathogens attack. It is nature's way.

Non-Woody Roots

Non-woody plant roots are organs that absorb water and elements dissolved in it. The two basic types of non-woody roots are:

1. Root hairs on non-woody roots are extensions of single epidermal cells. Common on seedlings, root hairs grow to maturity in a few days. They function for a few weeks and then begin to die.

On mature plants, they are usually not abundant. When they do form, they do so when soil conditions are optimum for absorption of water and elements. I have found root hairs growing in non-frozen soils beneath frozen soils in winter.

2. Mycorrhizae are the other type of non-woody roots. Mycorrhizae are organs made up of plant and fungus tissues that facilitate the absorption of phosphorus-containing ions and others essential for growth.

The fungi that infected developing non-woody roots to form mycorrhizae were very "biologically smart." Rather than competing with other microorganisms in the rhizosphere for exudates from the plant, the mycorrhizal-forming fungi went right to the source inside the plant. And, even more to their advantage, many of the mycorrhizal fungi grew thread-like strands of hyphae-long, vegetative tubes of fungi-out from the mycorrhizae. This inside and outside presence gave the fungi a distinct advantage over other microorganisms in the rhizosphere.

The plant gains efficiency with mycorrhizae in several ways.

1. With their extended hyphae, mycorrhizae not only greatly extend the absorbing potential into the soil, but the hyphae may connect with other hyphae on other plants. In this way, the mycorrhizae serve to connect plants of the same or a different species. This leads to the conjecture that the natural connections that developed over long periods in the natural forest may have some survival value. That is why forest types are often named for the groups of species commonly found growing together.

2. The mycorrhizae have been shown to provide some resistance against root pathogens. It may be that the pathogens would have difficulties in building their populations in the rhizosphere dominated by the mycorrhizal fungi.

Perhaps the most important feature of the mycorrhizal fungi is that their boundary material is mostly chitin. Chitin is slightly different from cellulose by the replacement of some cellulose atoms by a chain of atoms that contain a nitrogen atom. This slight change in some way makes chitin a material better suited for absorption of elements. Remember that the fungus hyphae gain all their essentials for life by absorption through their boundary substance.

There are other advantages to the chitin and the tube-like hyphae that ramify the soil in the rhizosphere and beyond. When the hyphae die, they add a nitrogen source for other organisms. Also, when the hyphae are digested, they leave tunnels in the soil that are about eight to 10 microns in diameter. For the bacteria, these small tunnels may mean the difference between life and death. The bacteria quickly colonize the tunnels. The survival advantage here is that the major threats to their survival are protozoa that are usually much larger than 10 microns. So the hungry amoebae are not able to get at the bacteria inside the eight-micron tunnels.

Who Was First?

I do not know if the fungi were the first to grow into the root to get first chance at exudates or whether it was the bacteria. Regardless, bacteria and their close relatives, the actinomycetes, also infect non-woody roots to form organs that serve for the fixation of atmospheric nitrogen. Fixation means that the nitrogen that makes up almost 80 percent of our air is converted to a soluble ionic form by the action of the bacteria and actinomycetes within the nodules on the roots. (Some free-living soil bacteria can also fix nitrogen.) An enzyme called nitrogenase is the catalyst for the reaction that will take place only under very exacting conditions. The mycorrhizae facilitate absorption of elements, and the nodules provide a nitrogen source. Many species of plants have actinorhizae, which are the nodules formed by the root infections by actinomycetes. Species of Alnus have very large nodules. The actinorhizae are common on tropical and subtropical plants, and especially on plants that have adapted to soils low in available elements essential for life.

On some subtropical and tropical plants, such as the macadamia, multi-branched clusters of non-woody roots called proteoid roots form. The proteoid roots alter the rhizosphere by acidification processes that facilitate the absorption of phosphorus-containing ions. Another type of nodule forms on species of cycads. These nodules harbor blue green algae, or cyanobacteria, that have the ability to fix atmospheric nitrogen.

My point is that many different synergistic associations have developed in, on and about non-woody roots that provide elements, not an energy source. These associations are of extreme benefit to all connected members. At the same time, the conditions that provide for the associations are very delicate and exacting. It does not take much to disrupt them.

It Does not Take Much to Disrupt Them

This statement deserves repeating and repeating. The delicate "threads" that hold these powerful associations together need to be recognized and respected. Plants in cities grow only so long as these "threads" remain connected.

Plants grow as large oscillating pumps, with the top trapping energy and pumping it downward. The bottom absorbs water and elements and pumps them upward. The pumps have developed over time to work on the basis of many synergistic associations that maximize benefits for all connected members and to minimize waste.

Many of life's essentials for the bottom associates come from the top of the plant. And, the top works only because the bottom works. Energy is required to move things, and elements and water are required to build things.

Compacted soil blocks air and water to the bottom and crushes all the microcavities where the microorganisms live. In nature, decomposing wood and leaves keep conditions optimal for the rhizosphere inhabitants.

Over-watering stalls the respiration processes in the roots. When respiration stops, carbonic acid is not formed. When carbonic acid is not formed, ions necessary for the absorption process do not form. When absorption is down, the plant system is in trouble. Fertilizers can be of great benefit to plants growing in soils low in or lacking elements essential for growth.

Elements or molecules made up of a few to many different atoms enter the roots as ions. An ion is a charged atom or molecule. Ions with a positive charge are cations, and those with a negative charge are anions. Each particle or granule of fertilizer is a salt made up of a lattice of anions and cations, just as ordinary table salt is made up of a grand lattice of connected sodium cations and chloride anions. When salt as sodium chloride dry granules is poured into water, the sodium and chloride ions separate. When they separate, they carry electrical charges and are called the sodium ion and the chloride ion. When a cation enters a root, another cation must exit. This is very important, as we will see. When nitrogen enters a root as nitrate anion, an anion of bicarbonate ion from carbonic acid exits. The bicarbonate ion is probably the second most important compound in nature, next to water, because it drives the absorption process. When a bicarbonate ion exits into the rhizosphere, the pH increases.

When urea is used in fertilizers as the nitrogen source, the pH in the rhizosphere could increase to 2 or more pH units. The chemistry behind this is complex, but here I present only the conclusion, because a common problem with plants in some high pH soils is chlorosis. There is no easy field method for measuring the pH of the one millimeter wide rhizosphere. The rhizosphere could be pH 8, and the bulk soil would measure pH 6. As pH increases, the availability of elements such as iron and manganese decreases. In soils, it is one thing to have an element present and another to have it in a form available to the plant as an ion. As pH increases, iron and manganese element, form molecules that precipitate in water rather than ionize. If they are not available as ions, they will not be absorbed. And, if they are not absorbed, several of the enzymes essential for chlorophyll formation and photosynthesis will not form. [If urea is first transformed to nitrates by nitrifying bacteria in an aerobic soil environment, then it does not present these problems. -Ed.]

When the energy flow from the top of the pump is blocked, then the bottom does not get enough energy for growth and defense. The pathogens invade, and the plant declines. This scenario does not mean that every time you use urea, plants will decline from chlorosis.

But the use of urea could be a contributing factor where plants with genetic codes for growth on low pH soils are planted in high pH soils. If fertilization is a desired treatment, then a fertilizer that has nitrogen in a positive charged ion, such as an ammonium ion, would help to reduce the rhizosphere pH. When the ammonium ion enters the root, a proton of positive charge will exit. The protons in rhizosphere water will bring about more acidic conditions, so there is a way out.

In summary, fertilizers can be very beneficial for healthy survival of plants planted outside their forest homes. How beneficial will depend greatly on an understanding of many of the points mentioned here and some basic chemistry.

Pumps and Food

Plants are oscillating pumps. When the pump begins to wobble, some parts will begin to weaken. When they weaken to the point where some other agent causes a part to break, the pump will stop.

It is very difficult to determine where problems start in an oscillating pump. Symptoms may be in the bottom, but the cause may have been in the top. Or, it could be the other way around.

I go back to two points that may be part of the answer: exudates and the self-thinning rule of ecology. All living things require food and water for growth. Leaves and photosynthesis provide the energy at the top of the pump. The nonwoody roots and the rhizosphere provide the elements and water at the bottom. Photosynthesis will not work without water and elements, and the absorption processes will not work without an energy source. Trees became plants growing in groups in forests where the self-thinning rule had strong survival value. Not only did exudates provide quick energy for the rhizosphere organisms, but the carbon in the wood of the plants that fell to the ground also provided a long-lasting energy source for a succession of organisms.

Reports from some countries indicate an abundance of soluble nitrogen compounds in runoff water and even in ground water. This is a strong indication that the carbon-nitrogen ratio has been disrupted in the soil. It is well established from studies of the physiology of fungal parasitism that the degree of parasitism is often determined by the carbon-nitrogen ratio. It is probably similar for other organisms.

The Carbon Connection

The requirements for carbon are much greater than what could be supplied by those sources alone. Carbon must come from the top of the pump [unless supplied by soaps, oils and conditioners through irrigation with graywater -Ed.]. When the energy source from the top begins to decrease, the rhizosphere organisms will begin to starve.

The oscillating pump model soon takes on the form of a circle, because now it could be said that the top did not work efficiently because the bottom had a problem first, and this could be so.

Some Sources

R.C Foster, A. D. Rovira, and T.W. Cock. 1983. "Ultrastructure of the Root-Soil Interface." The American Phytopathological Society, St. Paul, Minnesota.

Ken Kilham, *Soil Ecology* (Cambridge, Great Britain: Cambridge University Press, 1994).

Alan Wild, *Soils and the Environment: An Introduction.*(Cambridge, Great Britain: Cambridge University Press, 1994).

Creating a Urine Composter System

The nitrogen in urine will be quickly lost or de-nitrified, as it turns into the volatilized gasses of ammonia, nitrous oxide or molecular nitrogen. If you want to preserve it, you must immediately convert it to water-soluble nitrates—or *nitrify* it—before it is stored or used. This both preserves the nutrients and converts them to a plant-available form.

The easiest approach may be to direct urine to a subsurface-irrigated and aerated planter bed, where aerobic soil microbes nitrify the urine.

If you wish to convert it before applying it to plants, make a urine composter or urine-nitrifying system. One of the authors, Del Porto, is developing such a system, which also binds up urine's abundant sodium so it cannot harm plants and build up in the soil.

The system consists of a two-staged filter: The first stage is a constructed aerobic filter in which nitrifying bacteria grows. The filter is aerated*, and contains moist and porous media, such as coarse horticultural vermiculite, perlite or clay pellets. A good resident time for microbial conversion is 30 to 40 hours. For one person daily, a six-inch-diameter PVC pipe, 10 feet long and filled with coarse horticultural grade perlite, should convert a urine-sugar solution to nitrate solutions on a continuous basis at 87°F with adequate aeration (five mg/l dissolved oxygen).

It is now ready to use on plants.

The now-nitrified urine then undergoes a second stage, flowing to a container that contains a sodium-ion-sequestering material, such as granular or liquid gypsum. The size of the filter will depend on the strength, volume and design-flow rate. A simple or complex sugar carbohydrate such as sucrose or corn syrup is added to raise the C:N ratio of the urine from 0.8 to 25, which is required for the microbes to do their work.

The two bacteria primarily responsible for nitrifying urea are *Nitrobacter* and *Nitrosomonas.***

Where to find the microbes:

Aerobic *Nitrobacter* and *Nitrosomonas* bacteria can be found in soil or compost from a site where urine is already undergoing composting, such as urine-soaked sawdust or straw from animal pens and barnyards. To make your own urine composter inoculum, collect a gallon of soil or compost from these sites. Sift out larger stones and sticks and moisten it thoroughly with a pint or two of warm (100°F) rain or pond water (chlorine and fluorine that is found in tap water is toxic to most living organisms) to which you have added one-half cup of sugar or corn syrup and one-half teaspoon of liquid soap or dishwashing detergent. The former provides easily digested carbon for the organisms and the latter breaks the water's surface tension (makes the water wetter), allowing the microbes to get closer to the nutrients. Urine, too, needs surface area for aeration, so add sand, perlite, gravel, or even a fixed media, such as plastic balls which are later removed. Keep it warm—about 87°F is ideal. Pour off the liquid to use. The end-product is nitrates in a water solution.

*The system must maintain the urine at a dissolved oxygen (DO) of > 3 mg/l at the bottom of the Oxygen Sag condition and no more than 9 mg/l. Process temperature must be held at 80 to 87°F, the optimum temperature for the nitrifying bacteria.

**Nitrosomonas utilizes and converts organic nitrogen compounds to excreted nitrites (NO2) at a rate of 3.22 gram O2 per gram N. Then the Nitrobacter utilizes/converts the nitrites (NO2) from the Nitrosomonas to excreted nitrates (NO3) at a rate of 1.11 gram O2 per gram N.

Broad Approaches to Managing Wastewater Ecologically

In effluent management, the time has come to use a different model, a better paradigm:

1. PREVENTION. (Don't use and put toxic substances into your wastewater which cause pollution.)
2. CONSERVATION. (Use water efficiently to reduce the volume.)
3. Divert/Separate. (Don't mix it all together!)
4. UTILIZATION. (Emulate gardens and natural forested and wetland systems and use it up by irrigating plants. Design the size and plant selection of the garden based on the volume and characteristics of the effluent and use it ALL up!

GO TO THE BEGINNING OF THE PIPE, NOT THE END

What is required is to go to the source of the problem: that point in the process where we create effluent in our homes or buildings. This is the place to make changes, before the wastewater is created.

ENGINEERING ECOLOGICALLY

As witnessed in the highly evolved natural world, the ecological paradigm reveals how to safely utilize the polluting components of effluent that are unwanted residuals, or "wastes," to ultimately grow living plants that have intrinsic value.

Planning, designing and engineering with the principles of prevention and ecological engineering is the only truly sustainable way to resolve difficult pollution problems economically while maintaining an affordable quality of life.

Here are a few suggestions to consider when planning for effluent management be it septic tank (ST) and soil absorption systems (SAS) or other treatment modalities:

1.0 CONSIDER MODIFYING THE CHARACTERISTICS OF THE EFFLUENT TO BE TREATED.

1.1.0 Volume reductions

Permanent sewage conservation to reduce the volume and, consequently, the hydraulic component of the design flow, which affects type of treatment selected and the size and area required for soil absorption. Typical (without best available water conservation measures) daily volumes of indoor sewage generation is reported to be 55 gallons (208 liters) per person. Of that, 40%, or 22 gallons (83 l), is attributed to contributions from flushing toilets. The balance is washwater, and contributes 60% or 33 gallons (25 l) daily.

1.1.1. Washwater conservation of 50% or 16.5 gallons (62.5 l) can be achieved by installing best available water-conserving technology for plumbing fixtures and appliances. One such example is the front-loading washing machine which uses 16 gallons per cycle, versus 40-plus (151+ l) for a top-loading model (the savings in energy to heat hot water, water purchase costs, detergents and damage to clothing will more than justify this specification). A rinse-water saver collects the final and relatively clear rinse water and stores it for reuse a the first wash water

1.1.2 Indeed, up to 100% can be achieved by requiring that pre-treated washwater be safely used for subsurface irrigation of non-edible plants either in outdoor landscapes (if soils are suitable for proper drainage) or in sheltered greenhouse settings. Satisfying 100% of the watering requirements of specific plants with all of the washwater is the design goal. Then it is all used up.

1.1.3. After eliminating the black water from the effluent, an added benefit surfaces: changing the temperature of the effluent going into the system. Wash water alone leaves the house at approx. 89 ° F. The cold water from flush toilets cools the combined effluent. This can be important because above 5°C (41° F), for every 10 °C (18°F) the temperature increases , the microbial and chemical rates of reaction in the tank doubles, within certain upper limits. That means warmer effluent equals better performance!

1.1.4. Waterless excrement and toilet paper treatment systems, such as modern attractive composting systems can eliminate all of the flush water and eliminate 90% of the nitrogen contribution, most of the pathogens, suspended solids and all of the BOD attributed to urine, feces and toilet paper.

1.2.0 Removals of Solids:

Suspended and buoyancy-neutral solids, such as lint from the washing machines, will pass through the septic tanks and clog the soil absorption system (SAS) with non-biodegradable (polyester, nylon, etc.) lint particles, and cause the system to fail prematurely.

1.2.1 Specifying 70- to 160-micron or smaller lint filters for the washing machines, small under-sink grease traps and septic tank outlet filters will protect the soil absorption or irrigation system from failure.

1.2.2 A waterless toilet, as described in 1.1.3 above, can be considered a bio-filter, removing large quantities of feces and toilet paper in addition to being a water conservation device.

1.2.3 Eliminating kitchen garbage grinder effluent from entering the treatment system is a well-established method of educed solids and organics loading. The plumber can reconnect the kitchen sink grinder to an indoor or outdoor composter (where the organics are transformed to soil conditioner). This allows the homeowner the convenience of the food grinder without the negative effects on the treatment system.

1.3.0 Chemical modifications:

Many household chemicals cause significant trouble for STSAS or the receiving soils and waters that are ultimately connected to the STSAS via hydraulic gradients.

1.3.1 Reduce or eliminate toxic chemicals from entering the STSAS! If a chemical or house hold cleaner does not claim to be biodegradable, don't buy it! Changing the specification for bleach from toxic chlorine-based to oxygenating hydrogen peroxide-based, minimizes the death of beneficial microbes in the treatment tank and adds oxygen as well.

1.3.2 Waterless toilets as described in 1.1.3 above, will remove up to 90% of the nitrogen and table salt from the system. This recognizes that 90% of the nitrogen and table salt in residential sewage is from urine! We eat more protein than is necessary for cellular growth and maintenance, and the balance is excreted in urine as urea and uric acid (see discussion of urine management in Chapter 4). Sodium chloride (table salt) is troublesome for microbial nutrient transport in green plants (It gives them hypertension, too!).

1.3.3 Back-flush water from water softeners adds large amounts of microbe-toxic sodium chloride (salt) to sewage. Changing the water softener chemicals from a sodium-based to a potassium-based ion-exchange resin will reduce the toxic sodium and substitute potassium, a fertilizer for your garden! Direct it to a safe washwater irrigation system to use this nutrient instead of sending it to the STSAS.

1.4 Penetration of storm water an precipitation into the SAS is a major factor in system failures in recent years. Changes in the global climate have caused extraordinary rainfall in certain areas. Evaluate design changes to reduce rainwater penetration. This can be achieved by covering the system and managing storm water ponding on or around the SAS by crowning and planting the SAS to shed significant rain to drainage swales.

2.0 UPTAKE OF WATER AND NUTRIENTS WITH PLANTS

In nature, plants along with the bioplex called the rhizosphere (root system and its associated soil microbial population) are the mechanisms for the transformation of nutrients and water into beneficial and attractive landscapes. This is the ecological paradigm operating at its best.

2.1 Redesign the SAS to include attractive plants that will use up the effluent before it enters the deep receiving soils.

2.2 Zero-discharge and planted up-flow sand filter/evapo-transpiration beds (we call them Wastewater Gardens) with a holding tank can be employed in situations where SASs are not possible, such as where there is high ground water, impervious soils and sensitive areas, such as proximity to wetlands, surface water and sole-source aquifers. The plants will remove the nutrients and water that would normally be pumped and hauled to a treatment plant at great expense to the property owner.

Index

Plans for Composting Toilet Systems

Available from the Center for Ecological Pollution Prevention

Composting toilet systems are a proven way to reduce nitrogen and pathogen pollution of ground and surface waters, as well as supplement septic and sewage systems and provide remote toilet systems.

The Center for Ecological Pollution Prevention (CEPP) develops, promotes and demonstrates innovative lower-impact technologies and systems, with an emphasis on utilization and zero-discharge approaches.

Can be used with or without a toilet stool. Can be used with a urine-diverting toilet stool, dry toilet stool or a micro-flush toilet.

CEPP 55-Gallon Net Batch Composting Toilet System

Originally designed for an eco-tourism resort in Fiji, the 55-Gallon Net Batch system is an inexpensive solution that uses reused 55-gallon drums (or you can use rollaway trash bins) and an effective aeration mechanism that maximizes the composting material's surface to volume ratio. That means faster and more thorough composting. The capacity is as expandable as the number of reactors (drums) used. Plans include design for a simple Wastewater Garden to manage leachate and, expanded, gray-water, too!

Plans and maintenance manual . . .$30

BOTH SETS OF PLANS: JUST $50!

Clockwise from top left: A system in Yap, FSM, with an integrated Wastewater Garden leachate system. A system in Baja, Mexico. Skim-coating the removal doors. End-product will be removed for five years from start-up.

CEPP Twin-Bin Net Composting Toilet System

This design is an advancement of the classic two-vault composter system used worldwide. The difference: Improved aeration mechanisms that gets more air to the composting material and manage leachate. First used by Greenpeace for its appropriate technology and ecotourism projects, this system has proven successful in the South Pacific islands, Mexico and even cooler regions of the U.S., for its high efficiency, low maintenance and low construction cost.

Plans and maintenance manual $30

The Composting Toilet System Book
Features descriptions of manufactured and site-built composting toilet systems, compatible flush and waterless toilets, key maintenance and installation considerations, U.S. regulations, gray-water systems and much more. $29.95

How to Build a Washwater Garden
A simple step-by-step guide to constructing a Washwater Garden, a contained zero-discharge gray-water and leachate management system. $19.95

Add $3.20 for Priority Mail shipping. Order from:
Center for Ecological Pollution Prevention (CEPP)
P.O. Box 1330
Concord, MA 01742-1330 USA
Tel.: 978/369-3951 Email: EcoP2@hotmail.com